The Handbook of Environmental Chemistry

Volume 3 Part D

Edited by O. Hutzinger

Anthropogenic Compounds

With Contributions by
R. F. Addison, A. B. McKague, Å. Larsson, D. J. McLeay,
P. E. Ney, G. A. Parker, D. Rivin,
G. Sundström, M. Tarkpea, C. C. Walden

With 32 Figures

Springer-Verlag Berlin Heidelberg GmbH

Prof. Dr. Otto Hutzinger

University of Bayreuth
Chair of Ecological Chemistry and Geochemistry
Postfach 3008, D-8580 Bayreuth
Federal Republic of Germany

ISBN 978-3-662-16131-9 ISBN 978-3-540-39468-6 (eBook)

DOI 10.1007/978-3-540-39468-6

Library of Congress Cataloging in Publication Data.
Main entry under title: The Handbook of environmental chemistry.
Includes bibliographies and indexes.
Contents: v. 1. The natural environment and the biogeochemical cycles / with contributions by P. Craig ... [et al.] – [etc.] – v. 3. Anthropogenic compounds / with contributions by R. Anliker ... [et al.] – v. 4. Air pollution / with contributions by H. van Dop ... [et al.]
1. Environmental chemistry – Collected works. I. Hutzinger, O.
QD31.H335 574.5'222 [QHJ545.A1] 81-18272
ISBN 978-3-662-16131-9

Originally published by Springer-Verlag Berlin Heidelberg New York Tokyo in 1986

Softcover reprint of the hardcover 1st edition 1986

2154/3140-543210

Preface

Environmental Chemistry is a relatively young science. Interest in this subject, however, is growing very rapidly and, although no agreement has been reached as yet about the exact content and limits of this interdisciplinary discipline, there appears to be increasing interest in seeing environmental topics which are based on chemistry embodied in this subject. One of the first objectives of Environmental Chemistry must be the study of the environment and of natural chemical processes which occur in the environment. A major purpose of this series on Environmental Chemistry, therefore, is to present a reasonably uniform view of various aspects of the chemistry of the environment and chemical reactions occurring in the environment.

The industrial activities of man have given a new dimension to Environmental Chemistry. We have now synthesized and described over five million chemical compounds and chemical industry produces about hundred and fifty million tons of synthetic chemicals annually. We ship billions of tons of oil per year and through mining operations and other geophysical modifications, large quantities of inorganic and organic materials are released from their natural deposits. Cities and metropolitan areas of up to 15 million inhabitants produce large quantities of waste in relatively small and confined areas. Much of the chemical products and waste products of modern society are released into the environment either during production, storage, transport, use or ultimate disposal. These released materials participate in natural cycles and reactions and frequently lead to interference and disturbance of natural systems.

Environmental Chemistry is concerned with *reactions in the environment*. It is about distribution and equilibria between environmental compartments. It is about reactions, pathways, thermodynamics and kinetics. An important purpose of this Handbook is to aid understanding of the basic distribution and chemical reaction processes which occur in the environment.

Laws regulating toxic substances in various countries are designed to assess and control risk of chemicals to man and his environment. Science can contribute in two areas to this assessment; firstly in the area to toxicology and secondly in the area of chemical exposure. The available concentration ("environmental exposure concentration") depends on the fate of chemical compounds in the environment and thus their distribution and reaction behaviour in the environment. One very important contribution of Environmental Chemistry to the above mentioned toxic substances laws is to develop laboratory test methods, or mathematical correlations and models that predict the environmental fate of new chemical compounds. The third purpose of this Handbook is to help in the basic understanding and development of such test methods and models.

The last explicit purpose of the Handbook is to present, in concise form, the most important properties relating to environmental chemistry and hazard assessment for the most important series of chemical compounds.

At the moment three volumes of the Handbook are planned. Volume 1 deals with the natural environment and the biogeochemical cycles therein, including some background information such as energetics and ecology. Volume 2 is concerned with reactions and processes in the environment and deals with physical factors such as transport and adsorption, and chemical, photochemical and biochemical reactions in the environment, as well as some aspects of pharmacokinetics and metabolism within organisms. Volume 3 deals with anthropogenic compounds, their chemical backgrounds, production methods and information about their use, their environmental behaviour, analytical methodology and some important aspects of their toxic effects. The material for volume 1, 2 and 3 was each more than could easily be fitted into a single volume, and for this reason, as well as for the purpose of rapid publication of available manuscripts, all three volumes are published as volume series (see Preface to Parts C and D of the Handbook). Publisher and editor hope to keep materials of the volumes one to three up to date and to extend coverage in the subject areas by publishing further parts in the future. Readers are encouraged to offer suggestions and advice as to future editions of "The Handbook of Environmental Chemistry".

Most chapters in the Handbook are written to a fairly advanced level and should be of interest to the graduate student and practising scientist. I also hope that the subject matter treated will be of interest to people outside chemistry and to scientists in industry as well as government and regulatory bodies. It would be very satisfying for me to see the books used as a basis for developing graduate courses on Environmental Chemistry.

Due to the breadth of the subject matter, it was not easy to edit this Handbook. Specialists had to be found in quite different areas of science who were willing to contribute a chapter within the prescribed schedule. It is with great satisfaction that I thank all authors for their understanding and for devoting their time to this effort. Special thanks are due to the Springer publishing house and finally I like to thank my family, students and colleagues for being so patient with me during several critical phases of preparation for the Handbook, and to some colleagues and the secretaries for technical help.

I consider it a privilege to see my chosen subject grow. My interest in Environmental Chemistry dates back to my early college days in Vienna. I received significant impulses during my postdoctoral period at the University of California and my interest slowly developed during my time with the National Research Council of Canada, before I could devote my full time to Environmental Chemistry in Amsterdam. I hope this Handbook may help deepen the interest of other scientists in this subject.

Otto Hutzinger

Preface to Parts D of the Handbook

Parts D of the three series

- The Natural Environment and the Biogeochemical Cycles (Vol. 1)
- Reactions and Processes (Vol. 2)
- Anthropogenic Compounds (Vol. 3)

are now either available or in press. During their preparation it became obvious that further parts will have to follow to present the respective subject matters in reasonably complete form.

The publisher and editor have further agreed to expand the Handbook by three new series: Air Pollution, Environmental Modelling, and Water Pollution.

Again, I thank all authors as well as collaborators at the Springer Publishing House for their cooperation and help. Thanks are also due to many environmental chemists and reviewers in particular for their critical comments and their positive reception of the Handbook.

Bayreuth, December 1985 O. Hutzinger

Contents

List of Contributors

Dr. R. F. Addison
Marine Ecology Laboratory
Bedford Institute of Oceanography
Dartmouth, N.S. B2Y 4A2, Canada

Dr. Å. Larsson
National Environmental Protection Board
Emission and Product Control Laboratory
Brackish Water Toxicology Section
Studsvik
S-611 82 Nyköping, Sweden

Dr. A. B. McKague
Research Consultant
D. McLeay & Associates Ltd. and B.C. Research
4008 Quesnel Drive
Vancouver, B.C. V6L 2X2, Canada

Dr. D. J. McLeay
Research Consultant
D. McLeay & Associates Ltd. and B.C. Research
4008 Quesnel Drive
Vancouver, B.C. V6L 2X2, Canada

Prof. Dr. P. E. Ney
Tanzebengasse 1
D-8240 Berchtesgaden
Federal Republic of Germany

Dr. G. A. Parker
Department of Chemistry
University of Toledo
Toledo, OH 43606, USA

Dr. D. Rivin
Cabot Corporation
Concord Road
Billerica, MA 01821, USA

Dr. G. Sundström
National Environmental Protection Board
Emission and Product Control Laboratory
Special Analytical Section
Box 1302
S-171 25 Solna, Sweden

Dr. M. Tarkpea
National Environmental Protection Board
Emission and Product Control Laboratory
Brackish Water Toxicology Section
Studsvik
S-611 82 Nyköping

Dr. C. C. Walden
4008 Quesnel Drive
Vancouver, B.C. V6L 2X2, Canada

Cellulose Production Processes

C. C. Walden, D. J. McLeay, and A. B. McKague

Research Consultant, D. McLeay & Associates Ltd.,
and B.C. Research
4008 Quesnel Drive, Vancouver, B.C. V6L 2X2, Canada

Summary

The cellulose production industry utilizes large amounts and a wide variety of chemicals, operates various chemical processes and discharges large volumes of effluents. These effluents contain some unrecovered process chemicals and some wood extractives, as well as various extraneous chemicals which are formed during pulping and bleaching. Although the environmental significance of many of these compounds is unknown, cellulose plant effluent characteristics which potentially impact on the environment include abnormal pH, suspended/settleable solids content, oxygen demand, toxicity, color, persistent organics which are bioaccumulable, mutagenic chemicals and a fish-flesh tainting propensity. Toxicity is primarily attributable to resin and fatty acids, neutral wood extractives and chlorinated phenolics formed during bleaching. Mutagenic chemicals are primarily chlorinated molecules. Information linking chemical constituents of cellulose plant effluents to various other environmental effects, is sparse. Biotreatment of effluents reduces the concentration of most chemicals contained therein, and correspondingly reduces any potential environmental impact; excepting color. Concentrations of chlorinated phenolics discharged in cellulose plant effluents, which persist in the environment, are low and their environmental significance has not been demonstrated.

Introduction

The commercial production of cellulose normally utilizes a raw material or furnish of woody origin, involving very predominantly the coniferous softwoods, albeit the use of hardwoods and alternate sources is increasing. The cellulosic fraction of the wood is defibrated by a variety of either mechanical or chemical processes which require the use of substantial volumes of high quality water. The discharged effluents contain solubilized lignin, hemicelluloses and wood extractives, which may be modified in the chemical processes. In many instances, the pulp is brightened or bleached by a variety of chemical techniques, prior to its formulation into commercial and fine papers.

Groundwood or mechanical pulp production involves the wet refining of short wood bolts against a buhr-stone, or the disc refining of wood chips, previously prepared to narrow dimension tolerances. In the production of thermomechanical pulp (TMP) the chips are softened by the application of heat prior to disc refining. In still other processes, such as chemi-thermomechanical (CTMP) and neutral sulfite semi-chemical (NSSC) pulping, chemicals are added to facilitate defibration. Substantial washing of the defibrated pulp is essential to all these processes and the degree of internal water recycle can vary widely between processes and from plant to plant. The various groundwood pulps are generally utilized in the production of papers such as newsprint where colour and strength requirements are of lesser importance.

Kraft pulping is the major existing chemical process, in which the wood chips are digested under heat and pressure in strongly alkaline solutions of sodium sulfide. Recovery of pulping chemicals is an integral part of the kraft process, where the stronger liquors from pulp washing are concentrated first by evaporation and finally by incineration in the recovery furnace. This incineration destroys up to 95% of non-cellulosic organics solubilized during pulping. The recovered smelt is recausticized with burnt lime and sulfidity is adjusted by addition of sodium sulfate before re-use. Although water requirements may be similar for various kraft mills, particularly for newer mills designed to minimize chemical losses, water use and recycle patterns within individual plants are virtually unique.

In sulfite pulping, wood chips are digested under heat and pressure in mildly acidic bisulfite solutions. The corresponding cation or base, may be calcium (Ca), ammonium (NH_4), or magnesium (Mg). Excepting for the Mg-base processes, chemical recovery is not economically required for sulfite pulping. Indeed, it is not considered to be technically feasible for Ca-base pulping. Where chemical recovery is not practised, all the non-cellulosics solubilized during pulping are normally discharged in the plant effluents. Environmental regulatory pressures are forcing conversion of the older Ca-base plants to NH_4- or Mg-base operations, or even to kraft pulping, in which recovery is feasible. Notwithstanding, even for the NH_4- and Mg-base plants which practise recovery, incineration in the recovery stage seldom destroys more than 85% of the organics solubilized during pulping. The remainder is lost in the effluents.

Depending on their end use, a substantial proportion of pulps are processed further to improve brightness or clarity. Mechanical pulps are usually treated with dilute solutions of a reducing agent such as dithionite, whereas the chemical pulps are subjected to dilute solutions of oxidizing agents such as aqueous chlorine, hypochlorite and chlorine dioxide, either separately or in combination. Solubilized lignin is usually removed following each bleaching stage by tower washing with dilute caustic solutions. Chemical recovery of bleaching solutions is not practised commercially, because it would introduce inert sodium chloride into the recovery circuit. However, the increasing use of oxygen pre-bleaching permits diversion of the liquors from this stage to the recovery system.

Table 1. Typical volumes of some pulping discharges [4, 5]

Unit process	Water volume $m^3 \times 10^3$/t pulp
Groundwood	22– 34
Thermomechanical	22– 34
Neutral sulfite semi-chemical (with recovery)	10– 69
Kraft pulping	21– 69
Sulfite pulping (no recovery)	69–103
Sulfite pulping (with recovery)	69–103
Bleaching	69–137

Chemical pulps are used to augment the strength of mechanical pulps in the fabrication of lower-grade papers; or are processed further, usually involving additional defibration of the cellulosic fibres with the introduction of various additives prior to conversion into a variety of commercial products and fine papers. High purity kraft and sulfite (dissolving) pulps represent a source of α-cellulose for the production of various chemical derivatives, such as viscose rayon, cellophane films and others. Readers requiring additional detail concerning the various pulping processes should consult standard reference texts [1–3].

All pulp and paper processes utilize large quantities of high quality water; primarily for the various stages of pulp washing and for mat formation on the Fourdrinier machine. Some typical examples of water use are shown in Table 1 [4, 5]. The daily capacity of most pulp mills is in the range of several hundred tonnes, and chemical pulp mill capacities of more than 1,000 t/d are not uncommon. Inasmuch as water usages for pulping and bleaching are additive, the discharge volumes of a bleached chemical pulp mill frequently are of the order of 9,500 m^3/d, whereas large pulp and paper complexes can discharge 1,900–2,250 m^3/d. This latter volume is roughly equivalent to the domestic discharge of cities in the population range of 600,000 to 850,000 individuals. Entirely apart from the characteristics of the actual discharge, clearly a most significant feature of pulp and paper mill effluents is the single point release of exceptionally large volumes of effluent.

Pulp and paper mill effluent characteristics with a potential environmental impact are numerous and include: abnormal pH, suspended/settleable solids content, oxygen demand, toxicity, colour, persistent organics which are bioaccumulable, mutagenic chemicals, and a fish-flesh tainting propensity. Definitive knowledge pertaining to the chemical composition of pulp and paper effluents stems largely from investigations of toxicity.

Effluent Characteristics

Biochemical Oxygen Demand

The biochemical oxygen demand relates to the soluble organic constituents in an effluent which are fermented by an acclimated microbial population, under stan-

Table 2. Typical BOD_5 values of some pulping and bleaching discharges [1, 2, 11]

Unit process	5-day BOD kg/t pulp
Groundwood	7.5– 17.4
Thermomechanical	17.5– 27.5
Kraft and soda pulping	12.5– 25.0
Sulfite pulping (without recovery)	200 –300
Sulfite pulping (with recovery)	61.5– 63.3
Bleaching	15 – 27.5

dardized conditions, where the test sample is diluted so that the dissolved oxygen content is not limiting [6]. Complete biological oxidation requires 20–30 d, however the test is normally abbreviated to 5 d incubation, albeit in Sweden, for logistic reasons, this term is extended to 7 d. This total biochemical oxygen demand of pulping effluents, over 20–30 d, is 1.22–1.23 times greater than the value secured in the BOD_5 test [7].

The chemical characteristics of the components responsible for this oxygen demand have received scant attention. Rather, it has been assumed that they represent the wood extractives to be found in the particular wood species comprising the pulp mill furnish, together with the various hexoses and pentoses resulting from degradation of the hemicelluloses during chemical digestion, possibly with small quantities of D-glucose resulting from limited cellulose breakdown. On this basis, the principal sugars to be found in pulping effluents are D-glucose, D-galactose, D-mannose, D-arabinose, and D-xylose [1, 2, 8, 9]. The contribution of degraded solubilized lignin to BOD_5 is considered negligible and the fermentable organics in pulping effluents represent only about 25% of the total soluble organics [10].

Typical values for BOD_5 levels in effluents discharged from various types of pulp and paper plants are shown in Table 2 [1, 2, 11].

The acutal BOD_5 levels in pulping effluents are largely an inverse function of the efficacy of pulping liquor recovery; plants with newer, tighter recovery systems being characterized as having lower loadings per tonne. For Ca-base sulfite mills, where no recovery is feasible, the discharge of fermentable organics can be equivalent to that of the domestic sewage from cities with populations of several million individuals.

Toxicity

The toxicity of pulp and paper mill effluents is normally characterized by the response of one or more organisms to the time-concentration effects of the effluent, under highly standardized conditions. For the most part, the toxicity ascribed to various chemical compounds which are found in discharges of cellulose plants have been defined in terms of the acute (≤ 96 h) response of juvenile salmonid fish, usually rainbow trout (*Salmo gairdneri*), with lethality as a criterion [12]. A wide variety of acute lethal bioassay tests are available [13], although data therefrom are only crudely interconvertible [4, 14]. Isolation and identification of the responsible components must be undertaken concurrently with estimation of the corresponding toxicity.

There are essentially two classes of toxicants in cellulose plant effluents; the toxic extractives which are solubilized during processing, and those toxicants which are formed during processing, by the interaction of pulping and bleaching chemicals with the extractives and with degradation products of the non-cellulosic wood fractions. For the most part, the degradation products themselves are not toxic.

The principal toxicants in debarking and pulping effluents are various resin and fatty acids; together with juvabione derivatives in effluents corresponding to mill furnishes embracing wood from the true firs. Lesser amounts of related diterpene alcohols and corresponding neutral compounds have also been identified as minor contributors to effluent toxicity.

Resin acids have been recognized as toxic constituents of pulp and paper mill discharges (Fig. 1), since as early as 1931 [15]. Various demonstrations of their toxicity had been made [16–19] before their role was quantitatively defined in kraft pulping effluent [20]. In this quantitative approach, the identified toxicants, after being recombined in a synthetic mixture in the proportions found in the original effluent were shown by fish bioassay tests to produce a time-concentration lethal response, identical to that of the original effluent (Fig. 2).

The three major resin acids (isopimaric, abietic and dehydroabietic) accounted for over 80% of the total acute lethal toxicity according to fish bioassay tests. The same techniques have been extended to debarking effluents [21, 22], sulfite pulping effluents [23–25], and mechanical pulping effluents [26–28), demonstrating that the resin acids constitute the bulk of the toxicity in each of these effluent classes. Again, isopimaric, abietic, and dehydroabietic acids are the major toxicants in all these types of effluents, albeit palustric acid is frequently a significant contributor to the toxicity of debarking effluents [22]. The concentrations of the resin acids in various pulping effluents are illustrated in Table 3.

The principal resin acids, i.e., abietic, dehydroabietic, and isopimaric, are found in the discharges from bleached chemical pulping plants [23, 30, 31, 34, 35, 37] together with those of lesser importance vis-a-vis toxicity; i.e., neoabietic, palustric, and pimaric [27, 30, 31, 34, 35]. Similarly, resin acids are present in the discharges from paper plants [30, 31, 40–42]. In both instances, their presence is undoubtedly attributable to the inclusion of pulping sewers into the plant discharge.

Unsaturated fatty acids comprise the other major toxicity component of wood extractives to be found in cellulose pulping effluents [43, 44]; their contribution to acute lethal toxicity being estimated at 18% for a kraft pulping discharge [20]. These unsaturated fatty acids include oleic, linoleic, linolenic, and palmitoleic, with typical contents in pulping effluents, as shown in Table 3. Their presence has also been reported in discharges from bleached chemical pulp mills and paper plants [30, 31, 34, 35, 41]. Again, their presence in these discharges almost certainly stems from the inclusion of pulping sewers within the total plant discharge, since unsaturated fatty acids do not survive the bleaching processes, unchanged.

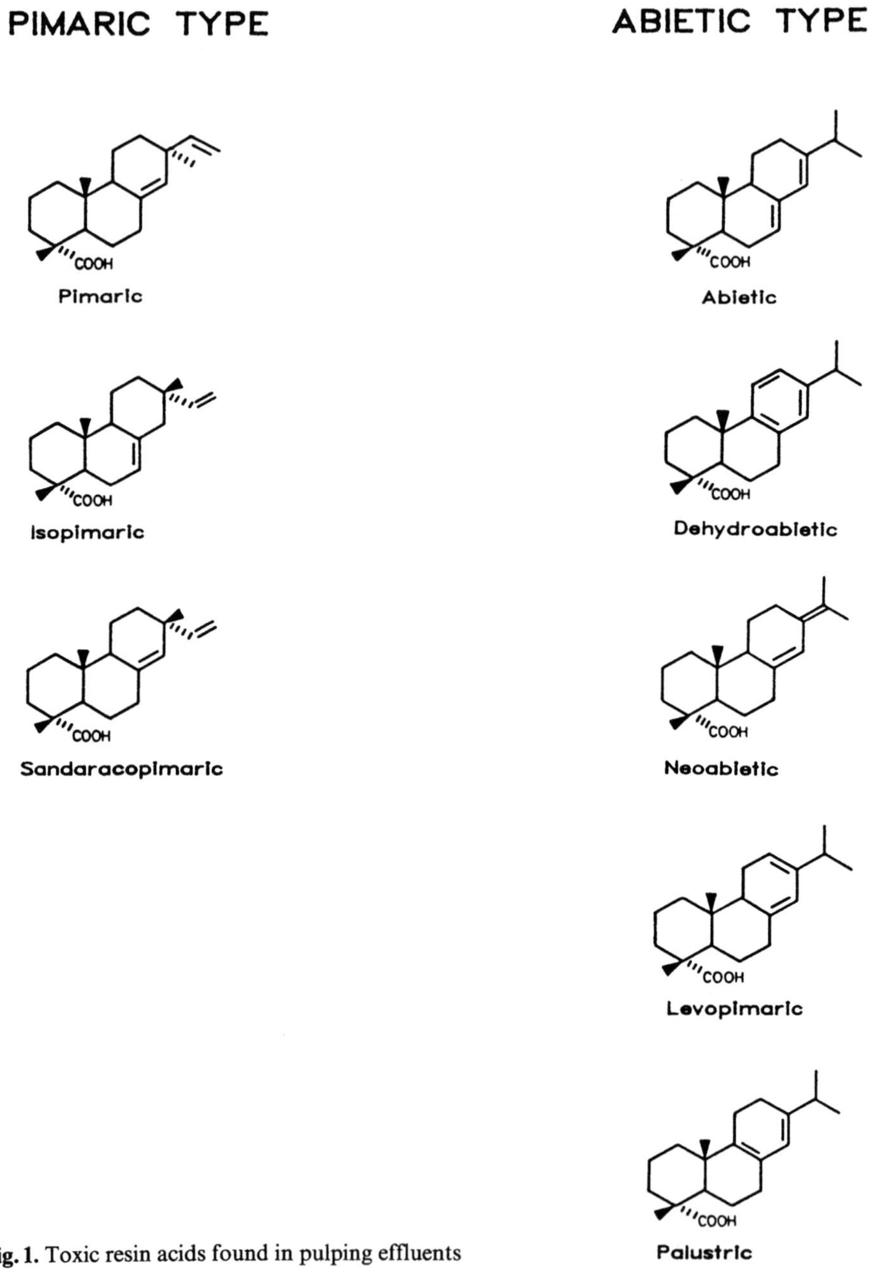

Fig. 1. Toxic resin acids found in pulping effluents

Various chemical species which are related to the resin acids, are found in effluents from mechanical pulping plants and can account for up to 30% of the total acute lethal toxicity [26]. These species include the diterpene alcohols; pimarol and isopimarol; dehydroabietal, abietal, abienol, 12E-abienol, and 13-epimanool [28, 45]. Miscellaneous chemical species also reported to be toxic to fish include

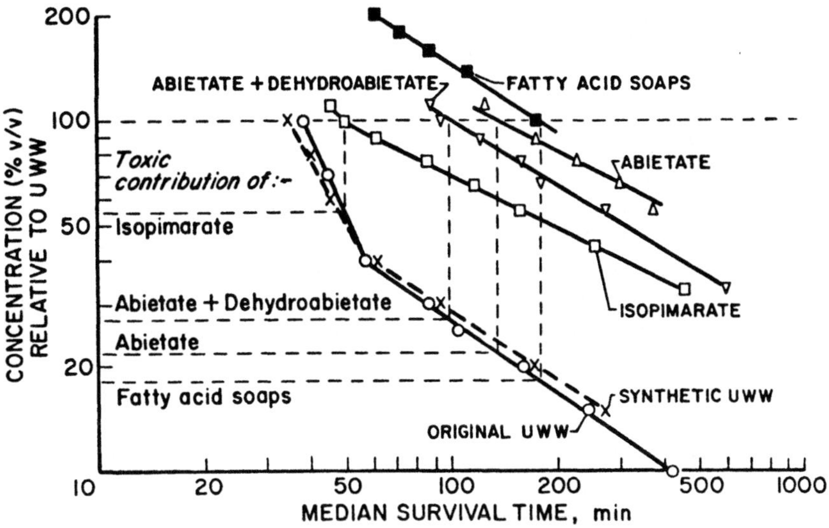

Fig. 2. Time concentration lethal response of juvenile coho salmon (*Oncorhynchus kisutch*) to effluent and to a mixed solution of resin and fatty acids recombined in the same concentrations as in the original unbleached white water effluent [20]

Table 3. Range of concentrations (μg/L) of resin and fatty acids in mechanical pulping and chemical pulping effluents

Acid	96-h LC50 (μg/L)	Kraft	Sulfite	Mechanical	References
Abietic	700–1,500	30–9,970	520– 4,840	210–16,000	[28–32]
Dehydroabietic	800–1,740	990–4,870	700–4,620	490–15,100	[25, 27, 28, 30–36]
Isopimaric	400–1,000	70–4,120	100– 5,070	150– 9,300	[25, 27–32, 35, 37]
Levopimaric	700–1,000	< 10–2,700	< 10– 2,400	< 10– 300	[27, 29, 31, 32]
Neoabietic	610– 730	< 50–1,200	–	30– 6,800	[31]
Palustric	500– 600	–	–	300– 7,700	[28, 29, 34, 38, 39]
Pimaric	60–1,200	–	–	20– 6,800	[28–30, 34, 35]
Sandaracopimaric	360	–	–	–	[37]
Linoleic	2,000–4,500	< 10–1,160	90–14,600	490– 9,000	[29–31, 36, 37, 39]
Linolenic	2,000–6,000	< 20– 110	270– 700	<100– 800	[28, 29, 37, 39]
Oleic	3,500–8,200	40–2,400	< 50– 6,780	230– 4,300	[29–31, 35, 38, 40]
Palmitoleic	2,000–6,000				[20]

sandaracopimaradiene, dihydroabietane, 4-*p*-tolyl-1-pentanol, and 13-abieten-18-oic acid [46, 47]. Also found in effluents from mechanical pulping of the true firs are juvabione ("paper factor"), juvabiol, $\Delta 1'$-dehydrojuvabione, $\Delta 1'$-dehydrojuvabiol, dihydrojuvabione [38, 48], and todomatuic acid [46]. Many of these toxic extractives do not survive chemical pulping, albeit they appear to persist in sulfite pulping effluents more so than in kraft effluents [48]. Other extractives which survive in sulfite pulping effluents include todomatuic acid and its derivatives [23, 46], as well as 3,4-divanillyltetrahydrofuran [48].

Two classes of extraneous chemicals with properties toxic to fish are formed by the interaction of pulping and bleaching chemicals with non-cellulosic degradation products. The volatile compounds include the group designated as total reduced sulfur (TRS) compounds and include hydrogen sulfide, methyl mercaptan, methyl disulfide, and dimethyl disulfide. The other major group includes chlorinated organics; derivatives of specific resin and fatty acids, simple phenol, guaiacols, and catechols.

The generation of TRS compounds is restricted to the kraft pulping process and their principal environmental impact normally stems from their malodorous properties, albeit their toxicity to fish is well recognized [17, 44, 49–51]. More recent evidence suggests that this toxicity is appreciably greater than earlier data indicated [52–54]. Nonetheless the contribution of TRS compounds to effluent toxicity is considered to be low, because these compounds are readily air-stripped from the effluents, or oxidized [55–58]. For example, it was estimated that for a group of 20 samples from three plants, TRS toxicity represented 5% or less of the total toxicity [59], prior to any form of treatment.

The toxic non-volatile compounds are the result of oxidation or chlorination of the unsaturated fatty acids, dehydroabietic acid and various phenol analogues resulting from lignin degradation. Unsaturated fatty acids form epoxystearic or dichlorostearic acid [29, 30, 60], whereas all the resin acids are destroyed, excepting dehydroabietic, which forms the mono- and dichloro-derivatives [27, 30, 31, 34, 35, 60, 61].

Toxic phenolic derivatives are shown in Fig. 3 and typical concentrations are shown in Table 4. Understandably these compounds, as well as the chlorinated derivatives of stearic and dehydroabietic acids, are all formed within the bleaching stages of the cellulose production process, albeit their concentrations are greatest following caustic extraction. Inasmuch as bleached pulp is almost always produced at the same physical location as the unbleached pulp, these compounds are always found in the discharges from a bleached cellulose pulping plant. The tri- and tetrachloroguaiacols represent the principal chlorinated phenolics in bleached kraft pulping discharges, whereas trichlorophenol is the principal chlorinated phenol in corresponding sulfite discharges. The dichloro-derivatives are present in considerably lesser concentrations. The chlorinated catechols have not been reported in sulfite effluents. Tetra- and pentachlorophenols are also found in cellulose plant discharges. However their presence is attributed to the use of slimicides within the plant [42, 74]; a practice still permitted in some jurisdictions.

Possibly excepting tetrachloroguaiacol and trichlorophenol, concentrations of individual clorinated phenolics in whole mill effluents do not exceed levels that are acutely lethal to fish or other aquatic life. Levels of chlorinated derivatives of simple phenol in sulfite effluents are generally less than in corresponding kraft effluents. Nonetheless the contribution of these derivatives of simple phenol to kraft or sulfite effluent toxicity is not great. Simple calculations suggest that the chlorinated derivatives of simple phenol account for only about 0.5% of the total acute lethal toxicity of bleached kraft mill effluents.

Toxicity attributable to the other non-cellulosic degradation products, by and large, has not been confirmed. No relationship has been found between phenolics

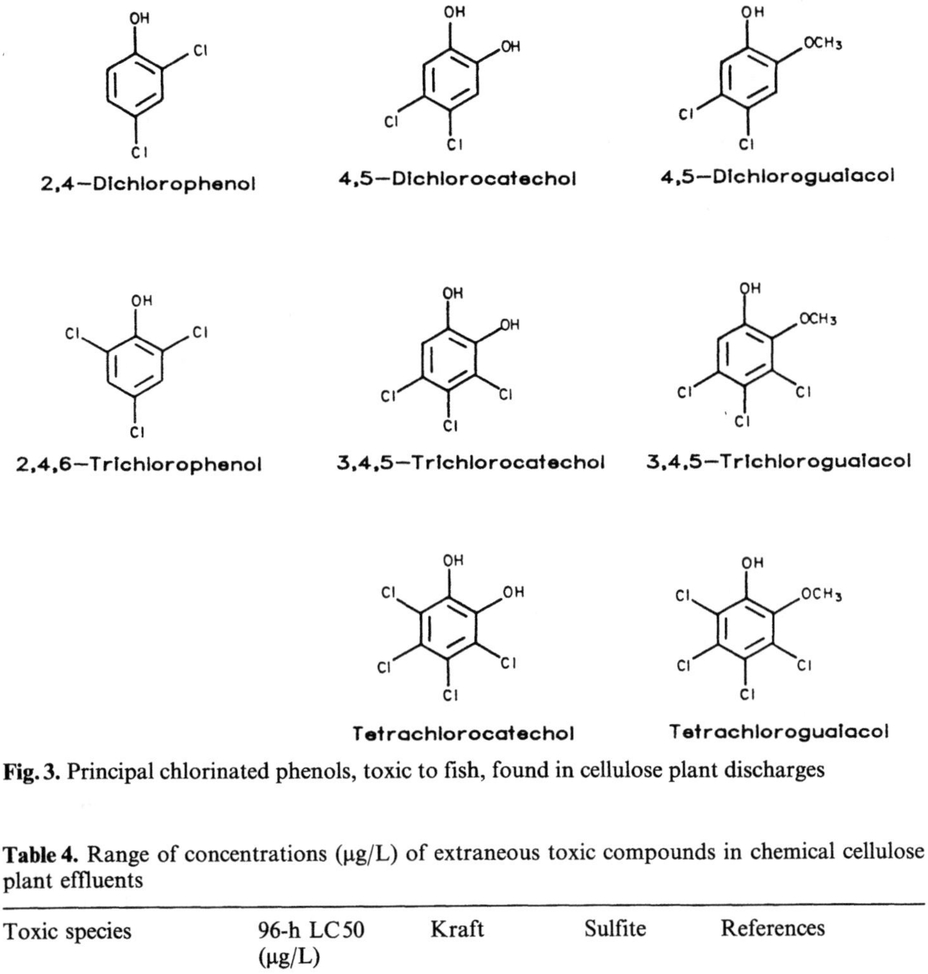

Fig. 3. Principal chlorinated phenols, toxic to fish, found in cellulose plant discharges

Table 4. Range of concentrations (μg/L) of extraneous toxic compounds in chemical cellulose plant effluents

Toxic species	96-h LC50 (μg/L)	Kraft	Sulfite	References
Dichlorocatechol	500–1,000	12– 90	–	[34, 65–67, 69]
Dichloroguaiacols	2,300	22– 100	6– 12	[34, 67–70]
2,4-Dichlorophenol	2,800	9– 15	4– 10	[35, 61, 65–67, 69, 70]
3,4,5-Trichlorocatechol	1,000–1,500	120– 270	–	[34, 65–67, 69]
Trichloroguaiacols	700–1,000	< 10– 340	16– 39	[29–31, 34, 35, 42, 60, 61, 66, 67, 69, 70, 72, 73]
2,4,6-Trichlorophenol	450–2,600	< 1– 51	3–764	[34, 35, 61, 66–70, 73, 74]
Tetrachlorocatechol	400–1,500	22– 420	–	[34, 65–67, 69, 71]
Tetrachloroguaiacol	200–1,700	< 10– 620	12–130	[29–31, 34, 35, 60, 61, 66–70, 72]
Chlorodehydroabietic acid	600– 900	< 10– 750	<10–900	[25, 29, 31, 34, 35, 37, 61]
Dichlorodehydroabietic acid	600–1,200	< 10– 410	<10– 40	[25, 29–31, 34, 35, 37, 61]
Epoxystearic acid	1,500–3,400	40–1,540	<40	[29, 30, 60]
Dichlorostearic acid	2,500	< 40– 552	<40	[30, 38]

and toxicity [75, 76], although eugenol, isoeugenol, and 3,3′-dimethoxy-4,4′-dihydroxystilbene have been implicated in the toxicity of unbleached sulfite pulping effluents [23]. In particular, no element of toxicity has been attributed to chlorinated lignin residues, other than the single nucleus derivatives already described. Many of the toxicity isolation and identification studies carried out in the seventies [20–22, 25, 27, 27, 36–39, 45, 60, 61, 65, 72] involved toxicity mass balances across the isolation-fractionation procedures [20], thus insuring that all major fractions responsible for acute lethal toxic effects were isolated for subsequent identification. It is safe to assume that no major toxic species, characterized as being acutely lethal to fish, remain unidentified.

Although specific investigations have quantitatively ascribed the acute lethal toxicity of individual effluent samples to various combinations of the responsible toxic compounds, efforts to develop a chemical assay have been less successful [31]. Attempts to predict toxicity of cellulose plant samples from analyses of these chemical constituents were for the most part, accurate [29, 31]. However the technique did not invariably yield correct results, and a chemical assay cannot reliably replace the acute lethal bioassay, for individual samples. Thus, our understanding of the interaction of the various toxic components is incomplete.

Present data permit some understanding of the influence of the different cellulose processes on toxic constituents of the effluents derived therefrom. The various mechanical pulping processes, as well as hydraulic debarking, solubilize the toxic wood extractives, i.e., resin and unsaturated fatty acids and various neutral compounds. Toxicity data are normally expressed on a concentration basis (v/v%) so that no assessment is feasible concerning the toxic loadings per tonne cellulose produced or the completeness with which the toxicants are extracted. The degree of water reuse within the mechanical pulping plants varies widely and, where the degree of recycle is high, the concentration of toxicants results in an acute lethal toxicity for these effluents, greater than that observed for effluents from chemical pulping [28]. A number of toxic chemical species, originally present in the wood furnish, undergo degradation in the kraft or sulfite processes. Palustric acid is a notable example. Neutrals such as juvabione epimers persist through sulfite pulping but not through kraft. However a large proportion of the total organics solubilized during chemical pulping are subsequently incinerated; about 85 and 95% respectively in the sulfite and kraft processes. On this basis, the lack of any appreciable difference in the toxicity of mechanical and chemical pulping effluents, and even between sulfite and kraft effluents, represents an apparent anomaly. Also anomalous is the similar toxicity levels for sulfite pulping effluents resulting from processes using different bases, inasmuch as chemical recovery is not part of the Ca-base process [31, 48].

It seems probable that the chlorinated dehydroabietic acids and the epoxy or chlorinated 18-carbon fatty acids are the chlorination (oxidation) products of these compounds carried over into the bleaching process, adsorbed on the unbleached pulp or in solution in the pulp slurry. Modified bleaching techniques have been examined, to ascertain their impact on the production of chlorophenols appearing in bleachery effluents. Maximum production, at 120% of the chlorine demand, occurred when the ratio of chlorine to chlorine dioxide was 1:1 [68, 71, 77, 78]. At higher levels of chlorine dioxide substitution for chlorine, levels of chlorophenols dropped, although reduction in toxicity was slight [79]. Prebleaching with oxygen, now practised commercially, reduces the quantities of but not the spectrum of compounds found in bleachery effluents [80–84]. Altogether, although a reduced environmental impact is a major benefit claimed for revised bleaching techniques, firm conclusions are not available.

Other chlorinated organics reported in pulp and paper plant discharges, albeit not present as significant contributors to fish acute lethal toxicity, include chloroform [61, 70, 74, 85–87] in levels up to 1.1 mg/L, 2,6-dimethoxy-3,4,5-trichlorophenol, chlorovanillins, protocatechualdehydes [34, 67], and di- and trichlorodimethyl sulfones [87]. The potential of these compounds and others for bioaccumulation and mutagenic activity is discussed subsequently. Notwithstanding, the concentration in bleaching effluents of chlorinated organics of known composition and concentration, is estimated to account for as little as 2% of the total chlorinated organics formed during the various bleaching processes [88].

Colour

Pulping and bleachery effluents possess a dark brown colour, which is usually measured by visual or spectrophotometric comparison to colour standards, based on 1 mg/L of cobalt and platinum present as sodium cobaltinitrite and sodium chloroplatinate [89]. In addition, colour levels are a function of pH, being 60% greater at pH 7.5 than at pH 5.0; both within the biological range [90]. The environmental impact of colour is two-fold: at a colour level of 60 APHA units and higher, diminution in light transmission may adversely affect photosynthesis by the aquatic flora (primary productivity) [91]; whereas visual acuity permits detection of 5–15 units, which may be esthetically undesirable [92].

Colour is attributed to lignin residues which have been solubilized during pulping and bleaching. The precise structure of these residues is unknown, although it is believed they are comprised of coniferaldehyde units, aryl ketones, quinone methides, and quinones. The quinones, which are responsible for the elevated colour at alkaline pH levels, are derived from the original lignin, but also from demethylation of guaiacyl and syringyl groups in the lignin and from autooxidation of catechol-type structures in polyphenolic extractives and sugar degradation products [93, 94]. The colour bodies can be divided into two groups according to their molecular weight. The lower molecular weight colour bodies (MW of 200–500) are less lignin-like in character, i.e., they posses a higher carboxylate to aromatic ring ratio, have a lower colour level per unit of total organic carbon and are not precipitated by calcium salts [90]. The higher molecular weight fraction, which is precipitated by lime, makes up the bulk of the colour and has a MW range of 1,000–3,500 [96, 97].

Colour levels are highest in bleached kraft pulping processes utilizing softwood furnishes, where colour values in excess of 2,000 APHA units are common [98]. The highest colour loadings are found in the caustic extraction effluent, where values can exceed 20,000 APHA units and contain 30–70% of the colour loading found in the total plant effluent [99].

Mutagenicity

Mutagenic activity has been indicated in cellulose plant effluents, particularly those originating from pulp bleaching [100–102]. Subsequent investigations, which confirmed these findings, also demonstrated mutagenic activity in various bleachery effluents, including those generated during direct chlorination, admixtures of chlorine with chlorine dioxide, and those involving prebleaching with oxygen [103, 104]. Low levels of mutagenicity were also identified in various individual process sewers, albeit in most total cellulose plant sewers mutagenicity could not be detected with the screening tests employed [103–107]; i.e., the mutagenicity had been diluted to below the detection threshold.

A number of compounds (chloroform, carbon tetrachloride, and safrole), which have been identified as procarcinogens (i.e., are metabolized in the liver to active carcinogens [108–110], also are found in small quantities in pulp and paper mill effluents [111]. Various attempts have been made to isolate and identify compounds in cellulose plant process sewers, effluents and environs which are responsible for their observed mutagenicity. These compounds, as shown in Table 5, embrace aromatic hydrocarbons, halogenated aliphatic hydrocarbons, various chlorinated aliphatic and aromatic aldehydes and ketones, and aliphatic and alicyclic acids. Brominated compounds are assumed to derive their bromine from process water [84] whereas the aromatic hydrocarbons (fluoranthrene, pyrene) originated in sediments adjacent to the mill outfall [115]. Some form of the Ames microbial test [105] was utilized to follow mutagenic activity in most of the isolation studies, although 4,5-dichlorocatechol, 7-oxodehydroabietic acid, and 3-chloro-*cis*-muconic acid were shown to be mutagenic to growing cells of common yeast [118]. Of the mutagenic compounds isolated and identified, no particular pattern emerges; although three-quarters of the compounds contained halogen atoms within their molecular structure.

Table 5. Compounds possessing mutagenic activity, known to be present in cellulose plant pulping and bleaching effluents

Compound	References	Compound	References
Carbon tetrachloride	[111]	1,3-Dichloroacetone	[114, 115]
Chloroform	[111]	1,1,3-Trichloroacetone	[114, 115]
Safrole	[111]	1,1,3,3-Tetrachloroacetone	[117]
1,1,2,3-Tetrachloro-2-propene	[112, 113]	Pentachloroacetone	[117]
1,1,2,3,3-Pentachloropropene	[112, 115]	Hexachloroacetone	[117]
Dichloromethane	[114, 115]	Acetovanillone	[118]
Trichloroethene	[114, 115]	2,4,6-Trichloroguaiacol	[115]
Tetrachloroethene	[114, 115]	Tetrachloroguaiacol	[115]
Bromodichloromethane	[114, 115]	4,5-Dichlorocatechol	[118]
Dibromochloromethane	[114, 115]	3-Chloro-4-dichloromethyl 5-hydroxy-2(5H)-furanone	[119]
Fluoranthrene	[116]		
Pyrene	[116]	Neoabietic acid	[112, 113]
Chloroacetaldehyde	[114, 115]	7-Oxodehydroabietic acid	[118]
2-Chloropropenal	[114, 115]	3-Chloro-*cis*-muconic acid	[118]

Off-Flavours in Fish

Most of those studies concerning off-flavours developed in commercially-relevant species of fish under controlled conditions relate to effluents from bleached kraft mills, alone or mixed with sulphite or mechanical effluents (Table 6). No specific test procedures have been universally accepted and data that have been secured have involved a variety of organoleptic evaluation techniques. Nonetheless exposure of test fish to cellulose plant effluents clearly does result in off-flavours in the fish flesh in exposures of 7 days or less at concentrations of $<1\%$ to about 5% (v/v) [120–125].

Flavour impairment can occur with exposure times as little as 4 h [125], whereas purging of the "taint" can occur in between 4 and 7 d. Thus although the flesh of fish in the immediate environs of cellulose plant outfalls may become tainted in a matter of hours, the effect is not permanent and the taint is cleared within days of the fish moving into an uncontaminated environment.

Condensates, produced within the chemical pulping processes, appear to be the process sewers primarily responsible for the off-flavours produced in fish flesh. Waste sulfite pulping liquor resulted in significant off-flavours in freshwater perch exposed for 7 d to 0.7% (v/v) [121, 122]. Kraft foul condensate caused tainting in rainbow trout at strengths as low as 0.05% [126]. Steam stripped foul condensates and other evaporator condensates have higher threshold tainting levels [126, 129]. Although recovery condensates are the most potent source of tainting materials within the chemical pulping processes, other contributory sources definitely exist, as shown in Table 7 [127].

Conclusive evidence as to the chemicals in cellulose plant effluents which are responsible for off-flavours in fish flesh is lacking. Chemicals known to cause fish flesh tainting, which occur in cellulose plant effluents include benzene, naphthalene, phenol, chlorophenol, and pentachlorophenol. However their concentrations in plant effluents are well below their respective tainting thresholds [128, 129]. The compounds 2,4-dichlorophenol and 2,4,6-trichlorophenol are found in bleached kraft effluents [67] in concentrations which overlap those concentrations reported to cause off-flavours in the flesh of exposed fish [129]. Nonetheless no direct proof exists that these specific compounds are responsible for tainting. Various phenolic compounds have been implicated at various times as being con-

Table 6. Induction of off-flavors in fish during exposure to cellulose plant effluents under controlled conditions

Type of effluent	Test fish	Exposure time (days)	Estimated tainting threshold (% v/v)	References
Unbleached kraft	Coho salmon	3 –4	1.5	[120]
Bleached kraft/sulfite	Perch	7	$>5<10$	[121, 122]
Bleached kraft/mechanical	Rainbow trout	2 –6	3	[123]
Bleached kraft	Perch	4 –6	1.5	[124]
Bleached kraft	Rainbow trout	0.2–1	0.2–0.8	[125]

Table 7. Fish flesh tainting propensities of certain major kraft mill process effluents [126]

Process effluent		Estimated tainting threshold (v/v %)
Source	% of total mill effluent	
Condensates + scrubber	8	0.007–1
Paper machine	24	>20
Bleach plant[a]	37	> 1
Whole mill	100	0.1

[a] Five consecutive stages, i.e., chlorine, caustic extraction, chlorine dioxide, caustic extraction, chlorine dioxide

tributors to off-flavours in fish caused by cellulose plant effluents [125, 127, 130, 131]; but no firm conclusions can be drawn.

Resin acids have also been suggested as contributing to off-flavours in fish flesh [16, 127] as have organosulfur compounds and monoterpenes [127, 132]; again without proof. Attempts to isolate specific chemicals from tainted fish flesh have encountered difficulties [130], albeit more recent attempts have been more successful. Positive correlations between tainting and concentrations of terpenes and their derivatives, with chlorinated compounds and with transformation products such as cymene and alkylbenzenes, have been demonstrated [131, 133].

Other Compounds

The cellulose industry utilizes a variety of raw materials together with a wide range of chemicals and large volumes of water. Apart from the compounds described previously, these various pulp and paper mill effluents additionally contain a wide variety of dissolved chemical constituents, usually in trace amounts, to which pollution characteristics may not have been specifically attributed. As shown in Table 8, these compounds embrace hydrocarbons and their chlorinated derivatives, aliphatic alcohols, carboxylic acids, phenols and related compounds, aldehydes and ketones, organosulfur compounds, together with terpenes, steroids and related compounds. It would seem probable that within this extensive list of chemical constituents of various pulping discharges will lie the answers to some of the environmental problems, associated with the cellulose industry, which are as yet not totally resolved.

Effect of Effluent Treatment

The only effluent treatments which are widely used by the cellulose processing industry involve primary treatment embracing some type of gravity separation of suspended solids and secondary treatment with microbiological oxidation of fermentable dissolved organic constituents. The technology has been developed over

Table 8. Miscellaneous chemical constituents, identified as being present in cellulose plant effluents

Chemical name	Concentration[a]	Reference
Hydrocarbons and derivatives		
1,2-Dichloroethane	8 μg/L	[134]
Dichloromethane	3–8 μg/L	[134, 135]
Dodecane	–[b]	[135]
Hexadecane	–	[135]
Pentadecane	–	[135]
1,1,2,2-Tetrachloroethane	–	[135]
Trichloroallene	0.1 g/t pulp	[136]
1,1,1-Trichloroethane	3–7 μg/L	[134]
Pentachloroethane	0.1 g/t pulp	[136]
Trichlorofluoromethane	74 μg/L	[134]
Acenapthene	1 mg/L	[41]
Chlorocymenes	–	[70]
Chlorobenzene	2 μg/L	[134]
Ethyl benzene	3 μg/L	[134]
Toluene	3 μg/L	[134, 135, 137]
Alcohols		
Butanol	–	[137]
Coniferyl alcohol	< 100 μg/L	[138]
Ethanol	500 mg/L (KC)	[137, 139]
Methanol	280–8,000 mg/L (KC)	[137, 139]
2-Methylpropanol	–	[137]
Propanol	–	[137]
Carboxylic Acids		
Acetic	300 g/100 Kg (BL)	[8, 139–142]
Arachidic	> 100 μg/L	[135, 138, 143]
Behenic	> 100 μg/L	[135, 138, 143]
Chloranilic	–	[144]
β-Chloromuconic	–	[144]
3,4-Dideoxopentonic	–	[8]
Formic	525 g/100 Kg (BL)	[8]
Gluco-isosaccharinic	963 g/100 Kg (BL)	[8, 140, 141]
Glyceric	2 g/100 Kg (BL)	[8]
Glycollic	171 g/100 Kg (BL)	[8, 140, 141]
2-Hydroxybutyric	–	[135, 140, 141]
Heptadecanoic	< 100 μg/L	[138]
Homovanillic	50–90 μg/L	[41, 138, 140, 141]
2-Hydroxypentenoic	103 g/100 Kg (BL)	[8]
Isosaccharinic	69 g/100 Kg (BL)	[8]
Lactic	302 g/100 Kg (BL)	[8, 140, 141]
Lignoceric	55 μg/L	[41, 138]
Malic	19 g/100 Kg (BL)	[8]
Malonic	3 g/100 Kg (BL)	[8]
Myristic	< 100 μg/L	[41, 138]
Oxalic	10 g/100 Kg (BL)	[8]
7-Oxodehydroabietic	< 100 μg/L	[138, 145]
Palmitic	< 100 μg/L	[41, 138, 139, 145, 146]

[a] Except as otherwise noted, concentrations represent raw effluent; KC – kraft condensates; BL – kraft black liquor; SWL – sulfite waste liquor

[b] Not measured quantitatively

Table 8 (continued)

Chemical name	Concentration[a]	References
Pentadecanoic	< 100 μg/L	[41, 138]
Stearic	25–100 μg/L	[20, 41, 135, 138–140, 143, 145, 146]
Succinic	39 g/100 Kg (BL)	[8]
Vanillic	5 μg/L	[41, 140, 141, 146]
Phenols and Related Compounds		
Acetosyringone	60–90 μg/L	[41]
Anethone	20–60 μg/L	[135, 140, 141, 147]
p-Benzoquinone	–	[144]
m-Cresol	–	
o-Cresol	–	[135, 137]
p-Cresol	–	
2,5-Dichloro-3,6-disulfo-hydroquinone	–	[143, 144, 148]
Dimethylphthalate	< 100 μg/L	[135, 138]
3,4-Dimethoxyacetophenone	6 μg/L	[135, 145, 146, 149]
Dioctyl phthalate	< 100 μg/L	[135, 138, 145]
Monochloropropiovanillone	–	[150]
Phenol	< 100 μg/L	[135, 137, 138, 140, 143, 148]
Safrole	< 100 μg/L	[135, 137, 138, 140, 143, 148]
Syringol	4–10 mg/L (KC)	[137]
Tetrachloro-*o*-benzoquinone	–	[150]
Trimethoxychlorobenzene	0.1 g/t pulp	[136]
Dimethoxy-2,4,6-trichlorophenol	–	[34]
Aldehydes, Ketones		
Acetone	3.5 mg/L (SWL)	[135, 137, 139]
Benzaldehyde	2–5 μg/L	[41, 135]
2-Butanone	–	[135, 137]
Chlorobenzaldehyde	0.1 g/t pulp	[136]
Furfural	200–270 mg/L (SWL)	[139]
p-Hydroxybenzaldehyde	0.6–0.9 mg/L	[41, 135, 140]
4-Hydroxy-3-methoxy-propiophenone	60 μg/L	[41]
3-Methyl-2-butanone	–	[137]
4-Methyl-2-pentanone	–	[137]
3-Pentanone	–	[137]
Veratraldehyde	1.5–2 mg/L	[8, 41, 149]
Organo-sulfur		
Thiophene	10–30 μg/L	[41, 135, 137]
Terpenes, Steroids and Related Compounds		
Borneol	10–100 μg/L	[41, 135, 138, 140, 141, 151, 152]
Camphene	30–40 μg/L	[135, 137, 138, 141, 151, 152]
Camphor	40–400 μg/L	[41, 135, 140, 144, 148]
Δ3-Carene	30–40 μg/L	[135–137, 141]
p-Cymene	–	[141, 147]
1,8-Cineole	–	[135, 141, 147]
d-Fenchone	40–250 μg/L	[41, 137, 140, 141, 147]
Fenchyl alcohol	0.1–0.6 mg/L	[41, 137, 140, 141, 147]
Limonene	20–200 μg/L	[135, 137, 141, 147]

Table 8 (continued)

Chemical name	Concentration[a]	References
l-Linolool	100–4,500 mg/L (KC)	[135, 137]
2-Methylfuran	100–4,500 mg/L (KC)	[137]
Myrcene	10–200 µg/L	[41, 135, 137, 140, 141, 147, 151, 152]
α-Phellandrene	30–300 µg/L	[135, 137, 141, 147]
β-Phellandrene	30–300 µg/L	[135, 137, 141, 147]
α-Pinene	30–500 µg/L	[135, 137, 138, 140, 141, 147, 149, 152]
β-Pinene	0.01–1.4 mg/L	[41, 135, 140, 141]
Sabinene	10–50 µg/L	[41, 135, 140, 141, 147, 151, 152]
β-Sitosterol	> 100 µg/L	[135, 138, 146]
α-Terpinene	100–4,500 mg/L (KC)	[137, 151, 152]
β-Terpinene	100–4,500 mg/L (KC)	[137, 151, 152]
δ-Terpinene	100–4,500 mg/L (KC)	[137, 151, 152]
Terpinolene	40 µg/L	[135, 137, 140, 141, 147]
Terpineol	30–50 µg/L	[41, 135, 140, 141]
4(p-Tolyl)-1-pentanol	100–4,500 mg/L (KC)	[137]

the last century for the treatment of domestic wastes; and virtually all variations of primary and secondary treatment developed therefore [153, 154] are applicable to cellulose plant wastes.

For secondary treatment to be effective, some specific considerations pertain. Excepting paper mill wastes, untreated cellulose plant effluents are seldom within the biological pH range, i.e., 5.0 to 8.0, and pH adjustment is normally necessary. Indeed, abnormal pH has been cited as responsible for as much as 75% of the observed acute lethality (to fish) of bleached kraft effluents [155]. Bleached chemical and mechanical effluents are acidic, ranging from pH values of around 2.0 for bleached cellulose plant effluents, to pH 4–5 for mechanical effluents. Industrial neutralization of pH is usually with burnt lime. In the bleached kraft process, two-stage neutralization is commonly practised, utilizing first waste lime mud (calcium carbonate) from the recaustizing section of the plant, followed by burnt lime. Unbleached kraft effluents are highly alkaline and where not admixed with bleachery wastes, commercial grade sulfuric acid is normally used for neutralization. The addition of microbiological nutrients may be required, depending on the residence time of the effluent in the secondary treatment section. For activated sludge systems with retention times of 4–8 h, nutrients are normally adjusted to a $\Delta BOD_5/N/P$ ratio of 100:5:1, whereas for retention times exceeding 5 d, nutrient adjustment is not required [156].

Secondary treatment of cellulose plant wastes is keyed to the removal of BOD_5. Well designed, well operated plants remove BOD_5 (or BOD_7) effectively, as is borne out by records maintained by the many mills and regulatory agencies involved (e.g., 157, 158]. However, secondary treatment systems do not oxidize the solubilized lignin residues and consequently do not reduce the color level of the effluent. For unbleached chemical pulping effluent, these unattached lignin residues can represent 75% of the total chemical oxygen demand of the effluent

[10]. The solubilized extractives and extraneous chemicals formed during pulping and bleaching are removed, in varying degree.

Evidence is limited concerning the persistence throughout secondary treatment, of extractives and extraneous molecules formed during pulping and bleaching. Although attention has focused primarily on those compounds implicated in some manner with an environmental impact, a few studies have been more generalized in nature [41, 140, 141]. Groups of simple chemicals reported as surviving industrial secondary treatment, to which no specific environmental impact is attributed other than constituting part of the residual BOD, include: aromatic hydrocarbons (acenapthene, thiophene); fatty acids (myristic, palmitic, pentadenoic, 2-hydroxybutyric); lignin degradation products (homovanillic acid, lignoceric acid, vanillic acid, acetosyringone, anethole, guaiacol, *p*-hydroxybenzaldehyde, 4-hydroxy-3-methoxypropiophenone, and veratraldehyde); and terpenoids and related compounds (borneol, camphene, camphor, Δ3-carene, *p*-cymene, *d*-fenchone, fenchyl alcohol, limonene, myrcene, α-phellandrene, α-pinene, sabinene, and terpinolene). Because of the highly variable nature of the wood furnish and of the operation and effectiveness of systems providing secondary treatment of cellulose plant wastes, it would not be logical that all of the above compounds would necessarily be found in any specific treated effluent, or conversely that these are necessarily the only compounds which may be present.

More specific information is available concerning the fate, during secondary treatment, of those compounds implicated in the acute lethal toxicity of cellulose plant effluents. Model fermentation systems, coupling microbiological oxidation of toxic molecular species with more readily assimilable molecules such as the simple sugars, indicated that all the toxicants are biodegradable [159–161], albeit with differing susceptibilities. For the major toxic compounds the specific degradation rates were as shown in Table 9. The two compounds most resistant to biooxidation were dichlorodehydroabietic acid and trichloroguaiacol. The biooxidation rate shown for tetrachloroguaiacol is that attained by the microorganisms after an induction period of about 5 d. By contrast, specific degradation rates for the BOD_5 loading (glucose, xylose, acetic acid) utilized in the model sys-

Table 9. Estimated removal of toxic compounds in cellulose plant effluents by secondary treatment systems [159, 160]

Toxic compound	Estimated effluent concentration (mg/L)	Specific degradation rate (µg/mg biomass/d)	Capacity[a] loading
Dehydroabietic acid	3	18	4.5– 6.0
Pimaric acid	1.4	10.5	5.5– 7.5
Tetrachloroguaiacol	0.3	8.4	21 –28
Pimarol	–	7.0	–
Monochlorodehydroabietic acid	0.6	4.7	5.9– 7.8
Dichlorodehydroabietic acid	1.1	1.7	1.2– 1.5
Trichloroguaiacol	0.4	0.8	1.5– 2.0

[a] Based on typical levels of microbiological solids in 6-h activated sludge and 5-d aerated stabilization basin systems, i.e., this ratio is independent of retention time

tems were more than 100-fold greater than the rate of removal of the toxic compounds [160].

On this basis, well designed, well operated secondary treatment systems have the potential to effectively remove all toxic compounds responsible for acute lethal toxic effects. Of the compounds listed in Table 9 the capacity of biotreatment systems to handle dichlorodehydroabietic acid and trichloroguaiacol is only marginally greater than their conventional loadings, i.e., capacity/loading ratio approaches unity. Thus, in actual treatment situations, dichlorodehydroabietic acid and trichloroguaiacol are the most likely compounds to "break through" into effluent discharges to receiving waters.

Ample evidence exists to demonstrate that, for various reasons, industrial biotreatment systems fail to detoxify cellulose plant effluents an appreciable proportion of the time [29, 31, 157, 162–164]. Table 10 shows concentrations of resin acids and of fatty acids, as found in the effluent from chemical and mechanical pulping plants, after biotreatment on an industrial scale. The upper range of concentrations corresponding to the individual resin and fatty acids in most instances exceeds the 96-h LC50 values reported also in Table 10. That is, some of the effluents were undoubtedly acutely lethal to test fish. However, the lower range of concentrations of these acids found in kraft, sulfite, and mechanical pulping effluents is extremely low, in many instances below the limit of detection of the analytical methods which were employed. Comparison of these data with those in Table 3 illustrate an overall drop in resin and fatty acid concentrations attributable to biotreatment. Although the data are not segregated according to the type of biotreatment system utilized, or with respect to operating effectiveness for BOD_5 removal, intrinsic differences in removal of toxicants by different systems (provided that they are well operated), is small [162, 163]. The lability of other toxic extractives is greater than that of the resin and fatty acids and, although it is assumed that they do not survive biotreatment, limited direct proof is available [29, 30].

The concentrations of extraneous chemical species, formed during pulping and bleaching, which persist through biotreatment, are shown in Table 11. In all

Table 10. Range of concentrations (μg/L) of resin and fatty acids, found in mechanical pulping and chemical pulping effluents after biotreatment

Acid	96-h LC50 (μg/L)	Kraft	Sulfite	Mechanical	References
Abietic	700–1,500	<10–3,630	<10– 500	14–4,200	[30–32, 34, 35]
Isopimaric	40– 100	<10–1,420	>10–300	12–7,900	[25, 30–32, 35]
Levopimaric	700–1,000	< 1–2,700	<10–310	11–1,800	[31, 32]
Neoabietic	610– 730	< 1–1,200	<10–200	< 1–3,800	[31, 34]
Palustric	500– 600	80	ND	ND	[34]
Pimaric	700–1,200	14–1,830	<20– 20	< 1–5,700	[30, 34, 35]
Linoleic	1,500–3,400	<20– 50	<10– 50	23–1,500	[30, 31, 34]
Linolenic	2,000–6,000	10– 30	20		[30, 34]
Oleic	3,500–8,200	20–2,340	10–370	24–1,400	[30, 31, 34, 35]

ND – Not detected

Table 11. Range of concentrations (μg/L) of extraneous toxic compounds formed as a result of pulping and bleaching, found in bleached chemical plant effluents after biotreatment

Chemical species	96-h LC50 (μg/L)	Kraft	Sulfite	References
2,4-Dichlorophenol	2,800	2– 51	2– 8	[35, 61, 66, 67, 69]
Dichlorocatechol	500–1,000	1–120	–	[34, 65–67, 69]
Dichloroguaiacols	2,300	12– 60	2– 7	[34, 67–69]
2,4,6-Trichlorophenol	450–2,600	< 1– 61	3– 30	[34, 35, 61, 66, 67, 69, 70, 73, 74]
3,4,5-Trichlorocatechol	1,000–1,500	2–280	4– 9	[34, 66, 67, 69]
Trichloroguaiacols	700–1,500	< 1–220	6– 60	[30, 31, 34, 35, 42, 61, 66, 67, 69, 73]
Tetrachlorocatechol	400–1,500	2–240	2– 5	[34, 65–67, 69, 71]
Tetrachloroguaiacol	200–1,700	1–120	1– 80	[30, 31, 34, 35, 61, 66, 67, 69]
Chlorodehydroabietic acid	600– 900	< 1–260	<10–450	[30, 31, 34, 35, 61]
Dichlorodehydroabietic acid	600–1,200	<10–152	<10– 30	[30, 31, 34, 35]
Dichlorostearic acid	2,500	<40–268	<40	[30, 38]
Epoxystearic acid	1,500–3,400	<40	<40	[29, 30, 60]

instances, the range of concentrations found in biotreated effluents is a small fraction of the acute lethal value (96-h LC50), and the lower end of the concentration range is virtually nil. Comparison of these data (Table 11) with those in Table 4 illustrates the efficacy with which these extraneous chemical species are removed by biotreatment. For the major toxicants, such as the chlorinated dehydroabietic acids, dichlorostearic acid, and the tetrachlorinated guaiacol (kraft) and catechol (sulfite) substantial removals were effected. The data indicate a lesser removal of the trichlorinated compounds, whereas removal of the dichlorinated molecular species was marginal albeit this latter class of compounds were minor toxicants.

The "Priority Pollutants" program, undertaken on behalf of the U.S. Environmental Protection Agency, included (in the analytical program involving the pulp and paper industry), analyses for most of the toxic compounds found in the mill discharges [165]. This massive program embraced 163 establishments, divided into 23 industry sub-categories, with 130 of the 163 plants operating biotreatment systems. A brief summary of the data from the bleached kraft and sulfite plants is presented in Table 12. All of the data pertain to effluents which received biotreatment. For virtually all constituent chemicals, the upper end of the concentration range pertains to values for the three dissolving kraft and four dissolving sulfite pulp mills, which were included in the study. Biotreatment resulted in marked diminution of the concentration of individual toxicant molecules, particularly for the major toxicants – the resin and fatty acids and their derivatives. Removals of the chlorinated phenols and guaiacols were almost as effective, although the concentrations of these compounds, before biotreatment, were generally much lower than for those studies reported in Tables 4 and 10. The overall data confirm the efficacy with which these molecules are degraded in well designed, well operated biotreatment systems. The data should not necessarily be

Table 12. Range of concentrations (μg/L) of various compounds found in effluent from bleached kraft and sulfite cellulose plants, before and after biotreatment [165]

Chemical	Bleached kraft					Bleached sulfite				
	No.[a]	Before		After		No.[a]	Before		After	
		Ave	Range	Ave	Range		Ave	Range	Ave	Range
Abietic acid	18	2,547	0–18,000	560	0–2,500	16	590	0–5,200	153	0– 940
Dehydroabietic acid	18	1,091	10– 5,200	292	0–1,000	16	598	2–1,870	227	0– 950
Isopimaric acid	18	240	0– 1,300	210	0– 590	16	240	0–1,760	42	0– 230
Pimaric acid	18	336	0– 1,900	273	0– 790	16	277	0– 450	21	0– 52
Oleic acid	18	1,326	0– 4,500	115	0– 810	16	387	14–1,860	77	13– 220
Linoleic acid	18	1,128	180– 3,900	46	0– 510	16	175	8–1,000	67	0– 160
Linolenic acid	6	70	0– 210	0	0	12	58	0– 130	0	0
Epoxystearic acid	–	–	–	–	–	15	494	0– 850	23	0
Chlorodehydroabietic acid	18	295	0– 1,300	98	0– 700	16	133	8– 360	56	0– 241
Dichlorodehydroabietic acid	15	25	0– 86	16	0– 65	16	25	0– 280	1	0– 30
Trichloroguaiacol	15	8	0– 21	0	0	16	5	2– 6	1	0– 2
Tetrachloroguaiacol	15	9	2– 23	1	0– 1	16	2	2– 4	1	0– 2
1,1-Dichloroethane	–	–	–	–	–	12	12	5– 22	0	0
2,4,6-Trichlorophenol	15	9	0– 26	3	0– 8	16	139	7– 370	81	1– 270
Chloroform	18	1,351	360– 4,000	18	0– 86	16	2,075	62–8,600	328	1–1,200
2-Chlorophenol	–	–	–	–	–	12	65	0– 120	27	21– 50
2,4-Dichlorophenol	15	3	0– 8	2	0– 8	16	78	0– 220	40	0– 130
Ethylbenzene	15	11	0– 82	0	0– 3	–	–	–	–	–
Fluoranthrene	3	2	0– 7	0	0	4	1	0– 4	1	0– 1

[a] Number of mills sampled

construed as demonstrating the biolability of such compounds as ethylbenzene and fluoranthrene, inasmuch as initially they are present in trace concentrations, at best.

The propensity of cellulose plant effluents to taint fish flesh is reduced by biotreatment [125]. Since limited data are available linking the tainting properties of effluents with the responsible chemical species, the effect of biotreatment thereon is not known.

Biotreatment of cellulose plant effluents showed a reduction in mutagenic activity in one instance [103, 104], albeit effluents are more typically non-mutagenic, even without treatment.

Environmental Fate of Organic Constituents

Monitoring of any deleterious effects of cellulose plant effluents on the flora and fauna within receiving waters is, by nature, primarily biological in function. Typical biological monitoring procedures require the retrospective examination of population numbers and diversity of biological communities in the path of the effluent plume and in a closely similar control area, away from any influence of the discharge [166, 167]. Crude discharge monitoring techniques such as the 96-h LC50 fish bioassay for toxicity are being replaced by a battery of short term, more sensitive sublethal and/or lethal techniques involving a variety of indigenous organisms, either *in situ* or in waters approximating receiving waters in their composition. Numerous reviews are available concerning these procedures which have been developed over the last twenty years, and in some instances are still under development [4, 56, 168–172].

Recent advances in analytical chemistry, particularly the widespread availability of coupled GC-MS systems, have permitted the detection of chemicals in environmental situations, at trace concentrations not heretofore considered feasible. With cellulose plant effluents attention has focused not only on the compounds identified as toxicants, but more especially on the chlorinated organics, because of the known persistence in the environment of chlorinated hydrocarbons such as DDT, PCB's, and the chlorinated phenols [173–175]. Also, the insertion of chlorine into an organic molecule enhances its fat solubility and thereby enhances its potential for bioaccumulation in animal tissues [176, 177].

The concentrations of various resin and fatty acids, and chlorinated phenolics downstream of two cellulose plants discharging into freshwater situations are shown in Table 13. In the lacustrine discharge, concentrations of resin and fatty acids were consistently lower than normal effluent concentrations (Table 3), at 0.8 km distance. In the riverine situation, the chlorinated phenols, although reduced substantially from normal effluent concentrations (Table 4) some 3 km downstream, were still detectable some 50 km distant. The results typify the concentrations associated with biotreated effluents and the continuing diminution in concentration resulting from naturally occurring biooxidation and from dilution in receiving fresh water situations.

In marine receiving waters, dilution of cellulose plant discharges is almost invariably greater than for corresponding fresh water situations. For this reason,

Table 13. Concentrations of resin and fatty acids and chlorophenols at freshwater sites distant from the outfall of the biotreated discharge from bleached kraft pulp mills [69, 178]

Resin/fatty acid	Receiving water concentration (μg/L)	
	0.1 km	0.8 km
Abietic	1–114	2– 7
Dehydroabietic	6–600	4–10
Isopimaric	2– 79	2– 3
Neoabietic	1– 10	<1
Pimaric	1– 67	2
Oleic	12–114	8–14
Linoleic	2– 54	3–10
Linolenic	< 3– 25	<2
Chlorophenols	**3 km**	**50 km**
2,4-Dichlorophenol	0.1	0.02
4,5-Dichloroguaiacol	0.08	<0.02
2,4,6-Trichlorophenol	0.09	0.04
Trichlorocatechol	0.4	0.009
Trichloroguaiacols	1.0	0.2
Tetrachlorocatechol	0.8	0.02
Tetrachloroguaiacol	0.3	0.09

Table 14. Composite data representing dispersion of chlorophenols in estuarine/marine waters at various distances from the untreated outfalls of three bleached kraft mills [69, 81, 179]

Chlorophenol	Range		Not detected Distance – km
	Distance km	Concn. μg/L	
4,5-Dichlorocatechol	0.25	0.06	0.72– 7.0
4,5-Dichloroguaiacol	0.16–0.25	0.02 –0.05	1.7 –12.5
2,4-Dichlorophenol	2 –8	0,002–0.28	8 –12.5
Tetrachlorocatechol	0.16–2.6	0.005–0.08	3.2 –12.5
Tetrachloroguaiacol	0.16–7.0	0.005–1.30	12.0 –15.0
Trichlorocatechol	0.16–7.0	0.004–1.77	12.5 –15.0
2,4,6-Trichlorophenol	0.16–7.0	0.003–0.90	12 –12.5

biotreatment of plant discharges frequently is not undertaken. Typical concentrations of chlorinated phenols, representing the more persistent chemicals to be found in bleached kraft discharges within the effluent plume and at various distances from the marine/estuarine outfall location, are shown in Table 14. Similar data have not been reported for the resin acids, chlorinated or otherwise. The concentrations of phenols are substantially reduced from normal discharge concentrations in virtually all instances (Table 14). However, the slow rate of biodegradation of these compounds and the initial concentrations of chemical groups such as the trichloro-catechols and guaiacols and tetrachloroguaiacol resulted in detectable concentrations at considerable distances from the mill outfall. Following the effluent plume is difficult in studies of this nature and in one study [179] the effluent plume was tracked by the dispersion of the chloroform levels contained therein.

Concentrations of the more persistent chemical entities found in cellulose plant effluents have been found in receiving water sediments in the environs of the mills. As the more stable of the resin acids, dehydroabietic acid has been found at levels 10 times above background (10 μg/g) at 15 km from a mill with a lacustrine discharge [180]. Abietic, pimaric, an abietanoic and an abietenoic acid were found in surficial sediment layers, adjacent to the same mill [145, 181]. Dehydroabietic acid has also been reported in sediments adjacent to marine/estuarine kraft outfalls [116, 118], but with few details.

Chlorophenols were found in riverine sediments 40–50 km downstream of the nearest mill [168, 183], ableit concentrations were in the 0.0007–0.01 μg/g range. Surprisingly, concentrations of tetrachlorocatechol were greater than those of the polychlorinated guaiacols, even though the concentration of the latter in secondary treated bleached kraft effluents is accepted to be the greater [69, 183, 184]. A similar situation existed with regard to concentrations of 2,4,6-trichlorophenol in the river sediment. Chlorophenol concentrations in marine/estuarine sediments varied with foreshore topography; higher values being reported by Swedish investigators [81, 179, 185] than by their Canadian counterparts [69]. Sediment concentrations in the one instance ranged from 0.001 to 0.08 μg/g over distances 2–10 km from mill outfalls and from 0.003 to 0.01 μg/g, 1.6–12.5 km from mill outfalls. Highest values were associated with the trichlorocatechols and guaiacols [69, 185].

Environmental concern about the persistence of chemical constituents in cellulose plant discharges in receiving waters and sediments stems primarily from their possible bioaccumulation in aquatic fauna, their potential for bioconcentration, their biomagnification in aquatic species at the upper end of the food chain, and any detrimental effects arising therefrom. Factors which affect the potential for bioaccumulation and biomagnification of a specific molecule include its resistance to microbial degradation and the partition of the chemical between water and animal fatty tissues. Lipophilic properties of organic molecules are enhanced by the substitution of chlorine into hydrocarbon molecules and by the absence of hydrophilic groups, whereas substitution of chlorine and the presence of aromatic nuclei detract from biodegradability. Thus, among the specific constituents in cellulose plant discharges, the chlorinated phenols offer the greatest intrinsic potential for bioaccumulation, albeit the phenolic grouping detracts from their partition to fatty tissues, and provides a potential point for microbial attack.

The concentrations of dehydroabietic acid and specific chlorinated phenols, known to be present in cellulose plant discharges, which were found in the tissues of fish exposed to solutions of these phenols under controlled conditions, are shown in Table 15. Also shown are data for some chlorinated benzenes and tetrachloroveratrole, reported to be a bacterial methylation product of the corresponding guaiacol [191]. The bioconcentration factor (BCF) is the ratio of the tis-

Table 15. Chemical bioaccumulation in aquatic organisms exposed to specific constituents of cellulose plant effluents and to some chemically related chlorinated compounds

Chemical	Organism	Tissue concentration µg/g (wet weight basis)	BCF	References
Dehydroabietic acid	Sockeye salmon[a]	19	30	[186]
		647[b]	996	[186]
		263[c]	404	[186]
2,4-Dichlorophenol	Brown trout	18	10	[118]
2,3,5-Trichlorophenol	Brown trout	6	12	[188]
4,5,6-Trichloroguaiacol	Bleaks	4	390	[189]
Tetrachloroguaiacol	Bleaks	4	400	[189]
3,5-Dichlorocatechol	Brown trout	6	2	[188]
Tetrachlorocatechol	Brown trout	10	4	[188]
1,2-Dichlorobenzene[e]	Rainbow trout	0.7	560	[190]
1,2,3,4-Tetrachlorobenzene[e]	Rainbow trout	0.3	12,000	[190]
Hexachlorobenzene[e]	Rainbow trout	0.2	20,000	[190]
Tetrachloroveratrole[e]	Zebra fish	2,300[d]	25,000	[191]

[a] For data on rainbow trout, see [187]
[b–d] Excepting these values which are µg/g bile, liver and fat respectively, all other tissue concentration values are µg/g whole body tissue
[e] Nor normal effluent constituents

Table 16. Concentrations of specific cellulose plant effluent[a] constituents contained in tissues of freshwater fish species taken at various distances from plant outfalls

Constituent	Organisms	Distance from outfall – km	Tissue	Concentration µg/g	References
Resin acids	Rainbow trout	0.8– 6	Bile	2–60	[192]
2,4,6-Trichloro-phenol	Northern pike, roach	< 5 –40	Muscle	0.001–0.056	[183, 193]
	Sucker, whitefish	50	Liver	0.015–0.033	[69]
Tetrachloro-guaiacol	Northern pike, pike	< 5 –40	Muscle	<0.001–0.007	[81, 193]
	Pike, sucker, whitefish	5	Liver	<0.019–≦0.57	[69, 81]
Tetrachloro-catechol	Northern pike	< 5 –40	Muscle	0.001–0.006	[193]
3,4,5-Trichloro-guaiacol	Pike	–	Muscle	≦0.008	[81]
	Pike, perch, whitefish, sucker	≦50	Liver	≦0.11 –≦0.36	[69, 81]
4,5,6-Trichloro-guaiacol	Roach	5	Muscle	0.047	[183]
	Pike, perch, whitefish, sucker	≦50	Liver	0.013–≦0.36	[71, 81]
Hexachloro-benzene	Pike	5 –40	Muscle	≦0.001	[193]
Chloroform	Fish (sp. ?)	≦ 2.5	Whole body	<0.001	[73]

[a] Virtually all the data pertain to bleached kraft discharges; some effluents were biotreated, the treatment (if any) of the remainder was not specified

sue concentration to that of the solution concentration. Excepting the tri- and tetrachloroguaiacol, bioconcentration of the specific constituents of cellulose plant effluents was minimal. Bioconcentration of the chlorinated benzenes was substantial, although these compounds are not normally considered to be constituents of cellulose plant effluents. The high bioconcentration factor for tetrachloroveratrole was derived on the basis of fatty tissue only. These data suggest a role for the hepato-biliary system in detoxification.

The concentration of various cellulose plant effluent constituents in fish captured in the environs of various Scandinavian and Canadian freshwater outfalls is shown in Table 16. Most discharges pertained to biotreated bleached kraft effluents, albeit in some instances no indication of treatment was provided. Concentrations in receiving waters from which the fish were taken were determined in only a limited number of instances, so that bioconcentration factors are not available. Muscle tissue concentrations for all the chlorinated phenols were low, i.e., less than 0.06 μg/g. The value for 3,4,5-trichloroguaiacol in fish muscle was only 0.047 μg/g. This chlorophenol has one of the higher concentrations in bleached kraft effluents [29–31, 165] whereas at the same time it is the most resistant to biodegradation [160]. In all instances bile and liver values are higher than muscle values, supporting a hepato-biliary mechanism for detoxification and elimination. Nonetheless, the highest value here is for the combined resin acids at 60 μg/g, whereas the highest chlorophenol value is for tetrachloroguaiacol at 0.57 μg/g. It is noteworthy that the whole-body concentration of chloroform was below the limit of detection, notwithstanding chloroform values in bleached kraft effluents of up to 1.1 mg/L [61].

Similar data for marine/estuarine outfalls are shown in Table 17, although it is assumed that effluents in these instances had not been biotreated. Concentrations of the bleached kraft effluent constituents in muscle or whole-body tissue are considerably lower than those shown in Table 16; not exceeding 0.055 μg/g and in most instances at or below the limit of analytical detection. Liver values were also low, for the most part, although individual values as high as 0.54 μg/g were recorded. Even though bioconcentration factors were not determined, in the absence of receiving water values, the absolute values suggest a minimal bioaccumulation of these compounds in fish tissues, at least at the distances from the discharging plants for which the data were obtained.

Evidence for detoxification in fish of the toxic compounds found in cellulose plant effluents support a hepato-biliary route, with the toxicants being eliminated as conjugated glucuronides. Fish exposed under controlled conditions to a resin acid mixture at a concentration equivalent to 0.08 of its 96-h LC50 value, showed elevated concentrations of the resin acids in blood plasma, bile, and in the brain, whereas at the same time liver UDP-glucuronyl transferase activity was impaired [194]. Three days after exposure more than 99% of the dehydroabietic acid found in the bile was conjugated as the glucuronide [192]. Labelled dehydroabietic acid, injected intraperitoneally, was found primarily in the bile and feces [195]. Clearing occurred rapidly. For example, zebra fish exposed to tetrachloroveratrole reached a steady state concentration in the liver of 3.50 μg/g, corresponding to a bioconcentration factor of 3,200 in this tissue. However, 80% of the chemical was lost within three days following return to uncontaminated water [191].

Table 17. Concentration of specific cellulose plant constituents contained in tissues of marine/estuarine organisms taken at various distances from plant outfalls

Constituent	Organisms	Distance from outfall – km	Tissues	Concentration µg/g	References
Chloroguaiacols	Mussels, salmon	? – 3	Muscle, whole body	ND – ≦0.055	[81]
	Fish, flounder, pike, rockfish		Liver	ND – ≦0.54	[69, 81]
3,4,5-Trichloroguaiacol	Sculpin, herring, mussels, shrimp	2 – 6	Muscle, whole body	ND – 0.04	[69, 81]
	Crab		Liver	0.03–≦1	[69, 81]
4,5,6-Trichloroguaiacol	Sculpin, bivalves	2 – 6	Whole body, liver, muscle	ND	[81]
2,4-Dichlorophenol	Sculpin, bivalves	2 – 6	Whole body, liver, muscle	ND – <0.02	[81]
2,4,6-Dichlorophenol	Sculpin, bivalves	2 – 6	Whole body, liver, muscle	ND – <0.02	[81]
	Salmon, mussels, shrimp, crab	3 –10	Muscle, whole body	ND – 0.02	[69]
	Rockfish	2.5–10	Liver	ND – 0.003	[69]
Tetrachloroguaiacol	Bivalves, sculpin	2 – 6	Whole body, liver, muscle	ND	[81]
	Herring, mussels, shrimp, crab	–	Muscle, whole body	ND – 0.008	[69]
	Salmon	3	Liver	0.01– 0.02	[69]
2,4,5-Trichlorophenol	Flounder	–	Liver	≦0.54	[81]
Chlorophenols	Bivalves, herring	–	Whole body, muscle	≦0.02–≦0.006	[81]
3,4,5-Trichloroguaiacol	Crab	2 – 3.5	Whole body	ND – 0.005	[69]
Dehydroabietic acid	Clam, tomcod	–	Whole body, muscle	0.3	[182]

ND – Not detected

Definitive data are lacking which link the design of individual cellulose plants with quantities of chemical constituents contained in their effluents; which document the survival of these effluent constituents through secondary treatment plants of differing design and variable efficiency of operation; which demonstrate the dispersion and dilution offered by various receiving water configurations; and finally which link all these factors to the uptake of the residual chemical constituents by aquatic fauna. However, available data do suggest that, apart from obvious overloading of the receiving water environment with respect to one or more parameters of a cellulose plant discharge, technology is available, within or external to the plant, for adequate environmental protection. Various chemicals shown to be present in cellulose plant discharges and known to have an environmental impact under certain time-concentration relationships have been identified in aquatic fauna. Nonetheless, the concentration of these chemical compounds in aquatic fauna is low in most instances and appears to be within the detoxifying capability of the organisms involved. Present data do not permit any firm conclusions relevant to undesirable environmental impacts in these situations. Certainly our technology has now advanced to the degree that we can detect these chemical constituents of cellulose plant discharges in an increasing variety of environmental situations. However, further studies are required before any firm biological relevance can be assigned to these chemicals, within the environment.

References

1. Casey, J.P. (ed.): Pulp and Paper Chemistry and Chemical Technology, 3rd ed., Vol. I–III, John Wiley & Sons, New York, NY (1980)
2. Macdonald, R.G., Franklin, J.N. (eds.): Pulp and Paper Manufacture, 2nd ed., Vol. I–III, McGraw-Hill, New York, NY (1965)
3. Beeland, G.V., Whitenight, D.K., Barnhill, K.G.: Environmental impact assessment guidelines for new source pulp and paper mills, EPA Publ. No. 13016-79-002, US Environmental Protection Agency, Washington, DC (1979)
4. Walden, C.C.: Water Res. *10*, 639 (1976)
5. McCubbin, N.: State-of-the-art of the pulp and paper industry and its environmental protection practices, EPS Publ. No. 3-EPS-84-2, Environment Canada, Ottawa, Ont. (1984)
6. Anon. Oxygen demand (biochemical), in: Standard Methods for the Examination of Water and Wastewater, 15th ed., American Public Health Association, Washington, DC (1980), pp. 483–489
7. Walden, C.C., Birkbeck, A.E., Mueller, J.C.: Pulp Paper Mag. Can. *73*, T153 (1972)
8. Lowendahl, L., Petersson, G., Samuelson, D.: TAPPI *59*(9) 118 (1979)
9. Roze, I.M., Kalnina, V.K., Vedernikov, N.A.: Khim. Drev. *4*, 66 (1976)
10. NCASI.: COD and BOD relationships of raw and biologically treated kraft mill effluents, Tech. Bull. No. 193, National Council Air Stream Improvement, Inc., New York, NY (1966)
11. Jackson, M., Ryberg, G., Dannielsson, O.: Pulp Paper *15*(10), 114 (1981)
12. Anon.: Bioassay procedures for fish (TENTATIVE), in: Standard Methods for the Examination of Water and Wastewater, 15th ed., American Public Health Association, Washington, DC (1980), pp. 723–743
13. Sprague, J.B.: Water Res. *3*, 793 (1969)
14. Walden, C.C., McLeay, D.J., Monteith, D.D.: Pulp Paper Mag. Can. *76*, T130 (1975)
15. Ebeling, G.: Vom Wasser *5*, 1982 (1931)
16. Hagman, N.: Paperi ja Puu *18*(1), 32 (1936)
17. Van Horn, W.M., Anderson, J.B., Katz, M.: Trans. Am. Fish. Soc. *79*, 55 (1949)

18. Mäenpää, R., Hynninen, P., Tikla, J.: Paperi ja Puu *4*(a), 143 (1968)
19. Rogers, I.H.: Pulp Paper Can. *74*, T303 (1973)
20. Leach, J.M., Thakore, A.N.: J. Fish. Res. Board Can. *30*, 479 (1973)
21. Leach, J.M., Gietz, W.C., Thakore, A.N.: Identification and treatment of the toxic materials in pulp and paper woodroom effluents, CPAR Rep. No. 148-2, Canadian Forestry Service, Ottawa, Ont. (1974)
22. McKague, A.B.: Identification and treatment of the toxic materials in pulp and paper woodroom effluents, CPAR Rep. No. 148-3, Canadian Forestry Service, Ottawa, Ont. (1975)
23. Wilson, M.A., Chappel, C.I.: Reduction of toxicity of sulphite effluents, CPAR Rep. No. 49-2, Canadian Forestry Service, Ottawa, Ont. (1975)
24. Nelson, P.J., Hemmingway, R.W.: TAPPI *54*, 968 (1971)
25. Howard, T.E., Leach, J.M.: Identification of toxic materials in sulphite pulp mill effluents, CPAR Rep. No. 407-3, Environmental Protection Service, Ottawa, Ont. (1978)
26. Row, R., Cook, W.H.: Resin acid soaps – toxicity and treatability, Proc. 6th Air Stream Improvement Conf., Tech. Sect., Canadian Pulp Paper Association, Quebec City, Quebec (1971)
27. Howard, T.E., Leach, J.M.: Identification and treatment of the toxic materials in mechanical pulping effluents, CPAR Rep. No. 149-1, Canadian Forestry Service, Ottawa, Ont. (1973)
28. Leach, J.M., Thakore, A.N.: TAPPI *59*(2), 129 (1976)
29. Chung, L.T.K., Meier, H.P., Leach, J.M.: TAPPI *62*(12), 71 (1979)
30. Easty, D.B., Borchardt, L.G., Wabers, B.A.: Removal of wood-derived toxics from pulping and bleaching wastes, EPA Publ. No. 600/2-78-013, U.S. Environmental Protection Agency, Cincinnati, OH (1978)
31. Leach, J.M., Chung, L.T.K.: Development of a chemical toxicity assay for pulp mill effluents, EPA Publ. No. 600/2-80-206, U.S. Environmental Protection Agency, Cincinnati, OH (1980)
32. Willard, H.K.: Toxicity reduction of pulp and paper mill waste water, EPA Contract Rep. No. 68-03-3028, WA20, U.S. Environmental Protection Agency, Cincinnati, OH (1983)
33. Davis, J.C., Hoos, R.A.W.: J. Fish. Res. Board Can. *32*, 411 (1975)
34. Holmbom, B.R., Lehtinen, K.-J.: Pap. Puu *62*, 673 (1980)
35. NCASI: Experience with the analysis of EPA's organic priority pollutants and compounds characteristic of pulp mill effluents, Tech. Bull. No. 343, National Council Air Stream Improvement, Inc., New York, NY (1981)
36. Walden, C.C., Howard, T.E.: Identification and treatment of toxic materials in mechanical pulping effluents, CPAR Rep. No. 149-2, Environment Canada, Ottawa, Ont. (1974)
37. Walden, C.C., Lockhart, R.W., Leach, J.M.: Identification of the toxic materials in sulphite pulp mill effluents, CPAR Rep. No. 407-1, Environment Canada, Ottawa, Ont. (1976)
38. Leach, J.M., Thakore, A.N.: Prog. Water Technol. *9*, 787 (1977)
39. Walden, C.C., Leach, J.M.: Identification and treatment of toxic materials in mechanical pulping effluents, CPAR Rep. No. 149-3, Environment Canada, Ottawa, Ont. (1975)
40. Richardson, D.E., Bloom, H.: APPITA *36*, 456 (1983)
41. Keith, L.: Environ. Sci. Technol. *10*, 555 (1976)
42. Ball, J., Priznar, F., Peterman, P.: Investigation of chlorinated and nonchlorinated compounds in the Lower Fox River watershed, EPA Publ. No. 905/3-78-004, U.S. Environmental Protection Agency, Chicago, IL (1978)
43. Bergstrom, H., Vallin, S.: The contamination of water by waste liquors of sulphate pulp mills, Medd. Statens Undersöken-Försöksants Sötvattenfisket, Kgl. Lantsbruksstyrelson No. 13 (1937)
44. Van Horn, W.M.: The toxicity of kraft pulping wastes to typical fish food organisms, Tech. Bull. No. 10, National Council Air Stream Improvement, Inc., New York, NY (1947)
45. McKague, A.B., Leach, J.M., Soniassy, R.N., Thakore, A.N.: Trans. Tech. Sect., Can. Pulp Paper Assoc. *3*, TR75 (1977)
47. Rogers, I.H., Mahood, J., Gordon, R.: Pulp Paper Can, *80*, T286 (1979)
47. Marvell, E.N., Wiman, R.: J. Org. Chem. *28*, 1542 (1963)
48. Lockhart, R.W., Leach, J.M.: Identification of toxic materials in sulphite pulp mill effluents, CPAR Rep. No. 407-2, Canadian Forestry Service, Ottawa, Ont. (1977)

49. Cole, A.E.: J. Pharmac. Expl. Ther. *54*, 448 (1935)
50. Van Horn, W.M.: A study of the toxic components of the waste waters of five typical kraft mills, Tech. Bull. No. 16, National Council Air Stream Improvement, Inc., New York, NY (1948)
51. Van Horn, W.M., Anderson, J.B., Katz, M.: TAPPI *33*, 209 (1950)
52. Adelman, I.R., Smith, Jr., L.L.: J. Fish. Res. Board. Can. *29*, 1309 (1972)
53. Smith, Jr., L.L., Oseid, D.M.: Water Res. *6*, 711 (1972)
54. Smith, Jr., L.L., Oseid, D.M.: Prog. Water Technol. *7*, 559 (1975)
55. Haydu, E.P., Amberg, H.R., Dimick, R.E.: TAPPI *48*, 136 (1952)
56. Walden, C.C., Howard, T.E.: The nature and magnitude of the effect of kraft mill effluents on salmon, Proc. Int. Symp. Ident. Measurement of Environ. Pollutants, National Research Council Canada, Ottawa, Ont. (1971), pp. 363–369
57. Seppovaara, O., Hynninen, P.: Paperi ja Puu – Papper och Tra *52*, 11 (1970)
58. Seppovaara, O.: Suom. Kemistilehri A, *44*, 5 (1971)
59. Ng, K.S., Mueller, J.C., Walden, C.C.: Study of foam separation as a means of detoxifying bleached kraft mill effluents, removing suspended solids and enhancing biotreatability, CPAR Rep. No. 233-1 (Part 1), Canadian Forestry Service, Ottawa, Ont. (1974)
60. Leach, J.M., Thakore, A.N.: J. Fish. Res. Board Can. *32*, 1249 (1975)
61. Claeys, R.R., La Fleur, L.E., Borton, D.L.: Chlorinated organics in bleach plant effluents of pulp and paper mills, in: Water Chlorination, Environmental Impact and Health Effects (R.J. Jolley, W.A. Brungs, R.B. Cumming, eds.), Vol. III, Ann Arbor Science, Ann Arbor, MI (1980), pp. 335–345
62. McKague, A.B., Walden, C.C.: Identification of the toxic constituents in kraft mill bleach plant effluents, CPAR Rep. No. 245-6, Environment Canada, Ottawa, Ont. (1979)
63. Rogers, I.H., Keith, L.H.: Organochlorine compounds in kraft bleaching wastes. Identification of two chlorinated guaiacols. Fish. Marine Service Res. Dev. Tech. Rep. No. 465, Pacific Environmental Laboratory, West Vancouver, B.C. (1974)
64. Rogers, I.H., Keith, L.H.: Identification of two chlorinated guaiacols in kraft bleaching waste waters, in: Identification and Analysis of Organic Pollutants in Water, (L.H. Keith, ed.), Ann Arbor Science, Ann Arbor, MI (1976), pp. 625–639
65. McKague, A.B.: Can. J. Fish. Aquat. Sci. *38*, 739 (1981)
66. NCASI: Experience with the analysis of pulp mill effluents for chlorinated phenols using an acetic anhydride derivatization procedure, Tech. Bull. No. 347, National Council Air Stream Improvement, Inc., New York, NY (1981)
67. Kovacs, T.G., Voss, R.H., Wong, A.: Water Res. *18*, 911 (1984)
68. Voss, R.H., Wearing, J.T., Mortimer, R.D., Kovacs, T., Wong, A.: Paperi Puu *62*, 809 (1980)
69. Voss, R.H., Yunker, M.B.: A study of chlorinated phenolics discharged into kraft mill receiving waters, Report prepared for Council of Forest Industries of British Columbia, Vancouver, B.C. (1983)
70. Leuenberger, C., Giger, W., Coney, R., Graydon, J., Molnar-Kubica, E.: Water Res. *19*, (in press) (1985)
71. Voss, R.H., Wearing, J.T., Wong, A.: Pulp Paper Can. *82*, T65 (1981)
72. McKague, A.B.: J. Chromatog. *208*, 287 (1981)
73. Anon.: Toxicity and environmental chemistry of wastewater from Proctor and Gamble Cellulose Ltd. (Grand Prairie), Res. Rep. AECV 84-86, Alberta Environmental Research Centre, Vegreville, Alta. (1984)
74. NCASI: Supplemental data reflective of available technological capability for separation of chlorinated organics from pulp and paper industry wastewaters, Special Rep. No. 82-1, National Council Air Stream Improvement, Inc., New York, NY (1982)
75. Brebion, B., Chopin, J., Humbert, F.: Chem. Ind. *77*, 1110 (1957)
76. Chopin, J.: Papetiere Bull. No. *3*, 147 (1959)
77. Voss, R.H., Wearing, J.T., Wong, A.: A novel gas chromatographic method for the analysis of chlorinated phenolics in pulp mill effluents, in: Advances in the Identification and Analysis of Organic Pollutants in Water, Vol. II (L.H. Keith, ed.), Ann Arbor Science, Ann Arbor, MI, (1981), pp. 1059–1095
78. Voss, R.H., Wearing, J.T., Wong, A.: TAPPI *64*(3), 167 (1981)

79. Kutney, G.W., Holton, H.H., Andrews, D.H., du Manoir, J.R., Donnini, G.P.: Pulp Paper Can. *85*, T95 (1984)
80. Wong, A., Le Bourhis, M., Wostradowski, R., Prahacs, S.: Pulp Paper Can. *79*, T235 (1978)
81. Kringstad, K.P., Stromberg, L.: Environmentally harmonized production of bleached pulp, Final Rep., Swedish Forest Products Research Laboratory, Stockholm, Sweden (1982)
82. Nikki, M., Korhonen, R.: J. Pulp Paper Sci. *9*, TR123 (1983)
83. Kringstad, K.P., Stockman, L.G., Stromberg, L.M.: J. Wood Chem. Technol. *4*, 389 (1984)
84. Kringstad, K.P., Lindstrom, K.: Environ. Sci. Technol. *18*, 236A (1984)
85. NCASI: Analysis of volatile halogenated organic compounds in bleached pulp mill effluent, Tech. Bull. No. 298, National Council Air Stream Improvement, Inc., New York, NY (1977)
86. NCASI: A study of methods of reducing chloroform concentrations in bleached pulp mill effluents, Tech. Bull. No. 399, National Council Air Stream Improvement, Inc., New York, NY (1982)
87. Voss, R.H.: Environ. Sci. Technol. *17*, 530 (1983)
88. Leach, J.M.: Loadings and effects of chlorinated organics from bleached pulp mills, in: Water Chlorination, Environmental Impact and Health Effects (R.L. Jolley, W.A. Brungs, R.B. Cumming, eds.), Vol. III, Ann Arbor Science, Ann Arbor, MI (1980), pp. 325–345
89. Anon.: Color, in: Standard Methods for the Examination of Water and Wastewater, 15th ed., American Public Health Association, Washington, D.C. (1980), pp. 60–70
90. Lee, E.G.-H., Mueller, J.C., Walden, C.C.: Color measurements of bleached kraft pulp mill effluents, Presented Annual Meeting, Canadian Pulp Paper Association, Pacific and Western Branches, Victoria, B.C. (1971)
91. Howard, T.E., Malick, J.G., Walden, C.C.: Pulp Paper Can. *80*, T291 (1979)
92. Whittemore, J.C., McKeown, J.J.: A study to define changes in pulpmill effluent contributed color in receiving waters detectable by human observers, Tech. Bull. No. 283, National Council Air Stream Improvement, Inc., New York, NY (1976)
93. Harkin, J.M.: TAPPI *55*, 991 (1972)
94. NCASI: The mechanism of color removal in the treatment of pulping and bleaching effluents with lime. I. Treatment of caustic extraction stage bleaching effluent, Tech. Bull. No. 239, National Council Air Stream Improvement, Inc., New York, NY (1970)
95. Dugal, H.S., Leekley, R.M., Swanson, J.W.: Color characterization before and after lime treatment, EPA Publ. No. 660/2-74-029, U.S. Environmental Protection Agency, Washington, D.C. (1974)
96. Bennett, D.J., Dence, C.W., Kung, F.L., Luner, P., Ota, M.: TAPPI *54*, 2019 (1971)
97. Marton, J., Marton, T.: TAPPI *47*, 471 (1964)
98. B.C. Research: Survey of color discharges in various pulp mills in British Columbia, Rep. submitted to Water Resources Branch, Victoria, B.C. (1973)
99. Soniassy, R.N., Walden, C.C., Serenius, R.: Examination of color removal techniques available to bleached kraft mills, Rep. submitted to Water Resources Branch, Victoria, B.C. (1975)
100. Ander, P., Eriksson, K.E., Kolar, M.-C., Kringstad, K., Ranning, V., Ramel, C.: Svensk Papperstidn. *80*, 454 (1977)
101. Eriksson, K.E., Kolar, M.-C., Kringstad, K.: Svensk Papperstidn. *82*, 95 (1979)
102. Bjorseth, A., Carlberg, E.G., Molner, M.: Sci. Total Environ. *11*, 197 (1979)
103. Lee, E.G.-H., Mueller, J.C., Walden, C.C.: Biological characteristics of pulp mill effluent (Part II), CPAR Rep. No. 678-2, Environment Canada, Ottawa, Ont. (1979)
104. Lee, E.G.-H., Mueller, J.C., Walden, C.C., Stich, H.: Pulp Paper Can. *82*, T149 (1981)
105. Ames, B.N., McCann, J., Yamasaki, E.: Mutation Res. *31*, 347 (1975)
106. San, R.H.C., Stich, W., Stich, H.F.: Intern. J. Cancer *20*, 181 (1977)
107. Stich, H.F., San, R.H.C.: Mutation Res. *10*, 389 (1970)
108. NIOSH: Suspected carcinogens. A subfile of the National Institute for Occupational Safety and Health toxic substances list, 2nd ed., U.S. Department of Health, Education and Welfare, Rockville, MD (1976)

109. Searle, C.E.: Chemical carcinogens, Monograph No. 173, American Chemical Society, Washington, DC (1976)
110. Fishbein, L., Morris, C.R.: Potential industrial carcinogens and mutagens, EPA Publ. No. 560/5-77-005, U.S. Environmental Protection Agency, Washington, DC (1977)
111. Lee, E.G.-H., Mueller, J.C., Walden, C.C.: Biological characteristics of pulp effluent (Part I), CPAR Rep. No. 678-1, Environment Canada, Ottawa, Ont. (1979)
112. Nestmann, E.R., Lee, E.G.-H., Mueller, J.C., Douglas, G.R.: Environ. Mutagenesis *1*, 369 (1979)
113. Lee, E.G.-H., Nestmann, E.R., Mueller, J.C., Douglas, G.R.: Mutation Res. *79*, 203 (1980)
114. Kringstad, K.P., Ljungquist, P.O., de Sousa, F., Strömberg, L.M.: Environ. Sci. Technol. *15*, 562 (1981)
115. Kringstad, K.P., Ljungquist, P.O., de Sousa, F., Strömberg, L.M.: Contribution of some chlorinated aliphatic compounds to the mutagenicity of spent kraft pulp chlorination liquors, in: Water Chlorination: Environmental Impact and Health Effects (R.L. Jolley, W.A. Brungs, Cotruvo, J.A., Cumming, R.B., Mattice, J.S., Jacobs, V.A., eds.), Vol. IV, Ann Arbor Science, Ann Arbor, MI (1983), pp. 1311–1324
116. Kinae, N., Hashisume, T., Makita, T., Tomita, I., Kimura, I., Kanamori, H.: Water Res. *15*, 17 (1981)
117. McKague, A.B., Lee, E.G.-H., Douglas, D.R.: Mutation Res. *91*, 301 (1981)
118. Nestmann, E.R., Lee, E.G.-H.: Mutation Res. *119*, 273 (1983)
119. Holmbom, B.R., Voss, R.H., Mortimer, R.D., Wong, A.: TAPPI *64*(3), 172 (1981)
120. Shumway, D.L., Chadwick, G.G.: Water Res. *5*, 997 (1971)
121. Cook, W.H., Farmer, F.A., Kristiansen, O.E., Reid, K., Reid, J., Rowbottom, R.: Effect of pulp and paper mill effluents on the taste and odor of water and fish, CPAR Rep. No. 12-1, Canadian Forestry Service, Ottawa, Ont. (1971)
122. Cook, W.H., Farmer, F.A., Kristiansen, O.E., Reid, K., Reid, J., Rowbottom, R.: Pulp Paper Can. *74*C, 97 (1973)
123. Whittle, D.M., Flood, K.W.: J. Fish. Res. Board Can. *34*, 869 (1977)
124. Liem, A.J., Naish, V.A., Rowbottom, R.S.: An evaluation of the effect of inplant treatment studies on the abatement of air and water pollution from a hardwood kraft pulp mill, CPAR Rep. No. 484-1, Environmental Protection Service, Ottawa, Ont. (1977)
125. Gordon, M.R., Mueller, J.C., Walden, C.C.: Trans. Tech. Sect., Can. Pulp Paper Assoc. *6*, TR2, March (1980)
126. Brouzes, R.J.P., Liem, A.G., Naish, V.A.: Protocol for fish tainting bioassay, CPAR Rep. No. 775-1, Environmental Protection Service, Ottawa, Ont. (1978)
127. Findlay, D.M., Naish, V.A.: Nature and sources of tainting in a kraft mill, CPAR Rep. No. 775-2, Environmental Protection Service, Ottawa, Ont. (1979)
128. Thomas, N.A.: Assessment of fish flesh tainting substances, in: Biological Methods for Assessment of Water Quality (J. Cairns, K.L. Dickson, eds.), ASTM STP 528, American Society for Testing and Materials, Philadelphia, PA (1973), pp. 178–193
129. Persson, P.E.: Water Res. *19*, (in press) (1985)
130. Blackwell, B.R., Mackay, W.B., Murray, F.E., Oldham, W.K.: TAPPI *62*(10), 33 (1979)
131. Paasavirta, J., Knuutinen, J., Tarhanen, J., Kuokkanen, T., Surma-Aho, K., Paukku, R., Kaariainen, H., Lahtipera, M., Veijanen, A.: Water Sci. Technol. *15*, 97 (1983)
132. Rogers, I.H.: J. Amer. Oil Chem. Soc. *55*, 113A (1978)
133. Berg, N.: Water Sci. Technol. *15*, 59 (1983)
134. NCASI: Abstracts of presentations at the 1977 NCASI West Coast Regional Meeting, Special Rep., National Council Air Stream Improvement, Inc., New York, NY (1977)
135. Shackelford, W.M., Keith, L.H.: Frequency of organic compounds identified in water, EPA Publ. No. 600/4-76-062, U.S. Environmental Protection Agency, Athens, GA (1976)
136. Lindström, K., Nordin, J.: Svensk Papperstidn. *81*, 55 (1978)
137. Hruitford, B.F., Johanson, L.W., McCarthy, J.L.: Steam stripping odorous substances from kraft effluent streams, EPA Publ. No. R2-73-196, U.S. Environmental Protection Agency, Corvallis, OR (1973)
138. Fox, M.E.: J. Fish. Res. Board Can. *34*, 798 (1977)
139. Eco Research Ltd.: Reduction of toxicity of condensates from sulphite waste liquor evaporators, CPAR Rep. No. 324-1, Canadian Forestry Service, Ottawa, Ont. (1975)

140. Hruitford, B.F., Friberg, T.S., Wilson, D.F., Wilson, J.R.: TAPPI *58*(10), 98 (1975)
141. Hruitford, B.F., Friberg, T.S., Wilson, D.F., Wilson, J.R.: Organic compounds in pulp mill lagoon discharges, EPA Publ. No. 600/2-75-028, U.S. Environmental Protection Agency, Corvallis, OR (1975)
142. Econotech: Identification and determination of the toxicity contribution of the toxic materials in bleach kraft and sulfite process effluents prepared in a pilot plant, CPAR Rep. No. 360-1, Canadian Forestry Service, Ottawa, Ont. (1975)
143. Leach, J.M., Mueller, J.C., Walden, C.C.: Process Biochem. *11*, 7 (1976)
144. Das, B.S., Reid, S.G., Betts, J.L., Patrick, K.: J. Fish. Res. Board Can. *26*, 3055 (1969)
145. Brownlee, B., Strachan, W.M.J.: J. Fish. Res. Board Can. *34*, 830 (1977)
146. Shimada, K.: Kama Pa Gikyoshi. *31*, 97 (1977)
147. Wilson, D., Hruitford, B.: Pulp Paper Can. *76*, T195 (1975)
148. Ericksson, M., Dence, C.W.: Svensk Papperstidn. *79*, 316 (1976)
149. Samuelson, D., Thede, L.: TAPPI *52*, 99 (1969)
150. Lindström, K., Nordin, J.: J. Chromat. Sci. *128*, 13 (1976)
151. Ryhage, L., von Syndow, E.: Acta Chem. Scand. *18*, 2025 (1963)
152. von Syndow, E.: Acta Chem. Scand. *19*, 1099 (1964)
153. Eckenfelder, W.W.: Biological waste treatment, Pergamon Press, New York, NY (1961)
154. Eckenfelder, W.W.: Industrial water pollution control, McGraw-Hill, New York, NY (1966)
155. Howard, T.E., Walden, C.C.: TAPPI *48*, 136 (1965)
156. Gehm, W.H.: Aerobic treatment of wastes high in BOD concentration, Purdue Univ. Engng. Bull. Ext. Ser. *38*, 346 (1953)
157. COFI: Brief submitted to the Provincial Government (BC) Public Inquiry on Pollution Control Practices in the Forest Products Industry, Council of Forest Industries of British Columbia, Vancouver, B.C. (1976)
158. PASS.: Computer file of Permit Application Sampling Systems, Waste Management Branch, Victoria, B.C.
159. Leach, J.M., Mueller, J.C., Walden, C.C.: Trans. Tech. Sec., Can. Pulp Paper Assoc. *3*, TR126 (1977)
160. Leach, J.M., Mueller, J.C., Walden, C.C.: Biodegradability of various toxic compounds in pulp and paper effluents, CPAR Rep. No. 408-1, Canadian Forestry Service, Ottawa, Ont. (1976)
161. Mueller, J.C., Leach, J.M., Walden, C.C.: TAPPI *60*(9), 135 (1977)
162. Mueller, J.C., Walden, C.C.: J. Water Poll. Contr. Fed. *48*, 458 (1976)
163. Walden, C.C., Howard, T.E.: Pulp Paper Can. *75*, T370 (1974)
164. Leach, J.M., Meier, H.-P.: Biodegradability of various toxic compounds in pulp mill effluents, CPAR Rep. No. 408-3, Environmental Protection Service, Ottawa, Ont. (1977)
165. Dellinger, W.: Development document for effluent limitations, guidelines and standards for pulp, paper and paperboard and the builders' paper and board mills, EPA Publ. No. 440/1-80/025-b, U.S. Environmental Protection Agency, Washington, DC (1980)
166. Anon.: Biological monitoring of water and effluent quality, Special Publ. No. 607, American Society for Testing and Materials, Philadelphia, PA (1976)
167. Anon.: Biological examination of water, in: Standard Methods for the Examination of Water and Wastewater, 15th ed., American Public Health Association, Washington, DC (1980)
168. Walden, C.C., Howard, T.E.: TAPPI *60*(1), 135 (1977)
169. Hutchins, F.E.: Toxicity of pulp and paper mill effluent. A literature review, EPA Publ. No. 600/3-79-013, U.S. Environmental Protection Agency, Corvallis, OR (1979)
170. Walden, C.C., Howard, T.E.: Trans. Tech. Sect., Can. Pulp Paper Assoc. *7*, T143 (1981)
171. Walden, C.C.: Biological effects of pulp and paper mill effluents, in: Advances in Biotechnology (M. Moo-Young, C.W. Robinson, eds.), Vol. II, Pergamon Press, Toronto, Ont. (1981), pp. 669–676
172. McLeay, D.J.: Laboratory monitoring for toxicity of pulp and paper mill effluents, in: Aquatic Toxicity of Pulp and Paper Mill Effluents. A Review, Can. Pulp Paper Assoc., Ontario Min. Environment, Fisheries Oceans Canada, Environment Canada, Ottawa, Ont. (in press) (1985)

173. Rao, K. (ed.): Pentachlorophenol: chemistry, pharmacology and environmental chemistry, Plenum Press, New York, NY (1978)
174. Jones, P.A.: Chlorophenols and their impurities in the Canadian environment, Econ. Tech. Rev. Rep. EPS3-EC-81-2, Environment Canada, Ottawa, Ont. (1981)
175. Jones, P.A.: Chlorophenols and their impurities in the Canadian environment: 1983 supplement, Econ. Tech. Rev. Rep. EPS3-EP-84-3, Environment Canada, Ottawa, Ont. (1984)
176. Neeley, W.B., Branson, D.R., Blau, G.E.: Environ. Sci. Technol. *8*, 1113 (1974)
177. Mackay, D.: Environ. Sci. Technol. *16*, 274 (1982)
178. Oikari, A., Holmbom, B., Anas, E., Bister, H.: Paperi Puu *62*, 193 (1980)
179. Xie, T.-M., Abrahamsson, K., Fogelqvist, E., Josefsson, B.: The distribution of chlorophenolics in a marine recipient, Appendix Rep., in: Investigation of Chlorophenolic Compounds from the Paper and Pulp Industries (Xie, T.-M., Ph.D. thesis), University of Goteborg, Goteborg, Sweden (1984)
180. Fox, M.E.: Fate of selected organic compounds in the discharge of kraft paper mills into Lake Superior, in: Identification and Analysis of Organic Pollutants in Water (L.H. Keith, ed.), Ann Arbor Science, Ann Arbor, MI, (1976), pp. 641–659
181. Brownlee, B., Fox, M.E., Strachan, W.M.J., Joshi, S.R.: J. Fish. Res. Board Can. *34*, 838 (1977)
182. Bacon, G.B., Silk, P.J.: Bioaccumulation of toxic compounds in pulp-mill effluents by aquatic organisms in receiving waters, CPAR Rep. No. 675-1, Environment Canada, Ottawa, Ont. (1978)
183. Paasivirta, J., Sarkka, J., Leskajarvi, T., Ross, A.: Chemosphere *9*, 441 (1980)
184. Salkinoja-Salonen, M., Saxelin, M.-L., Pere, J.: Analysis of toxicity and biodegradability of organochlorine compounds released into the environment in bleaching effluents of kraft pulping, in: Advances in the Identification and Analysis of Organic Pollutants in Water (L.H. Keith, ed.), Vol. II, Ann Arbor Science, Ann Arbor, MI (1981), pp. 1131–1163
185. Xie, T.-M.: Chemosphere *12*, 1183 (1983)
186. Kruzynski, G.M.: Some effects of dehydroabietic acid (DHA) on hydro-mineral balance and other physiological parameters in juvenile sockeye salmon (*Oncorhynchus nerka*), Ph.D. thesis, University of British Columbia, Vancouver, B.C. (1979)
187. Oikari, A., Holmbom, B., Anas, E., Bister, H.: Ann. Zool. Fennice *19*, 61 (1982)
188. Hattula, M.L., Wasenius V.M., Reunanen, H., Arstila, A.U.: Bull. Environ. Contam. Toxicol. *26*, 2951 (1981)
189. Renberg, L., Svanberg, O., Bengtsson, B.E., Sundstrom, G.: Chemosphere *9*, 143 (1980)
190. Oliver, B.G., Nümi, A.J.: Environ. Sci. Technol. *17*, 287 (1983)
191. Nielson, A.H., Allard, A.S., Reiland, S., Ramberger, M., Tarnholm, A., Viktor, T., Landner, L.: Can. J. Aquatic Sci. *41*, 1502 (1984)
192. Holmbom, B.R.: Studies on resin acids and chlorinated phenolics in Finnish pulp mill effluents and on their bioaccumulation in fish, Presented at the 89th National AIChE Meeting, Portland, OR (1980)
193. Paasivirta, J., Sarkka, J., Aho, M., Suma-Aho, K., Tarhanen, J., Roos, A.: Chemosphere *10*, 405 (1981)
194. Oikari, A., Nakari, T., Holmbom, B.: Ann. Zool. Fennici *21*, 45 (1984)
195. Oikari, A., Anas, E., Kruzynski, G., Holmbom, B.: Bull. Environ. Contam. Toxicol. *33*, 233 (1984)

Asbestos

P. E. Ney

Tanzebengasse 1, D-8240 Berchtesgaden
Federal Republic of Germany

Summary

After description of the structures of asbestos minerals, their formation and occurrence in the lithosphere is outlined and their physical and chemical properties are listed. Mining and processing is dealt with briefly and figures are given for production. The most important technical applications are indicated. Not exclusively, but largely through its unusually versatile utilization, asbestos is led by many, partially unexpected, paths into the human environment. There, bundles of asbestos may unravel very easily to fine respirable "asbestiform" fibres which after having been inhaled cause serious diseases. The pathology, diagnosis and epidemiology of these diseases and their role in modern industrial society is explained. By a method described it is possible to determine 1 asbestiform fibre/ml (= 1 million fibres/m^3) in the air at working places and by suitable protection measures it is possible to maintain or even reduce this concentration. Owing to the lack of an effective therapy measures of prevention have high priority.

Introduction

"Asbestos" is a collective name for six silicates with different composition but two main types of crystal structure, which have in common that they are resistant to alkalies and that their aggregates are mechanically separable into flexible and incombustible fibres of different length and fineness. The combination of these properties is unique and distinguishes them, on principle, from numerous other fibrous minerals such as wollastonite, palygorskite, sillimanite, gypsum or halloysite. They are formed only naturally and occur not only as normal constituents of wide-spread, mostly metamorphic rocks, but also locally in mineable deposits of ores or non-metallic minerals.

Asbestos, whose greek name means "indestructible", was known already in antiquity. It was used at that time and during the Middle Ages only in a very small volume. Its industrial utilization began 1860 in Italy and 1870 in Canada and Germany. After the invention of asbestos cement by the Austrian Hatscheck in 1900, usage started in most countries and grew to big volumes in some. In 1977, production amounted to 5.8 million t. In some of its more than 3,000 applications it is still indispensable at present and in some others it will remain so in future. The comparative frequency of asbestos in nature and still more the versatility of technical applications imply a wide distribution of it within the human environment.

Asbestos occurs only in minimal amounts as dust in nature. Ambient air contains 10^3 fibres and less/m^3, owing to natural erosion of asbestos-bearing rocks and blowing. During mining, processing, manufacturing, usage of asbestos and asbestos-bearing products and during the intended destruction or natural weathering of composites made from asbestos, the aggregates and bundles of fibres become mechanically divided. The same is true for asbestos as constituent of mined ores or quarried rocks. In every case the result is fine respirable, "asbestiform" fibres of dust size. Whereas compact asbestos or firmly bound component of composites does not exert any detrimental physiological effect, its asbestiform fibres resp. all asbestos-containing dusts independent of the particular type of asbestos are definitely hazardous for human health. After having been inhaled they cause serious diseases, whose dependency from dose and period of exposition is not yet fully understood. These diseases comprise above all "asbestosis", an irreversible and at the time of diagnosis already incurable damage of the lungs, and less often, the formation of carcinoma (malignant tumours) of the bronchi, pleura, and peritoneum (mesothelioma). Owing to the extremely long latency period, any therapy of these diseases is practically ineffective. There are no indications that asbestiform fibres in water and beverages are also pathologically efficient.

Especially in the major industrial countries asbestos-linked diseases have caused invalidity and untimely death in thousands of cases. They are now acknowledged and compensated as occupational diseases. Since asbestiform fibres remain dangerous to all human beings ("We all share the same air"), every suitable measures should be applied to avoid their generation, distribution, and respiration.

Sophisticated instruments and measurements enable the detection of much less than 1 fibre/ml in air, and by suitable protection measures concentrations of 0.1 fibre/ml may be maintained in the air at working places, but there are efforts to further lower the concentration of respirable fibres considerably.

The hitherto proposed and in their prices comparable substitutes for asbestos still do not possess equally favourable combinations of properties, particularly most of them are not alkali-resistant, which is crucial for the durability of asbestos cement. However, present medical knowledge, does not consider these substitutes harmful. The search for better substitutes, the displacement of asbestos from as many applications as possible and the further reduction of asbestiform fibres in breathing and ambient air remain urgent tasks for the future.

Literature on asbestos, especially that concerning environmental problems, is widely scattered in scientific and technical journals. The author attempted to list as many relevant references as possible for each of the chapters. General publications and books on asbestos are listed by [1–30].

Mineralogy and Crystallography of Asbestos Minerals and Their Genesis in the Earth's Crust

There are only two structurally, but 6 chemically different types of asbestos:

a) Sheet silicates ("serpentine asbestos")
   The sheets are composed of two layers. One layer is a pseudohexagonal network of linked SiO_4-tetrahedra, all tetrahedra pointing one way. It is joined by a layer in which divalent cations (mainly Mg^{2+}) are surrounded by octahedra of oxygens and hydroxyl-groups. On one side only, two out of three hydroxyl-groups are replaced by apical oxygens of SiO_4-tetrahedra. Figure 1 shows this structural arrangement [31–39]. The sole representative of this

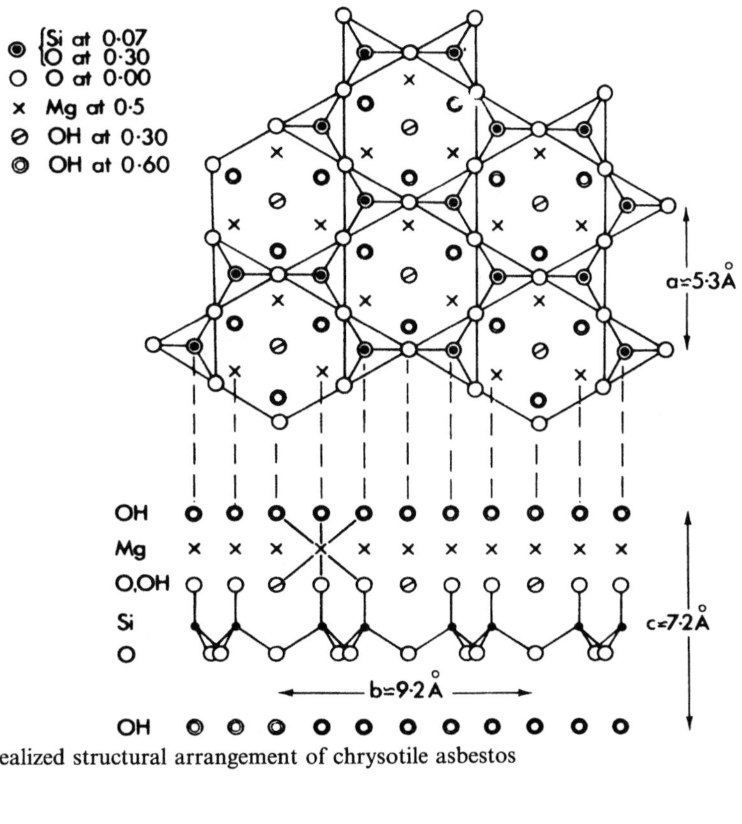

Fig. 1. Idealized structural arrangement of chrysotile asbestos

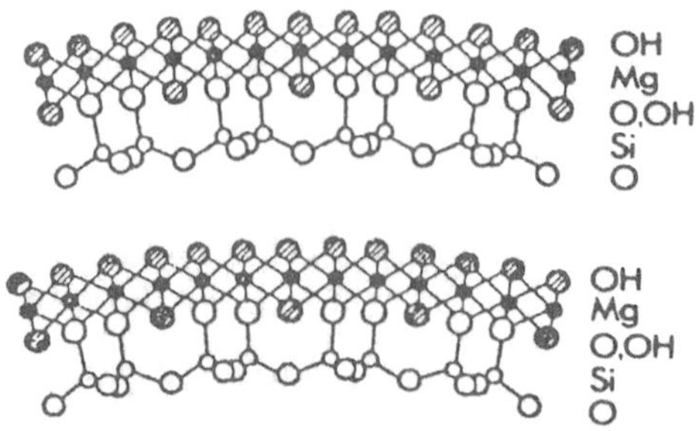

Fig. 2. Curvature of composite sheets in the fibre of chrysotile asbestos

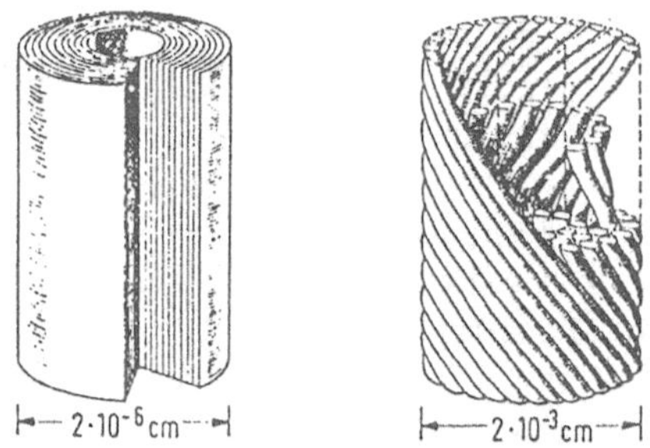

Fig. 3. Comparison between the architecture and proportion of a fibril of chrysotile asbestos (left side) with that of cotton [20]

group is (clino-)-chrysotile asbestos, $Mg_6(OH)_8Si_4O_{10}$, occasionally with little Fe^{2+}, Ni^{2+} or Al^{3+} for Mg^{2+} in the centers of the octahedra. A critical consideration of the two layers mentioned shows, that their joining involves mismatching. As may be seen from Fig. 2, a better matching is achieved by curvature of the composite sheet with its tetrahedral component on the inside of the curve [40–43]. Mostly the sheets are curved about the x axis of the structure, forming either concentric cylindrical hollow tubes or spiral rolls (scrolls), elongated parallel to x. However, in all these "fibrils" only one sheet will be at the ideal radius of curvature, that is the number of sheets in a fibril is limited. Yada has been able to demonstrate by high-resolution electron microscopy on transverse sections of chrysotile asbestos, that the fibrils are built prevailingly by about 10 successing sheets, and that the tubes or rolls have in average a diameter of 20 nm.

Thus, a sliver of chrysotile asbestos with cross-section 0.1 mm^2 contains about 20×10^6 tubular fibrils in more or less parallel orientation. Therefore, the very fine threads which result from unraveling an asbestos bundle still contain thousands of fibrils. Owing to the varying degree of orientation of fibrils within a bundle dissimilar values of tensile strength with chrysotile asbestos of different origin may be explained. Interesting is a comparison between the architecture and proportion of a fibril of chrysotile with that of cellulose (cotton) (see Fig. 3).

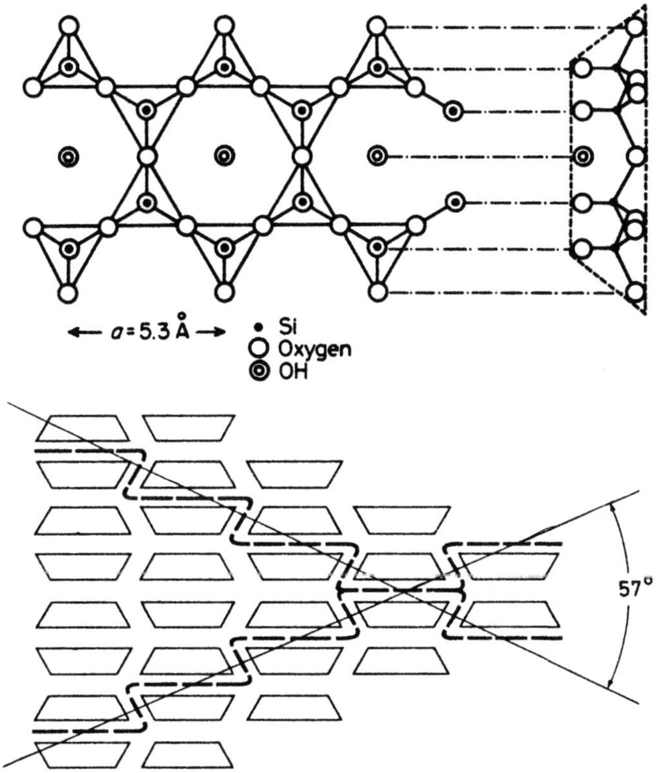

Fig. 4. *Above:* Structural arrangement of one anionic double chain or band of amphibole asbestos. *Below:* Arrangement of anionic bands, bonded to each other by cations (not indicated), seen perpendicular to the direction of the bands. The indicated probable paths of weakness mark the cleavage in the length direction of fibres of amphibole asbestos

b) (Double-)Chain silicates (amphibole asbestos) [44–53].

(Si, Al)O_4-tetrahedra are linked to form double chains or bands. The chains are generally four tetrahedra wide and of very great length. The fibres of the amphibole asbestos run parallel to the chain length (see Fig. 5).

The double chains are bonded to each other laterally by planes of cations which carry additional hydroxyl- or fluoride-ions to balance the charge. The sizes of the cations determine their coordination by oxygens of the double chains and this in turn determines the positions of the double chains relatively to one another. In most amphibole asbestos minerals the stacking of double chains is such as to produce a monoclinic cell, but in anthophyllite it results in a orthorhombic cell (see Fig. 5).

In contrast to chrysotile asbestos, the cross-section of amphibole asbestos is not rounded but irregularly angular. In the plane perpendicular to the fibre axis a bundle of fibres resembles a polycrystalline aggregate of anhedral grains.

The following compositions of amphibole asbestos minerals should be regarded as ideal, in fact most amphiboles are non-stoichiometric:

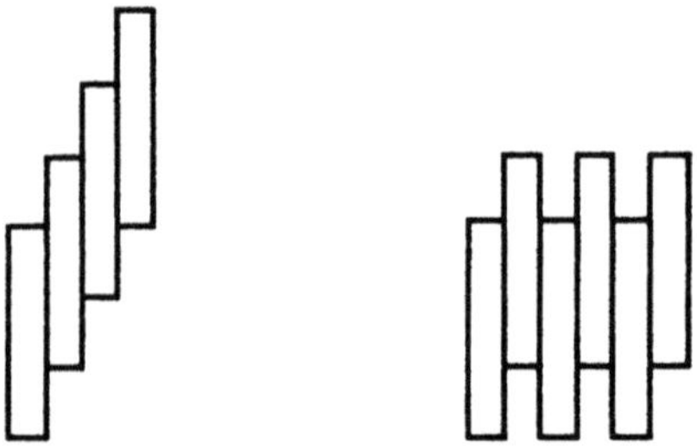

Fig. 5. Different arrangement of double chains in the structures of amphibole asbestos, leading to monoclinic or orthorhombic symmetry of their crystals

Tremolite (grammatite):

$Ca_2(Mg, Fe^{2+})_5(OH, F)_2(Si_4O_{11})_2$ with 0–20 mol% FeO for MgO

actinolite:

$Ca_2(Mg, Fe^{2+})_5(OH, F)_2(Si_4O_{11})_2$ with 20–80 mol% FeO for MgO

crocidolite (riebeckite, 'blue asbestos'):

$Na_2Fe_3^{2+}(OH)_2(Si_4O_{11})_2$

amosite:

$MgFe_6^{2+}(OH)_2(Si_4O_{11})_2$ with up to 2.9% Al_2O_3

anthophyllite:

$(Mg, Fe^{2+})_7(OH, F)_2(Si_4O_{11})_2$ with up to 1.5% Al_2O_3.

Minerals of the asbestos group do not occur in large coherent masses in nature like many other raw materials, but are contained in their host rocks prevailingly as two-dimensional bodies and are mineable only from those. The term "veins" is used only in connection with chrysotile asbestos, the bodies of crocidolite and amosite are called "bands" or "seams", and tremolite, actinolite, and anthophyllite occur most frequently as irregular lenses. The mineable thickness of the veins or seams begins at about 3 mm and exceeds 15 cm only very rarely. The mineable content begins at about 3%–5% by volume and exceeds 10% only occasionally.

The most valuable type of asbestos forms "cross-fibres", that is, fibres perpendicular to the side walls of their host rocks and strictly parallel to each other. However, the fibrous state of a mineral and the formation of cross fibre-veins ore -bands is not confined to the asbestos group; it may be found with numerous other minerals, although their fibres are not separable like those of asbestos. Obviously, the formation of cross-fibres must be dependant on very similar factors, independant from the compositions or minerals involved, namely formation of many closely spaced nuclei that grow parallel to one another in a slowly opening fissure. In most cases the ultimate cause for this opening seems to be strain operating on the individual body of host rock. Further indispensable conditions are availability of water, and continuous supply of the components, a favourable range of temperature and pressure, a steady and slow heating and cooling during a long period of tectonic quiescence [54–99].

Each type of asbestos occurs not only in deposits of very different shape and size within some few host rocks, but also as a normal constituent of these host rocks and sometimes in some other rocks. However, with regard to the entirety of rocks, asbestos-bearing rocks are rather scarce and, with some exceptions, dont occupy larger areas on the earth's surface. Rocks bearing the economically most important chrysotile asbestos are about equally frequent as rocks bearing the economically meaningless actinolite. The host rocks of tremolite and anthophyllite are distinctly less frequent and those of amosite and crocidolite are quite rare. Some asbestos-bearing rocks are extensively used as ballast for roads and rails, some other scatter their finegrained products of weathering continuously to the water running through them or to the wind passing over them.

The most important host rocks for chrysotile, which accounts for 95% of the world production are serpentinites. These are compact, tough but not very hard rocks, mostly greenish to black coloured. They consist of the compositionally equal or very similar, but in their micro-shapes quite different minerals chrysotile (fibrous), antigorite (platy), and lizardite (massive). Generally one of these minerals prevails. The other constituents of serpentinites (magnetite, brucite) are not environmentally relevant with exception of the very rare and scarcely occurring heazlewoodite, Ni_3S_2, which should be as strongly cancerogenous as its synthetic counterpart.

Serpentinites occur in two very different geological environments: Either in the lower parts of layered complexes of ultramafic rocks in Precambrian cratons or as steeply dipping plates in the most strongly folded orogens. In both cases very extended rock-bodies may accrue. While serpentinites occur in many countries, deposits of chrysotile asbestos with a significant production (more than 10,000 t/year) are confined to only a few. In order of production figures they are: USSR, Canada, P.R. China, South Africa, Zimbabwe, Brazial, USA, Australia, Italy, Mexico, Cyprus, CSSR, Swaziland, and Turkey. Asbestos-bearing rocks and unprocessed asbestos ore belong to the many bulk raw materials with which most of the readers never will probably come in contact.

A brief introduction to the terms "metamorphism" and "metamorphic rocks" is given below, because the host rocks for all types of amphibole asbestos, and according to the opinion of many geologists, even the serpentinites belong to the group of metamorphic rocks. Metamorphism means an alteration of the mineral association of a given rock in the solid state and in most cases almost without alteration of the bulk chemical composition (but with addition or removal of water and/or CO_2) by increasing pressure and heat, e.g. by displacement of the rock in deeper parts of the earth's crust or in contact with hot intruding magma. Metamorphic rocks result from metamorphism of magmatic or sedimentary rocks, and even of metamorphic rocks with a lower degree of a first metamorphism. The number of publications on the genesis of deposits of amphibole asbestos is small [100–109].

Actinolite is an essential constituent of actinolite schists and of many "green schists" which formed by a low-grade metamorphism from so widespread rocks as basalts, basaltic tuffs, diabases or gabbros. Deposits of actinolite occur only in the rocks mentioned. Tremolite appears especially in metamorphic rocks, which formed from siliceous dolomites or dolomitic limestones or from olivine diabases or olivine gabbros. Anthophyllite-bearing rocks or schists emerged by metamorphism from serpentinites, peridotites, pyroxenites, and talc schists. Crocidolite and amosite have been found almost exclusively in rocks which were formed by metamorphism of "banded iron formations". Their deposits are confined to the Precambrian.

Generally speaking, asbestos minerals as constituents of rocks or in deposits occur only in metamorphic rocks. They are absent from magmatic and sedimentary rocks which together constitute more than 90% of the earth's crust. This statement is important not only for prospecting for new deposits of asbestos minerals but also with regard to possible natural source of asbestos and the resulting danger for human health. From it follows necessarily that in regions not influenced by industrial activity the concentrations of asbestos in the ambient air and in water should be significantly higher in regions where metamorphic rocks occur near the surface and without a protecting cover of soil than in corresponding regions of magmatic or sedimentary rocks. It follows further that a latent danger exists for workers and for the environment of the facilities in the following operations: Winning of rocks like serpentinites, peridotites, pyroxenites, diabases, green schists in quarries; mining of asbestos minerals, chromite, diamond, talc, gold, ores of iron, nickel, copper, platinum within asbestos-bearing host rocks.

Physical and Chemical Properties of Asbestos Minerals

The application of an asbestos mineral or its suitability for a certain purpose, its identification and measurement in the environment, and its attitude towards the human organism is closely related to its physical and chemical properties. These are summarized in Tables 1–3.

Most characteristic is the divisibility of every compact asbestos mineral into very fine and mostly flexible fibres (see [111–137]. The necessary forces for sepa-

Table 1. Chemical composition of asbestos

	Chrysotile	Crocidolite	Amosite	Anthophyllite	Actinolite	Tremolite
SiO_2	38.80–41.5	49 –53	49 –53	56 –58	51 –56	55 –56
Al_2O_3	0.04– 4.7	0 – 0.2	0	0.5– 1.5	1.5– 3	0 – 2.5
Fe_2O_3	0.04– 1.6	17 –20	0	0	0 – 3	0 – 0.5
FeO	0.3 – 2.0	13 –20	34 –44	3 –12	5 –15	0 – 4
MgO	38.2 –42.6	0 – 3	1 – 7	28 –34	15 –20	21 –26
CaO	0.35– 2.0	0.3– 2.7	0	0	10 –12	11 –13
Na_2O	0.04– 0.1	4 – 8.5	0	0	0.5– 1.5	0 – 1.5
K_2O	0.02– 0.2	0 – 0.4	!0 – 0.4	0	0 – 0.5	0 – 0.6
H_2O^+	11.40–12.9	2.5– 4.5	2.5– 4.5	1 – 6	1.5– 2.5	0.5– 2.5
H_2O^-	0.6 – 0.9					

Table 2. Physical and chemical properties of asbestos essential for application

	Chrysotile	Crocidolite	Amosite	Anthophyllite	Actinolite	Tremolite
Texture	Soft–harsh also silky	Soft–harsh	Coarse, pliable	Harsh	Harsh	Harsh
Tensile strength N/mm^2	1,000–2,300	2,300–6,900	400–2,100	Up to 50	Up to 12	12–100
Spinnability	Verry good	Fair	Fair	Poor	Poor	Poor
Filtration prop.	Slow	Fast	Fast	Medium	Medium	Medium
Mohs's hardness	2½–4	4	5½–6	5½–6	±6	5½
Resistance to heat	Good; brittle at high temp.	Poor, fuses	Good; brittle at high temp.	Very good	–	Fair–good
Temp. at maximum loss of ignition	980 °C	650 °C	870–980 °C	980 °C	–	980 °C
Fusion point	1,520 °C	1,195 °C	1,400 °C	1,470 °C	1,395 °C	1,315 °C
Electric charge in aqueous suspension	Mostly positive	Negative	Negative	Negative	Negative	Negative
Resistance to acids	Poor	Poor	Good	Very good	Fair	Good
Resistance to alkalies	Good	Good	Good	Very good	Fair	Fair
Magnetite content	0.5–2%	3.0–5.9%	0	0	0	0

Table 3. Optical and physical properties of asbestos essential for identification

	Chrysotile	Crocidolite	Amosite	Anthophyllite	Actinolite	Tremolite
Colour	White, gray, green, yellowish	Lavender-or metallic blue	Ash-gray or brown	Grayish-white, brown-gray, green	Greenish	Gray-white, greenish, yellowish, bluish
Refractive indices						
nα	1.535–1.548	1.68	1.64	1.59	1.61	1.60
nγ	1.545–1.558	1.70	1.69	1.63	1.65	1.63
Optical character of main zone	+	–	+	+	+	+
Extinction	Parallel fibre axis	Parallel fibre axis	Parallel fibre axis	Parallel fibre axis	Oblique to fibre axis ≠/c=12–18°	Oblique to fibre axis ≠/c=12–18°
Density (g/cm^3)	2.4–2.6	3.2–3.3	3.1–3.25	2.85–3.1	3–3.2	2.9–3.2
d-values of the 3 strongest X-ray diffraction lines	7.36 (100) 3.66 (100) 2.456 (80)	8.42 (100) 2,72 (100) 3.09 (80)	8.33 (100) 3.06 (70) 2.766 (70)	3.05 (100) 3.24 (100) 8.26 (55)	8.38 (100) 3.12 (100) 2.71 (90)	8.38 (100) 3.12 (100) 2.71 (90)

ration are small and hence respirable fibres will be generated very easily. Related to the divisibility is the spinnability, and the most important physical property of asbestos, its tensile strenght. While all types of amphibole asbestos become brittle after heating up to 450 °C within a short time or after slow heating up to 300 °C, chrysotile asbestos retains one third of its original tensile strength even after heating up to 650 °C or 500 C, respectively.

Much unproductive labor has been focussed to determination of the content of poisonous trace elements and cancerogenic organic substances in asbestos as a possible source of toxic effects (see [138–148]). Chrysotile and especially crocidolite frequently contain inclusions of magnetite between and at the ends of the fibres which make them ferrimagnetic and thus give troubles in electrical applications.

For many purposes the behavior of asbestos against acids or alkalies is crucial, e.g. the fibres in asbestos cement must resist to high alkalinity (pH > 12), whereas the low resistance of chrysotile and crocidolite even to weak acids is the main reason for the quick weathering of these asbestos minerals. Amosite and actinolite are seriously attacked by strong acids [149–169].

In most cases chrysotile shows a positive electrical charge (zeta potential) in aqueous suspensions, but there are types of it with a negative zeta potential. The isoelectric point or pH for zero zeta potential of chrysotile varies from 10.5 to 11.3, depending from the origin of the fibres (see [170–176]).

The different types of asbestos, in most cases, can be easily distinguished by the polarization microscope owing to their characteristic optical properties. Chrysotile has a unique morphology and pattern with selected area electron diffraction (SAED), which make its identification by electron optical methods rather easy. The differences in composition of the asbestos minerals establish in most cases their identification by the energy-dispersive X-ray spectrum (EDX). From IR spectra it is possible to extract characteristic frequencies for the individual types of asbestos [177–187]. Methods for identification of asbestos types and particles will be treated also in chapter 8 where numerous further references on this item are given.

Production, Mining, and Processing of Asbestos

After a long period (late 1960's and early 1970's) of oversupply, asbestos was much in demand in 1973 – a combination of industrialized countries enjoying a boom and developing countries entering asbestos product manufacture on a large scale. The result was shortages in 1974 which have become even more acute in 1975. Since then, on the one hand asbestos consumption declined drastically in the USA, the U.K. and other industrialized countries owing to recession and to environmental concern. On the other hand, new mines came into production in Colombia, Greece, and in the USSR, but apparently exploration of new asbestos deposits occurs only on a very small scale at present. Table 4 shows figures on production of asbestos and on estimated production capacity for 1980 [202].

Table 4. Figures on production of asbestos and on estimated production capacity for 1980

Country	Capacity (t)	Production (t)
USSR	3,050,000	2,150,000
Canada	1,510,000	1,202,511
Rep. of South Africa	420,000	276,734
Zimbabwe	300,000	250,949
P.R. China	300,000	250,000
Italy	200,000	157,794
Brazil	185,000	170,000
USA	112,000	80,079
Greece	100,000	–
Australia	95,000	83,466
Fed. Rep. Germany	88,500	80,000[a]
Cyprus	40,000	35,535
Swaziland	35,000	32,833
India	35,000	31,253
Japan	23,000	3,362
Yugoslavia	20,000	12,106
Colombia	20,000	–
Turkey	20,000	8,724
South Korea	14,000	9,854
Bulgaria	10,000	8,800
Others	20,000	8,007
Total		4,852,200

[a] From imported Canadian crude ore

The total production comprises approximately 95% chrysotile, 3% crocidolite, 1.5% amosite, and 0.5% other varieties of asbestos. The preceeding list does not mean that environmental problems with asbestos are only relevant to the countries mentioned. Virtually all industrial and developing countries produce asbestos cement building materials. Beyond that, in most countries serpentinites and/or other asbestos-bearing metamorphic rocks occur which at least partially are used as aggregates, gravel, dimension stone or are mined for ores or non-metallic minerals.

Mining of asbestos will be considered here mainly from the viewpoint of its environmental concern. Generally, surface mining is more advantageous in recovery, grade, control, economy, and safety. The method of mining depends from the shape and depth of the ore-body and from the thickness of ore and waste. A waste-to ore ratio up to 3:1 or even more is permissible. Many ore-bodies are tabular in shape with a pronounced dip, with the result that the economic limit for quarrying is reached at a comparatively early stage. According to the different shape of the deposits, the mining and quarrying methods and equipment vary in different countries and even in the same country. They comprise the following steps: Drilling, blasting, digging or loading, haulage, crushing, and stockpiling or immediately processing. Generally, on each of these steps, dust, carrying respirable asbestos fibres, will be generated unless care will be taken by spraying water for suppression of the generation and spreading of dust. Asbestos mining gets its

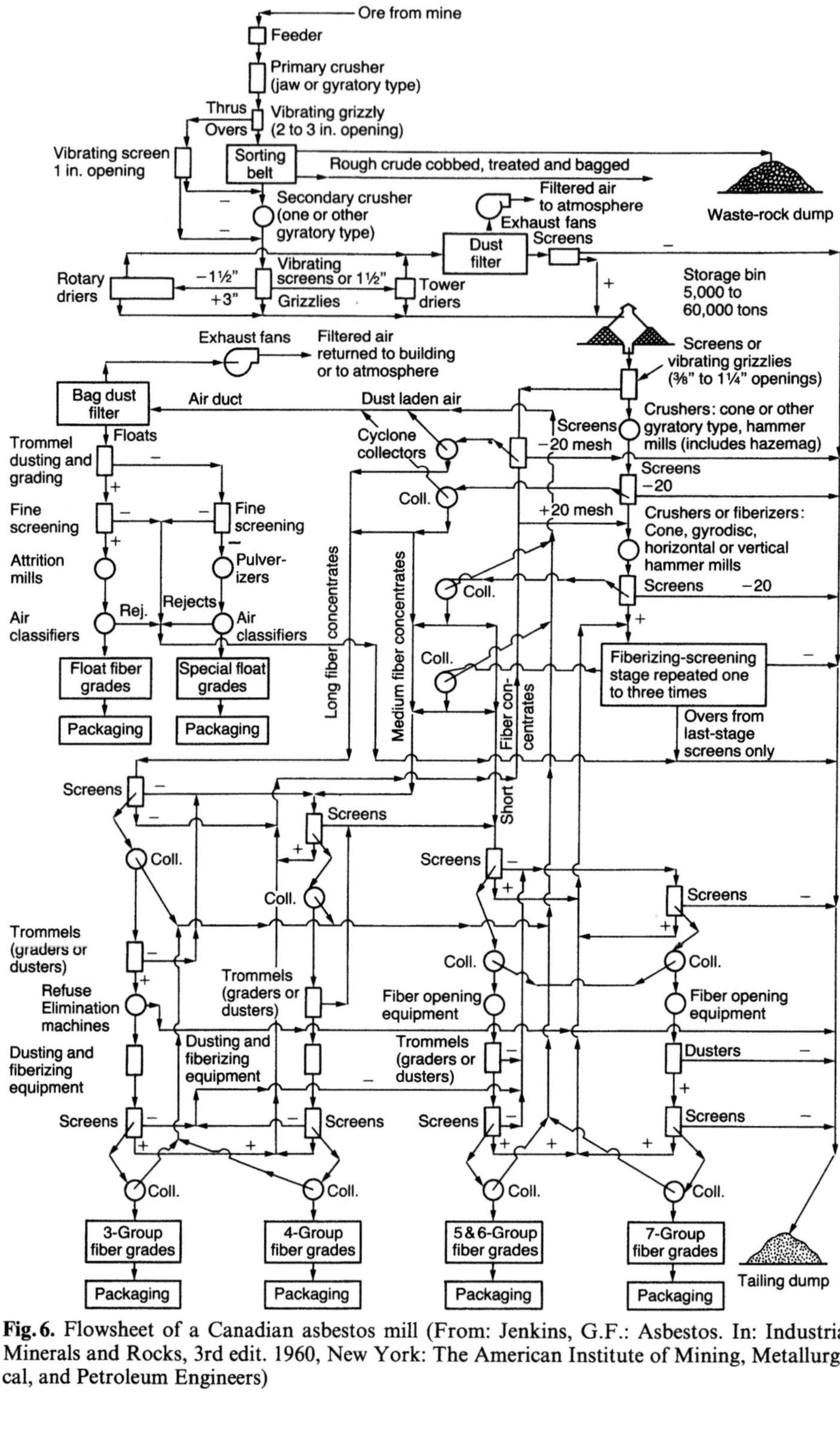

Fig. 6. Flowsheet of a Canadian asbestos mill (From: Jenkins, G.F.: Asbestos. In: Industrial Minerals and Rocks, 3rd edit. 1960, New York: The American Institute of Mining, Metallurgical, and Petroleum Engineers)

true dimension when considering that the mass of total waste amounts 10- to 20-times the production of crude ore. In most asbestos mining areas the dumps of this waste form characteristic barren hills, which are difficultly to restore because of the almost total lack of essential plant nutrients such as calcium, phosphorus, and potassium in the host rocks. Many of these hills remain for a long time a source of asbestos-bearing dust in windy regions. Special references on methods and equipment for mining of asbestos proved to be extremely scarce, (see [188, 189]), but in many other [54–109] relevant informations may be found.

For an outsider, processing or milling of the crude asbestos ore looks quite complex. To give an example, Fig. 6 shows the flowsheet of a Canadian asbestos mill. The main purpose of such a complicated process is to separate asbestos and waste impurities as completely an possible and, at the same time, to save the length of the fibres or bundles of fibres as well as possible. Pieces of crude ore with particularly long and correspondingly valuable fibres still are sorted by hand and the barren rock is knocked off by a hammer. Separation is achieved by very careful further comminution of the ore coming from the primary or secondary crusher, by a strict control of humidity (by wetting or drying), by manyfold screening of the intermediate size fractions and by an extensive removal of dust. In Canadian mills the average consumption of air per ton of ore treated is in the order of magnitude of 6,000 to 12,000 m^3. This requires that all air has to be filtered in extensive bag filter units before discharging from the mill into the ambient air [24].

Only at the later stages of processing the bundles of fibres are "opened" by special machines ("fiberizing"). In the last stages, several standardized "groups" are sorted out by sieves, chiefly according to the length of fibres. These "groups" are named after their main application, whereby the "groups" 3–5 still comprise some different "grades":

Group 1: Crude cross-fibre veins, having $^3/_4$ inch staple or longer
Group 2: Crude cross-fibre veins, having in staple $^3/_8$ up to $^3/_4$ inch
Group 3: "Spinning" ("textile", shipping, long shingle fibres)
Group 4: "shingle fibres" for asbestos cement
Group 5: "Paper stock"
Group 6: "Stucco"
Group 7: "Refuse" ("shorts" and "floats")
Group 8: "Sand".

Utilization of Asbestos Fibres

The selection of asbestos fibres for a particular application depends on the desired properties of the composite end product and on the "group" of asbestos. The different uses of asbestos fibres have been in a continuous change in the last years and differ in various industrialized countries. An overview is given in Tables 5 and 6.

It is likely that only few people are aware of the hundreds of products of which asbestos is a part, or of its broad contribution to everyday life. In Table 7 some

Table 5. Estimated consumption of asbestos fibres in the United Kingdom for the years 1973 and 1980 (in 1,000 t) (from [200])

Application	Consumption	
	1973	1980
Asbestos cement building products	32.2	34.4
Asbestos cement pressure pipes	5.2	3.3
Fire resistant insulating board	13.0	–
Insulation products	2.3	–
Jointings and packings	6.6	8.9
Friction materials	9.9	13.3
Textile products	4.8	4.4
Floor tiles	9.4	11.1
Moulded plastics and battery boxes	1.6	0.2
Fillers, reinforcements	14.9	24.4
Total asbestos usage	172.5	90.2

Table 6. Application of asbestos fibres in the USA in the years 1973 and 1980 (in 1,000 t) (from [202])

Application	Consumption	
	1973	1980
Flooring products	198	100
Asbestos cement pipes	151	160
Roofing products	73	28
Friction products	72	48
Asbestos cement sheet	58	8
Packing and gaskets	24	12
Insulation	23	8
Paper products	16	1
Textiles	16	4
Other	158	32
Total demand	795	401

of the more important uses of asbestos are listed. * Denotes uses in which reinforcement for strength, impact or abrasive resistance is an essential function of the asbestos content.

The references [198–216] on utilization of asbestos represent only a very small part of the existing literature on this subject, but especially the more general publications in "Industrial Minerals" give an excellent insight into the versatility of applications of asbestos and its economic background.

Table 7. Important uses of asbestos

*Acoustical ceilings	*Linings (stoves, heaters, furnaces)
*Adhesives	*Marine construction materials
*Adhesive wall sprays	*Millboard
Annealing blankets	*Noise dampeners
*Atomic energy shielding	Oil-well muds
*Auto undercoats	*Packing
Bags (for filtering dust without fire hazard)	Paints
*Battery boxes	*Paper
Blankets (insulating and fire-smothering)	*Pipe insulation
*Brake linings	*Pipes and piping
Caulking (Chemical containers)	*Plastics
*Cements	*Polishing cloths
Clothing (protective)	Press covers (for drycleaning)
*Cluch facings	Radiator sealing compounds
*Coatings	Rocket discharge shielding
Conveyor belts (for carrying hot objects)	*Roofing felts
Cord (for electrical pads and blankets)	*Roofing tiles and slates
Decorative "snow"	Rope (caulking and gasketing steam and air lines)
Diaphragms (electro-chemical)	Safety clothing
*Draperies	*Sheets (structural)
*Dryer felts (for paper-making machines)	*Shingles (roof and siding)
Filters (lye, sugar, corrosive liquids, beer, wine, oils, dust)	Table pads and mats
*Fire-proofing cement	Theater curtains
*Floor tile	*Tubing (insulation for cables and pipes)
Furnace hoods	*Vibration dampeners
Gas filters	*Wall coverings
*Gaskets	*Wall tile
Greases	*Welding rod coatings
*Instrument shielding	Wicks
Insulation	*Wiping pads
Ironing board covers	*Wrappings (thermal and electric insulation)

Paths of Asbestiform Fibres into the Human Environment

There is now general agreement about the size limits of pathogenic asbestiform fibres: Fibres with a length of 5–100 μm and a thickness of 3 μm and below are considered as particularly dangerous for the human health. Asbestos-bearing dusts are generated continuously, periodically or episodically, but in every case necessarily during the following processes:

- Mining of asbestos minerals
- Mining of asbestos-bearing host rocks of metallic or non-metallic ores
- Quarrying of asbestos-bearing rocks
- Haulage, loading, and transhipment of crude asbestos ore
- Haulage and loading of crude or crushed asbestos-bearing ores or rocks
- Processing of asbestos ore by comminution, milling, screening, fiberizing, sizing, bagging
- Procesing of asbestos-bearing rocks and emptying silos with such rocks
- Filling of bags or containers with processed asbestos at the mine site and emptying them at the site of the trader or consumer

- Manufacturing of commercial products, manually or with the aid of machines, from asbestos
- Machining (Roughing, fashioning) of asbestos-bearing products
- Repairing or patching of asbestos-bearing installations or insulations
- Cutting, sawing or facing of asbestos cement building products or asbestos cement pipes by craftsmen or do-it-yourself homeworkers
- Normal or accelerated wear of asbestos-bearing products during usage
- Weathering of asbestos-bearing building materials outdoor in a normal climate or in a much more aggressive urban or industrial climate
- Disposal of tailings from mining or processing of asbestos ore or from asbestos-bearing other ores and rocks
- Disposal of asbestos-bearing industrial waste
- Removal of asbestos-bearing insulations or building materials
- Removal of whole asbestos-containing buildings by demolishing or blasting without a protective spray of water
- Filtration of beverages with the aid of asbestos
- Usage of fillers like tremolite-bearing talcum
- Continuous usage of asbestos garments
- Air-conditioning with an asbestos-containing equipment
- Burning of asbestos-containing paper, plastics, and textiles

The above surely incomplete list allows conclusions as follows:

1. In view of the fact that asbestos minerals occur in so many countries and environments and find numerous applications, there is a great chance for everybody to get in contact with asbestos dust. Even persons who are not at all engaged in mining, processing, manufacture or application of asbestos in any form may be exposed.
2. Consumers, manufacturers and users of asbestos products may enlarge the general risk considerably by ignorance of dangers, by lacking precaution or by unsuitable handling. Smoking aggravates the hazardous action of asbestos dust.
3. Certain groups of persons will be in contact with asbestos dusts during the whole period of their professional activity. They, particularly, need protection and supervision.

Owing to its positive electrical charge, the dust of chrysotile tends to adhere on most natural or artificial surfaces, which prevailingly are negatively charged, and may accumulate there locally or temporarily. It is interesting that there are almost no publications on weathering of asbestos minerals. But, as outlined already, the mostly used chrysotile and crocidolite do not resist acids. Owing to its small size, especially these minerals weather rather quickly, and therefore don't become accumulated or enriched in soils or fluvial and aeolian sediments. Baris et al. [218] reported 1976 from an epidemy of the asbestos-linked pleural mesothelioma in the village of Karain near Ürgüp, Central Anatolia, Turkey. All 37 patients were peasants who in cultivating their fields have been exposed to dust from soil. A critical examination of the soil proved that it did not contain any asbestos mineral but the fibrous mineral erionite, a rare mineral of the zeolite group [219].

For faster information the references to this chapter are arranged in the following manner:
[220–227]: General statements on asbestos in the environment
[228–232]: Exposure to asbestos in mines and quarries
[233–243]: Air pollution by asbestos
[244–249]: Asbestos in water and beverages
[250–256]: Building materials as source of asbestos
[257–262]: Asbestos-bearing dust from brake linings
[263–275]: Various sources of asbestos-bearing dusts
[276–281]: Asbestos as constituent of some commercial talcum
[282–287]: Asbestos in foods, drugs, and dental materials
[288–289]: Asbestos in waste disposals.
As far as cases of illness by reception of asbestos from these sources or the determination of asbestos fibres in them are concerned, some references of the following chapters shall give further information.

Impact of Fine Asbestiform Fibres on Human Health: Pathology, Diagnosis, and Epidemiology of Asbestos-Linked Diseases

The impact of asbestiform fibres on the health of modern humanity does not come by chance: Asbestos-linked diseases have been known since 1907, the effects of asbestos dust on the lungs have been reported already in 1930 [402], the medical risks of exposure to asbestos have been discussed intensively in the last 18 years, and today in industrialized countries ten thousands of people occupationally and millions of people by pollution of air and water come in contact with asbestos (see [290–301]). The amount of asbestos manufactured in numerous branches of industry is quite considerable. Table 8 shows the imports of asbestos by its main consumers in 1979, whereby some of the importing countries additionally used asbestos from own deposits.

Table 8. Imports of asbestos by main consumers in 1979 (after [202])

Country	Import (t)
United States of America	513,084
Federal Republic of Germany	386,118
Japan	291,531
France	126,476
United Kingdom	116,669
Italy	77,151
Belgium and Luxembourg	50,754
Mexico	50,120
India	34,760
Spain	29,399
Netherlands	28,235
Denmark	26,865
Austria	23,912

From the very beginning, the fibrous nature of asbestos has been suspected as ultimate cause of the diseases linked with it. Central importance to the risk of health, to diagnosis, and to all counteractions is the question for the size of the "asbestiform" fibres which induce the diseases. Most experts today consider fibres with a diametre below 3 μm and a length between 5 and 100 μm as "dangerous", and those with a length of 20 μm as "most dangerous". Inhaled fibres of these sizes may arrive at the alveoli, are retained there and penetrate the tissue of the lungs, where they become coated with a ferro-protein complex. These "asbestos bodies" can be found by microscope in the sputum of persons who have been exposed to asbestos dust after as short a time as 4 months, but are found even still 30 years after exposure. Shorter fibres may be eliminated by the clearance mechanism within the respiratory system. The longer the fibres the longer the time they will be retained within the bronchi and the lungs, and the more difficult is their elimination. References [302–349] deal with the many-sided aspects of this subject.

During the exploration of the interplay between retained asbestos fibres and the tissue some possible causes of disease have been taken into consideration:

- Proliferation of cells, owing to mechanical irritation and continuous local processes of inflammation.
- Effects of chemical factors, e.g. trace elements or organic contaminations, as shortly mentioned on page 10. Some further references on this item are given by [350–356].
- Adsorption of additionally inhaled carcinogenic substances (e.g. from smoking or aromatic hydrocarbons) or of pathogenically effective molecules from the metabolism of the lungs at the surface of the fibres.
- Intensification of the effect of genotoxic carcinogenes.
- Repeated percussion and change of permeability of the membrane of cells and changed metabolism by that, leading to degeneration and death of cells.

It lasted decades before a causal relation was generally accepted between the inhalation of asbestiform fibres and diseases of the bronchi, lungs, pleura, and peritoneum. Inconceivably it lasted again decades or years before the unambiguous medical findings led to consequences in industrial hygienics and brought the legislator in the main asbestos-producing and asbestos-consuming countries to decisive measures of prevention and compensation. This interval may be well seen from Table 9, which shows the date of detection of asbestos-linked diseases and the date of their acceptance as occupational diseases, liable for compensation, on the example of Germany.

Table 9. Date of detection of asbestos-linked diseases and date of their acceptance as occupational diseases liable for compensation (in Germany) (after [337, 548])

Kind of disease	Date of detection	Date of acceptance
Asbestosis	1907	1936
Asbestos-linked bronchial carcinoma	1933	1943
Pleural mesothelioma	1938	1977
Peritoneal mesothelioma	1954	1977

Indeed, it is to the credit of the German asbestos-manufacturing industry that it introduced consistent dust prevention early. Schoellmann comments in his book from 1925 [21]: "All machines developing dust during action are connected with a dust removal plant. These plants are prescribed by trade inspectorates in order to keep the air clean and free of dust in the working rooms in the interest of the employees." The regrettable fact that within the past decades so many workers in the asbestos industry fell ill and died untimely shows that the danger of asbestos dust has been underestimated and has not been recognized to the full extent by the industry, by health authorities, and by the individual workers. Indeed, the information about possible risks has been quite inadequate.

In contrast to many other poisonous inhaled substances, dusts of all types of asbestos possess unusually long periods of latency. The time between the beginning of exposure and the first diagnosis is on the average 25 years with asbestosis, the most frequent disease, and asbestosis emerges (in the F.R.G.) at an average of 56 years. With the asbestos-linked types of cancer, the period of latency is about 30 years and more.

The long time of latency and the time-consuming development of methods for simpler and more secure diagnosis explain, at least partially, the terribly long interval between scientific recognition and official actions. It makes a shortest possible detection of asbestos-linked diseases a quite urgent task. This task is practized in the Federal Republic of Germany systematically since 1974, whereby in the last years computer-assisted tomography has proven as especially effective. It is still unknown which dose of asbestiform fibres (expressed as number of fibres per cm^3 of respiratory air) is necessary for causing a disease. However, it is suspected that at least with tumours a linear relation exists between dose and frequency, whereby the very important question of the existence of a threshold value has not been adequately resolved. This question is still object of research. For a better elucidation of dose-response relationships, Robock [227] proposes the following modifications of definition and presentation of "dose":

- Instead of multiplication of the average concentration with time of exposure in years should be accepted the summing up over total exposure time the actual dose (number of fibres per m^3 air) inhaled per shift.
- Instead of the linear scale for the dose should be accepted a logarithmic one which will allow to mark concentrations as occurring in the general environment as in certain branches of the asbestos industry.

More general publications on asbestos-linked diseases are listed in [357–378]. The references concerning the four most important of them are grouped as follows:

[379–416]: Asbestosis
[417–436]: Pleura plaques
[437–473]: Mesothelioma
[474–486]: Carcinoma of the bronchi, lungs, pleura, peritoneum
[487–491]: Relations between smoking and asbestos-linked diseases.

"Asbestosis" or parenchymal "fibrosis" of the lung is a direct consequence of the fact that asbestiform fibres, once inhaled, tend to remain in the lung air spaces. Even at the now generally accepted threshold value of 1 asbestiform fibre per cm^3 of air at the working place, an asbestos worker during an 8 h working day breathes about 3–5 million of such fibres [18], which of course do not all remain in the lungs.

The firstly formed "asbestos bodies" which envelop the fibres break down in the course of time, damage the lung tissue and create a peribronchial and perivasale diffusively distributed neoplasm of connective tissue. This leads to a progressive coarsening of the alveoli and by that to an increasing reduction of gas-exchange in the lungs. Therefore, breathlessness, a troublesome cough, and perpetual tiredness are most frequent signs of an advanced asbestosis. Given an appropriate occupational history, the complaints mentioned, and "asbestos bodies" in the sputum, the diagnosis will be accomplished by a chest radiograph. However, early cases may be difficultly to ascertain.

The diagnosis and the syndrome of asbestosis may be considerably complicated, and the illness may be additionally aggravated by a contemporary or subsequent occurrence of any of the other asbestos-linked diseases. After Buchanan [18, p. 397] currently over half of the British death certificates recording asbestosis also record a lung cancer. He also notes that owing to the destruction of lung tissue, an increasing strain is put on the heart to pump blood through the lungs and thus, heart failure is the other common cause of death in victims of advanced asbestosis. Unfortunately, and in contrary to other types of pneumoconiosis, asbestosis remains progressive even after the contact with asbestos has ceased. But, if exposure ceases at an early stage the progression may be slowed down.

"Pleura plaques" are mostly bilaterally and fairly symmetrically arranged yellow-white areas of thickening (some millimetres) on the parietal pleura which lines the chest cavity. They consist of bundles of collagen, and contain calcium in fine and irregularly distributed granules. Asbestos fibres have been identified in them too. Well calcified plaques therefore may be detected radiologically during life. Although especially many cases of pleura plaques have been recorded in areas where anthophyllite, the hardest and chemically least reactive asbestos mineral is mined or quarried, any of the other asbestos minerals may lead to them, and besides that, asbestos seems to be not the sole cause, but the most common. In absence of other disorders pleura plaques are not clinically harmful, in contrary, they often provide a suitable radiological indication of asbestos exposure. The period of latency is usually 20 years.

A "mesothelioma" is a primary serosal tumour or cancer affecting the lining membrane of the lung surface or inner chest wall (pleura), or a similar membrane lining of other organs, e.g. the peritoneum. Histologically very characteristic is the variability of structures within the same tumour. This, and other features complicate the distinction from a peripheral carcinoma of the lung or from a secondary carcinoma of the peritoneum, and have contributed to the late recognition of mesothelioma as an asbestos-linked, occupational and compensatable disease. A detailed description of the macroscopic, histological, and cytological features of mesothelioma, and of its diagnosis, may be taken from the references given, especially from Jones [18, pp. 425–441].

In the context of environmental risks it is important that about 85%–90% of the cases of mesothelioma are associated with occupational exposure to asbestos from which crocidolite carries the greatest danger. Wagner et al. [467] studied the histories of 33 patients who died of mesothelioma, and found that all but one had either worked in crocidolite mines in South Africa or had lived near these mines. On the other hand, after McDonald [539], from 4,547 deaths which occurred between 1910 and 1975 in 11,000 Canadian chrysotile miners and millers, only 10 (0.24%) were due to mesothelioma. Although in the lungs of about 40% of all city-dwelling males after prolonged search "asbestos bodies" have been found, the risk of mesothelioma for the general public should be considered as very small.

The "bronchial carcinoma" produced by asbestos and mostly arising in the epithelium of the major bronchi, does not differ in any of its essential features from that of other causes, and may be initiated by all commercial types of asbestos. Very remarkable is the finding that the combined carcinogenic properties of asbestos and smoking cigarettes produce a multiplicative effect on incidence of bronchial carcinoma. Selikoff et al. [489] have shown that asbestos workers smoking regularly cigarettes had an 8-fold greater risk of bronchial carcinoma deaths than cigarette smokers who did not work with asbestos, and a 90-fold greater risk than men who neither smoked cigarettes nor worked with asbestos.

While the references [492–523] deal with the diagnosis of all asbestos-linked-diseases, the references [524–550] relate to cases in special industries or locations or for certain products, [551–575] to pronouncedly epidemiological publications, and [576–596] to the mortality. Finally, the references [597–622] concern the experimental pathology. In this area in most cases it was possible only to investigate short-time effects and not any of the experiments gave results which were applicable for therapy.

At present it is neither possible to quantify the actual risk for the health of people working in the asbestos industry nor for the whole population, for the following reasons:

- Owing to the long period of latency, statistical investigations have to extend over at least 40 years, and have to be based on equal or comparable methods of diagnosis and on complete occupational and residential histories of all persons concerned – conditions which are realized nowhere.
- The long period of latency also defeats precautionary investigations.
- In the last decades in virtually all industralized countries more effective measures of prevention have been introduced which certainly diminished the danger for many people working with asbestos. On the other hand, the world production of asbestos has almost continuously grown in the last 25 years, as Table 10 shows, and its cumulative value from 1954 to 1984 amounts to about

Table 10. World production of asbestos in the last 25 years

Year	Production (1,000 t)
1955	1,768
1960	2,202
1964	3,208
1970	4,320
1975	4,240
1979	5,278
1980	4,852

100 million tons. Even if the world production should decline further in the next years or decades, additional quantities of asbestos certainly will enter our environment by repair or destruction of existing buildings. Therefore, more recent figures for frequency distribution of cases of asbestos-linked diseases cannot be compared with older ones, and even the cessation of asbestos production will not mean the happy end of the menace by asbestos to human health in this century.

- Contrary to asbestos, there is no credible evidence of malignant or chronic progressive non-malignant disease in man resulting from exposure to man-made vitreous fibres. In spite of this, the harmlessness of synthetic fibres, especially those used as substitute for asbestos in building materials, have to be monitored (see [875–934].

The present situation of occupational asbestos exposure will be briefly outlined below. The many years of well organized registration and regulation of occupational diseases in the Federal Republic of Germany, will allow epidemiological considerations.

The percentage of all occupational diseases with unprecedented compensation for working accidents in 1979 amounts to only 12, and for asbestos-linked diseases to 1.7%. At the „Zentrale Erfassungsstelle asbestgefährdeter Arbeitnehmer“ (Central registration office of asbestos-endangered employees) at the end of 1979 altogether 28,476 persons has been registered. Of them 14,759 persons have been employed continuously or for a longer time "asbestos-endangered". The following Table 11, taken from Versen [548], shows, according to different branches of industry, the distribution of compensated diseases caused by asbestos for the end of 1979.

Table 11. Compensated asbestos-linked diseases in the Federal Republic of Germany in 1979 (after Versen [548])

Branches of industry	Asbestosis	Asbestosis + bronchial carcinoma	Mesothelioma	Total
Insulation	49	13	9	71
Shipyards	13		4	17
Asbestos cement	137	28	6	171
Rubber/Plastics	26	1	1	28
Brake linings	9		1	10
Spinning-mills	101	9	4	114
Weaving-mills	54	5		59
Other textiles	27	5		32
Others	49	14	11	74
Total	465	75	36	576
Percentage	80.7%	13.0%	6.3%	100%

Identification, Measurement, Evaluation, and Monitoring of Asbestiform Fibres in Different Matrices

In view of the diagnosis, pathology, and epidemiology of diseases linked with asbestos, first of all a detection in affected organs or tissues is desirable. For this task enough methods are now available (see [623–632]).

Dependent on concentration, size, and matrix, the mere identification of asbestos minerals (or their distinction from other fibrous materials) is difficult and needs sophisticated instrumentation or separation. In industrial products and often also in dusts, asbestos is accompanied by cement, calcium or sodium silicates, clays, chalk, diatomite, talcum, resins, thermoplastics, organic fibres, and many other substances. For a direct visual identification the following methods are available (instrumentation and methods are listed roughly according to increasing cost and complexity):

- On heating of picked fibres in a crucible to dull red heat may be recognized wether the fibres are chrysotile (or ceramic fibres!) (they appear unchanged, will neither burn nor fuse) or amphibole asbestos (will become brown by oxidation) or of organic origin (will fuse before charring or burning) or man-made glass fibres (will fuse to a bead).
- With a stereo-binocular microscope even traces of asbestos fibres may be discovered in and picked from relatively large samples of quite heterogeneous composition.
- By virtue to their optical properties asbestos fibres embedded in a liquid of known refraction index may be determined by aid of a polarization or phase-contrast microscope very precisely and distinguished from other types of fibres, provided that the bundles have been picked into pieces and the fibres have been set free as complete as possible from adhering substances [639–659].
- The usage of the transmission or scanning electron microscope (TEM respectively SEM), selected area electron diffraction (SAED) or energy dispersive X-ray diffraction (EDX) for identification or determination of number and dimension of the particles is expensive and time consuming but indispensable if their size is below the optical resolution [660–691].

The Figs. 7–14, taken from Miller [675] show the morphology, electron diffraction pattern, energy-dispersive X-ray spectra, and X-ray diffraction pattern of chrysotile and crocidolite.

An indirect (non-visual) identification may be achieved by:

- X-ray diffraction analysis. Asbestos minerals can be determined in a mixture qualitatively down to 2%, but quantitatively just from 15%–20%, provided that any dry grinding of the mixture or of the fibres has been avoided strictly, that the mounted sample is free of texture, and that all interferences will be taken into account [692–704].
- Infrared spectrophotometry. Samples of only 250 µg suffice, but they must have been purified particularly careful from all accompanying materials [705–714].
- Differential thermal analysis (DTA) and differential gravimetric analysis (DGA). In spite of their wide application as heat-insulating materials, the asbestos minerals decompose thermally. The reactions involved are partially en-

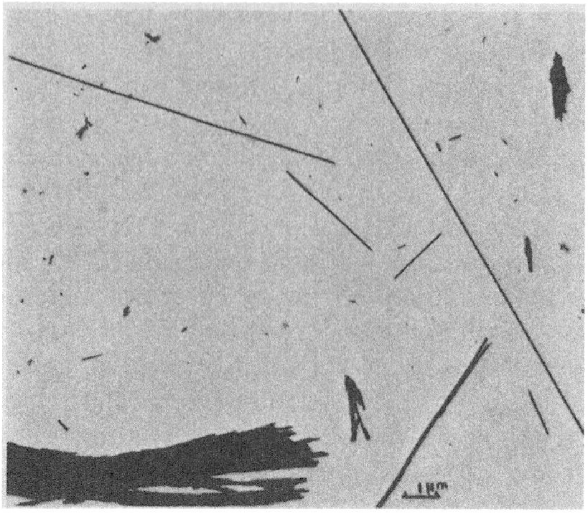

Fig. 7. Morphology of chrysotile asbestos, showing fibrils, fibres, and fibre bundles (electron micrograph)

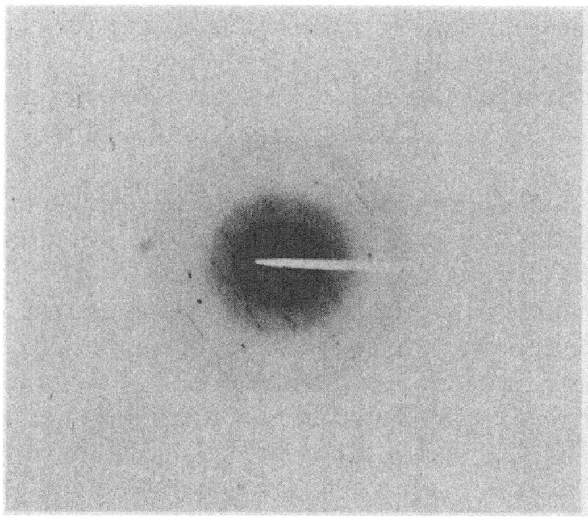

Fig. 8. Selected area electron diffraction pattern of chrysotile asbestos

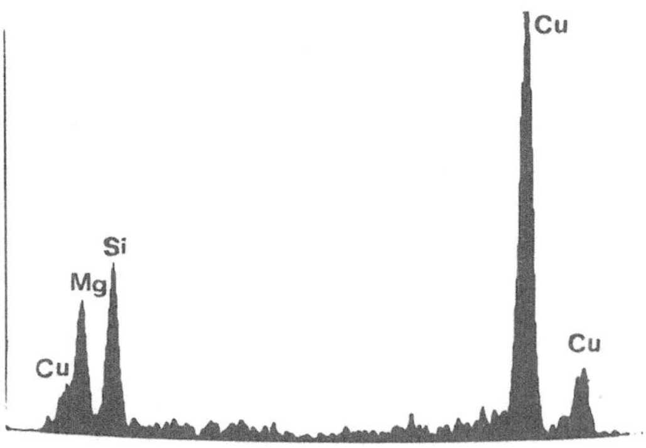

Fig. 9. Energy-dispersive X-ray spectrum of chrysotile

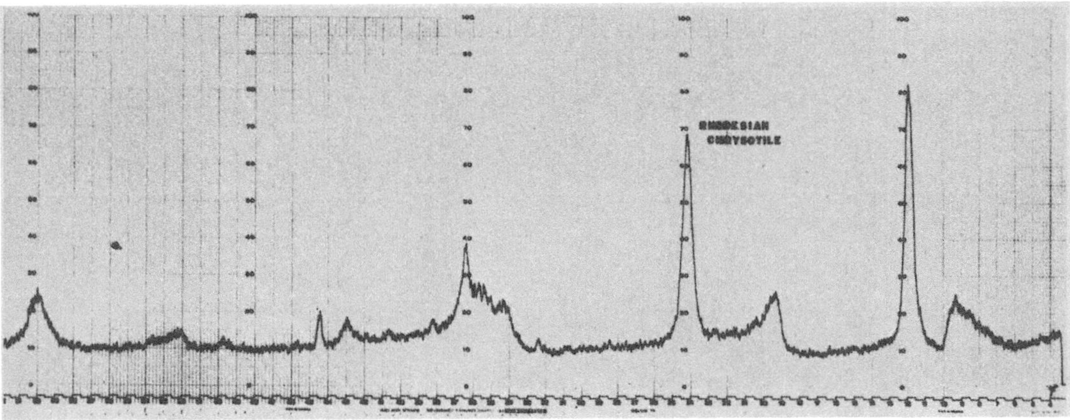

Fig. 10. X-ray diffraction pattern of chrysotile asbestos, using copper radiation. (Figures 7–10 after Miller [675]

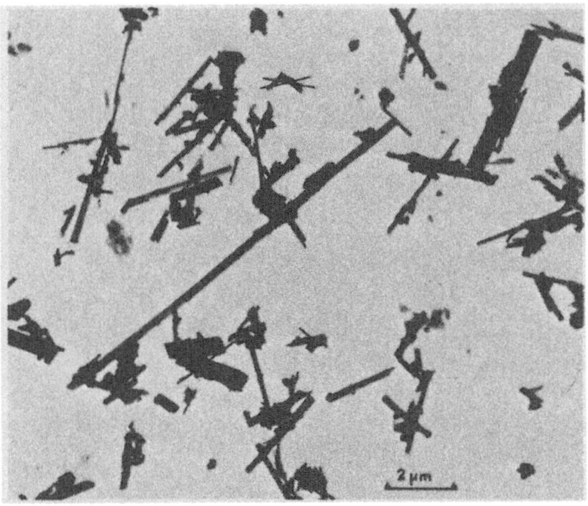

Fig. 11. Morphology of crocidolite asbestos, ground in a ball mill. (Electron micrograph)

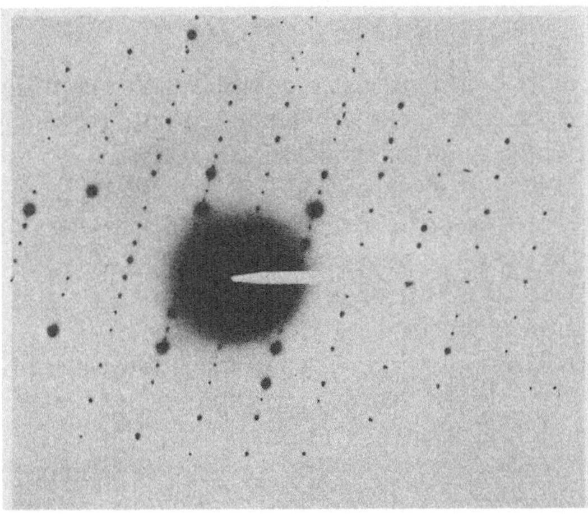

Fig. 12. Selected area electron diffraction pattern of crocidolite asbestos

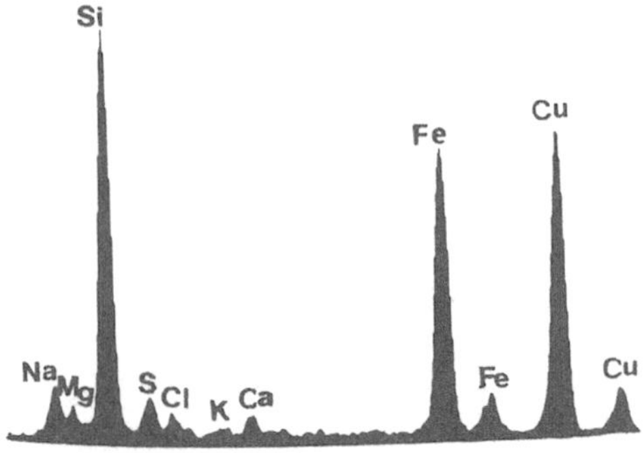

Fig. 13. Energy-dispersive X-ray spectrum of crocidolite

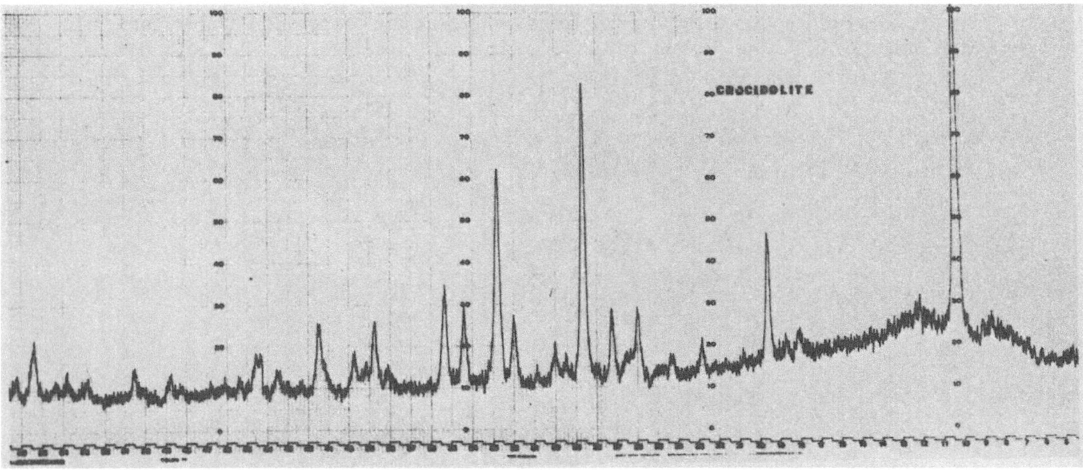

Fig. 14. X-ray diffraction pattern of crocidolite asbestos, using copper radiation. (Figures 11–14 after Miller [675])

dothermic, partially exothermic and coupled with weight losses. These effects provide characteristic, temperature dependent curves which enable the determination of different types of asbestos, especially in not too diluted mixtures with thermally inert materials [132–137, 712].

There are different reasons for measuring the concentration of asbestos and asbestiform fibres in dust:

- determination of the dust situation at the workplace
- personal protection of workers
- systematic control of regulations or equipment for protection
- control of emissions and of dedusting equipment
- long-term epidemiological investigations.

There should be only one aim: The measurement must be performed uniformly in all countries, from sampling until evaluation, in order to get comparable and useful values on an international bases. Generally, the following require-

ments will not be met by a single apparatus:
- The concentrations should be measurable over a wide range of magnitudes with equal precision.
- The measurement should be possible directly at the workplace during a whole shift and as near to the respiratory organs of the person, concerned, as possible.
- The data should become available as quickly as possible, e.g. for engineering or dust suppression purposes.
- During continuous measurement an alarm system should indicate warning or an intensified dust separation should be put in motion if a certain limit value has been exceeded.

In the past decades different methods have been developed in some countries as Fig. 15 shows (Robock [852]).

Instrument	Country (year of introduction)					
	USA	UK	SA	CAN	D	AIA
K			1916		1951	
TP		1935/1951	1940			
MI	1935			1948		
GS		×			1973	
FS	1965	1965	1970	1974	1976	1979

Fig. 15. Measuring methods for asbestos in air in various countries and year of introduction. (AIA = Asbestos International Association) (after Robock [852]). K: Konimeter; TP: Thermal Precipitator; MI: Midget-Impinger; GS: Gravimetric Sampler; FS: Filter Sampler (Membrane Filter Method)

The "konimeter" is a portable instrument which allows to take snap samples of dust. In it, about 5 cm^3 dust-laden air are impinged with high velocity through a small hole on to a movable plate, which is coated with an adhesive. The dust spots are examined and evaluated by a microscope. The readings are particles/cm^3.

In the "impinger" a high-velocity jet of air is directed to a liquid. The dust particles are wetted and retained in the liquid which is then examined under a microscope. The parameter obtained is million particles per cubic foot (mppcf).

The "thermal precipitator" is a relatively heavy instrument which makes use of the thermophoretic effect: The dust particles in a very slow stream of air are heated by a wire and deposit in a thin line across a cold glass plate. The precipitate is evaluated by an optical microscope. Again, particle/cm^3 are measured.

All types of gravimetric samplers merely allow to determine the mass of inhalable fine dust (in mg/m^3), and do not discriminate between fibrous and non-fibrous particles.

Reliable and simple information on the concentration of pathogenous asbestiform particles in dust-laden air as number of particles per volume unit air is best obtained by the membrane filter method. Like with the other methods mentioned, the knowledge of the aerodynamic properties of asbestiform particles is essential for function (see [633–638]). In this method all details have to be standardized and the application of uniform equipment, auxiliary substances and calibration standards must be warranted to allow international comparison.

At present, internationally a "reference method" is used which was developed and is recommended by the "Asbestos International Association" (AiA) [639]. In this method a measured volume of air is sucked with a battery-driven sampling pump through a rastered membrane filter with a pore size of 1.2 μm and a diameter of 2.5 cm, made from mixed esters of cellulose nitrate. The opaque filter is transformed into a transparent homogeneous membrane by treatment with acetone vapour and triacetin. Then the fibres are microscopically classified according to their length/thickness-ratio (at present still 3/1), to their length (greater than 5 μm) and diameter (less than 3 μm), and are counted by means of a phase-contrast microscope at a magnification of about 500 X. A distinction between different types of asbestos or of asbestos from other fibrous materials is impossible. The result is indicated as "fibres per ml air".

At present, the dust standards are not yet the same in different countries, but an international agreement can be expected. The results depend to a certain extent on the contrast differences in the microscope and on the skill of the microscopist. The membrane filter reference method is now the most widely used method for monitoring asbestos dust at workplaces.

References on all methods described and some others which are only seldom applied or have become obsolete are listed under [715–796]. The concentration of pathogenous asbestos dust or asbestiform particles respectively in the air fluctuates within a range of 6–7 magnitudes. The following graph (Fig. 16), drawn from Robock [227] shows the average asbestos fibre concentrations and short term peak concentrations (darts) in various branches of the asbestos industry without suppressing measures as well as maximum concentrations in the environment. One should be aware that the limit value of 1 fibre/ml air, now widely accepted and maintained, still means 1 million (!) respirable and pathogenous fibres per cubic-meter air at the workplace. About 30 years ago a limit value of 180 fibres/ml air were considered as exceptionally low. This demonstrates (compare Fig. 16) the impressive progress in protection measures. But these still need improvement.

As in air, the determination of asbestiform fibres in surface or drinking water and in beverages requires first a separation, which is mainly done with appropriate membrane filters. The next step is counting and evaluating, mostly done with the aid of an electron microscope. But there are considerable difficulties in such determinations, and the results vary by more than one magnitude. Estimates [18, p. 199] indicate, that the ingestion of electron-microscope-visible fibres in the USA amounts to

6.9 million/year with 50% of the population,
2,190 million/year with 10% of the population,
and 20,260 million/year with 1% of the population,

assuming a person consumes 1.5 liter of water per day. The references [797–813] deal with methods involved in these investigations.

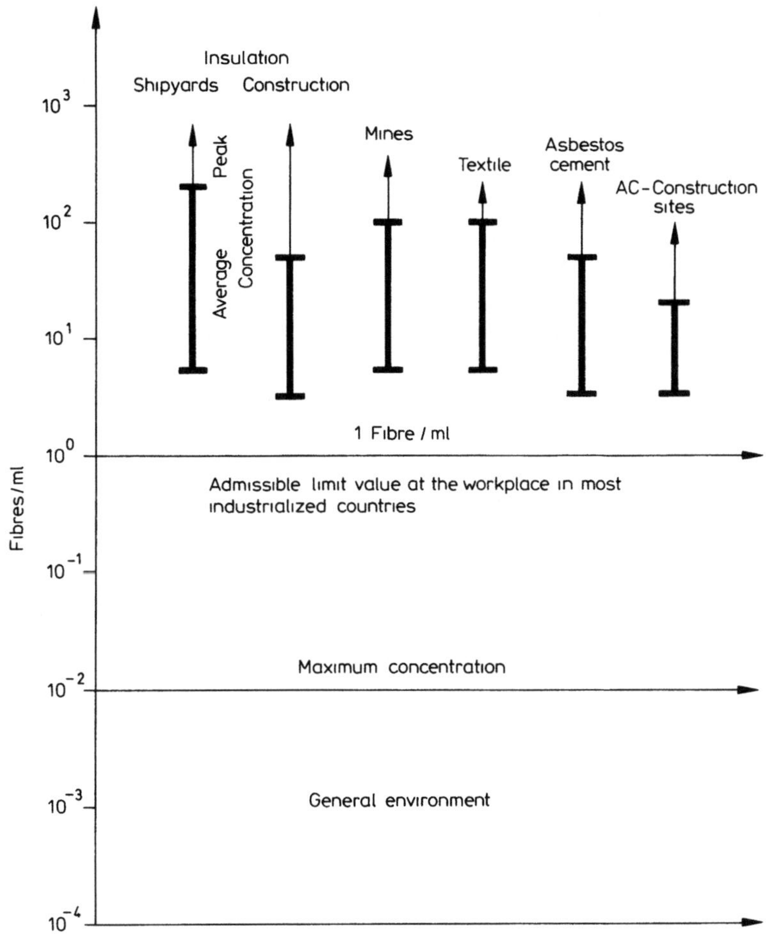

Fig. 16. Average concentrations (between crossbeams) and short term peak concentrations (darts) of asbestos fibres in air in various branches of the asbestos industry (without suppressing measures) and maximum concentrations in the environment (after Robock [227])

Prevention of Health Hazards by Asbestiform Fibres During Continued Usage of Asbestos

In addition to persons who are in contact with asbestos or products made from it occupationally, also the general population may be threatened by asbestos dust in our environment. Therefore, everyone should be interested in effective protection. It is important to consider the following facts:

1. Asbestos, owing to its unique combination of properties, will continue to be used in the future to some degree, because it fulfills indispensable functions.
2. Asbestos should be substituted increasingly in as many applications as possible by synthetic fibres. The synthesis of new and still better fibres should be intensified.

3. The serious danger of fine respirable dust of asbestos, even at a rather short exposure, has been recognized beyond doubt by world-wide medical findings and by the sickness and death of thousands of victims. Owing to the anomalous long period of latency, the mostly delayed diagnosis, and the total lack of any therapy with asbestos-linked diseases the sole chance lies in an early and consistent prevention.
4. The concentration of respirable fibres in air is measurable now within the whole range from 10^3 fibres/m^3 air until 10^9 fibres/m^3 and so we have an objective control of exposure.
5. A reduction of the concentration down to 1 fibre/ml (= 10^6 fibres/m^3 air) is already possible at many workplaces. A further considerable reduction has been realized in some cases and should be striven for in the future.

Some desirable and possible measures of protection now will be discussed related to the entry of asbestos into our environment:

a) Mining or quarrying of asbestos-bearing ores and rocks:
   - Rocks and ores containing more than 1% asbestos must be drilled wet.
   - Asbestos-bearing debris must be wetted before hauling, loading and transport by trucks or conveyor belts.
   - Stored asbestos-bearing rocks, ores or tailings should be effectively covered or continuously wetted.
   - Mining, particularly in open pits, should become still more mechanized in order to keep the number of exposed workmen low.
   - Dust-producing equipment for loading must be provided with an enclosure and exhaust ventilation.

b) Processing of asbestos-bearing raw materials:
   - All dust-producing machines (for crushing, milling, fiberizing, screening, bagging) should be provided with the maximum enclosure and with an adequate exhaust ventilation, and should be maintained under negative pressure.
   - All possible combinations of process technology which promise the avoidance of formation and spreading of dust should be used.

c) Manufacturing of asbestos-bearing commodities:
   - The application of asbestos should be reduced generally. It should be replaced by harmless substitutes as far as this is economically admissible.
   - The usage of crocoidolite should remain prohibited.
   - Fibre-emitting operations should be isolated and the access to them should be restricted.
   - Dust should be caught with the suction hole of the exhaust ventilation immediately at the spot of its generation.
   - Cleaned air should be returned into the working rooms only if the increased efficiency of the dust separator has been proven.
   - Depositions of dust should be avoided and removed regularly.
   - The time of exposure for employees should be limited per working day.
   - The control of exposure by measurement of fibre concentration should be intensified. (A realistic aim in the Federal Republic of Germany seems to be 4–5 measurements per workplace and per year).

- Smoking, eating, drinking in asbestos-endangered rooms or places should be strictly prohibited.
- In emergency and in certain workplaces personal protective equipment must be available and must be used, if necessary.
- The usage of asbestos in installations for air-conditioning must be prohibited.

d) Working with asbestos-bearing products:
- All asbestos-bearing products must be labelled clearly as "cancerogenic", especially for the "do-it-yourself workers".
- Places of dust formation in workshops should be isolated and carefully exhausted.

e) Application of asbestos or asbestos-bearing products:
- The application of asbestos as filter aid, especially for beverages or medicine, must be strictly prohibited.
- Asbestos-bearing drinking water should be purified painstakingly by a combination of filtration and flocculation.

f) Disposal of asbestos and asbestos-bearing materials:
- All wastes which may form dust on loading, transport or disposal should be tested beforehand for asbestos. Asbestos-bearing wastes should be separated from non-hazardous wastes.
- People employed with disposal of asbestos-bearing wastes should be informed sufficiently about dangers and suitable treatment of the wastes.
- Asbestos-containing wastes should be deposited in a landfill site with adequate covering layer of non-asbestos-bearing waste or soil.
- Asbestos-bearing parts of buildings should be demolished only under a dense film of water.

The references to this chapter [814–874], are incomplete because many national directions indicated in literature have been declared obsolete now and have, therefore, not been listed. Information on the actual status may be obtained from the following authorities:

Australia:
National Health and Medical Research Council
P.O. Box 100, Woden, Canberra, A.C.T. 2606

Belgium – Netherlands – Luxembourg:
Benelux Asbestos Information Committee
B-1000 Brussel, Boite 32, Boulevard E, Jacqmain 162

Canada:
Institute of Occupational & Environmental Health (Association)
AMAQ, Suite 320, 580 est. Grande Allée, Québec, Qué. GIR 2K2

Occupational Health and Saftey Unit and Department of Epidemiology and Health
McGill University, Montréal, Québec H3A 2B4

France:
Association Francaise de l'Amiante
9 Rue de Téhéran, F-75008 Paris

Laboratoire d'Hygiène et de Contrôle des Fibres Minérale
10, Rue de la Pépinière, F-75008 Paris

Federal Republic of Germany:
Asbest-Institut für Arbeits- und Umweltschutz e. V.
Görlitzer Str. 1, D-4040 Neuss 1

Berufsgenossenschaftliches Institut für Arbeitssicherheit
Lindenstraße 80 (Postfach 2043)
D-5205 St. Augustin-2

Umweltbundesamt
Bismarckplatz 1, D-1000 Berlin 33

India:
Asbestos Information Centre
401 Padma Palace, 86 Nehru Place, New Delhi 110019

Italy:
Centro Nazionale Amianto
Corso Europa 12, I-20122 Milano

Republic of South Africa:
South African Asbestos Producers Advisory Committee
P.O. Box 10505, Johannesburg 2000

Sweden:
Swedish National Board of Industrial Safety
Fack, S-100 26 Stockholm 39

United Kingdom: Asbestos Information Centre
Sackville House, 40 Picadilly, London W1V 9PA

Asbestos International Association
68 Gloucester Place,
London W1H 3HL

United States of America:
Asbestos Information Association/NA
1745 Jefferson Davis Highway, Crystal Square, Suite 509, Arlington, Virginia 22202

U.S. Environmental Protection Agency, Office of Research and Development
Research Triangle Park, North Carolina 010919

Occupational Safety and Health Administration (OSHA), Department of Labor
Constitution Avenue NW, Washington, D.C. 20210

Acknowledgement

My best thanks go to Prof. Dr. K. Robock, Asbest-Institut für Arbeits- und Umweltschutz e.V., D-4040 Neuss for providing so much of invaluable literature.

References

1. Autorenkollektiv: Der Rohstoff Asbest und seine Verwendung. Freib. Forsch.-H. C *289*, 1–285 (1973)
2. Badollet, M.S.: Asbestos. In: Encyclopedia of Chemical Technology, Vol. 2, 734–747 (ed. Kirk-Othmer). Interscience 1963
3. Becker & Haag: Asbest, seine Fundstellen, Gewinnung, Aufbereitung, Verarbeitung und Anwendung in Industrie und Technik. 99 pp. Berlin 1927
4. Bellhouse, G.: Report on Conferences between employers and inspectors concerning methods for suppressing dust in asbestos textile factories. HMSO, London, 35–214 (1931)
5. Berger, H.: Asbestfibel. Gentner, Stuttgart 1961
6. Berger, H.: Asbestos Fundamentals. 171 pp. Chemical Publ. Co., New York 1963
7. Bobeth, W., Böhme, W., Techel, J.: Anorganische Textilfaserstoffe. Verlag Technik, Berlin 1955
8. Bogovski, P. et al. (ed.): Biological effects of Asbestos. Proceedings of a working conference at IARC, Lyon, Oct. 2–6, 1972. IARC Scientific Publ. No. 8, 1–346, Lyon 1973
9. Bohlig, H., Otto, H.: Asbest und Mesotheliom. Fakten, Fragen, Umweltprobleme. Thieme-Verlag, Stuttgart 1975
10. Bowles, O.: The Asbestos Industry. Bull. 552, U.S. Bur. Mines, Washington, D.C. 1955
11. Carroll-Porczynski, C.Z.: Asbestos. From Rock to Fabric. The Textile Institute, Manchester 1956
12. Carroll-Porczynski, C.Z.: Inorganic Fibres. Natl. Trade Press Ltd., London 1958
13. Frank, H.: Asbest. Becker & Haag, Hamburg 1952
14. Green, A.K., Pye, A.M.: Asbestos. Characteristics, applications and alternatives. Fulmer Spec. Rep. No. 5, Reedprint (1976)
15. Jenkins, G.F.: Asbestos. In: Industrial Minerals and Rocks, 3rd ed., 23–53. Amer. Inst. Mining, Metall., Petrol. Eng. Inc., New York 1960
16. Ledoux, R.L. (ed.): Short course in mineralogical techniques of asbestos determination, Quebec, May 1979. 279 pp. Min. Ass. Can. 1979
17. Levine, R.J.: Asbestos: An informational resource. U.S. Dept. Health, Educ., Welfare, Natl. Cancer Inst., Divis. Cancer Control and Rehab., Prevention Branch 1978
18. Michaels, L., Chissick, S.S.: Asbestos. Properties, applications and hazards, Vol. 1, 553 pp. Wiley 1979
19. Ney, P.: Asbeste. Arten, Entstehung, Bedeutung, Problematik. Naturwissenschaften *68*, 597–605 (1981)
20. Noll, W.: Asbest. In: Ullmanns Encyklopädie der Technischen Chemie, 4. Aufl., Bd. 8, 67–79. Verlag Chemie GmbH, Weinheim 1974
21. Schoellmann, W.: Das Ganze der Asbest-Verarbeitung. 3. Aufl., 69 S. Union Deutsche Verlagsgesellschaft, Berlin 1925
22. Selikoff, I.J., Churg, J.: Biological effects of asbestos. Proceed. Conf. New York, Oct. 19–21, 1964. – Ann. N.Y. Acad. Sci. *132*, 1–766 (1965)
23. Sinclair, W.E.: Asbestos: its origin, production and utilization. Mining Publ. Ltd., London 1959
24. Winson, R.W.: Asbestos. In: Industrial Minerals and Rocks, 4th ed., 379–425. Amer. Inst. Mining, Metall., Petrol. Eng., Inc., New York 1975
25. Campbell, W.J. et al.: Selected silicate minerals and their asbestiform varieties: Mineralogical definitions and identification-characterization. U.S. Bur. Mines, Inform. Circ. 8751 1977
26. Hodgson, A.A.: Fibrous Silicates. Lecture series No. 4. Royal Institute of Chemistry, London 1965

27. Kramer, J.R.: Fibrous and asbestiform minerals. In: Nat. Bur. Stand. Spec. Publ. No. *506*, 19–32 (1978)
28. Ross, M.: The "asbestos" minerals: definitions, description, modes of formation, physical and chemical properties, and health risk of the mining community. – In: Nat. Bur. Stand. Spec. Publ. *506*, 49–63 (1978)
29. Tröger, W.E. (Hrsg. O. Braitsch): Optische Bestimmung der gesteinsbildenden Minerale, Teil 2, Textband, 429–431, 438–443, 451–453, 611–622. E. Schweizerbartsche Verlagsbuchhandlung, Stuttgart 1967
30. Zussman, J.: The mineralogy of asbestos. – In: Michaels, L., Chissick, S.S. (eds.): Asbestos: Properties, Applications and Hazards., Vol. 1, 45–66, Wiley 1979
31. Deer, W.A., Howie, R.A., Zussman, J.: Serpentines. – In: Sheet Silicates, Rock-forming minerals, 4th impress., Vol. 3, 170–190 (1965)
32. Krstanović, I., Pavlović, S.: X-ray study of chrysotile. Amer. Mineral. *49*, 1769–1771 (1964)
33. Middleton, A.P.: Crystallographic and mineralogical aspects of serpentine. D. Phil. Thesis. Oxford Univ.: Dept. of Geol. and Miner. (1974)
34. Middleton, A.P., Whittaker, E.J.W.: The structure of povlen-type chrysotile. Canad. Mineral. *14*, 301–306 (1976)
Middleton, A.P., Whittaker, E.J.W.: The nature of para-chrysotile. Canad. Mineral. *17*, 693–697 (1979)
35. Warren, B.E., Bragg, L.W.: The structure of chrysotile, $H_4Mg_3Si_2O_9$. Z. Krist. *76*, 201–219 (1930)
36. Whittaker, E.J.W.: The structure of chrysotile. II. Clinochrysotile. Acta Cryst. *9*, 855–862 (1956)
37. Whittaker, E.J.W.: The structure of chrysotile. III. Orthochrysotile. Acta Cryst. *9*, 862–864 (1956)
38. Wicks, F.J., Whittaker, E.J.W.: A reappraisal of the structures of the serpentine minerals. Canad. Mineral. *13*, 227–243 (1975)
39. Wicks, F.J.: Mineralogy, chemistry and crystallography of chrysotile. In: Ledoux, R.L. (ed.): Short course in mineralogical techniques of asbestos determination, 35–78. Quebec. Mineralogical Society of Canada (1979)
40. Bates, T.F. et al.: Tubular crystals of chrysotile asbestos. Science *3*, 512–513 (1950)
41. Yada, K.: Study of chrysotile asbestos by a high resolution electron microscope. Acta Cryst. *23*, 704–707 (1967)
42. Yada, K.: Study of microstructure of chrysotile asbestos by high resolution electron microscope. Acta Cryst. A *27*, 659–664 (1971)
43. Yada, K.: Microstructures of chrysotile and antigorite by high resolution electron microscopy. Canad. Mineral. *17*, 679–691 (1979)
44. Finger, L.W.: Refinement of the crystal structure of an anthophyllite. Carnegie Inst. Year Book *68*, 283–288 (1970)
45. Franco, M.A. et al.: Structural imperfection and morphology of crocidolite (blue asbestos). Nature *266*, 520–521 (1977)
46. Rabbitt, J.C.: A new study of the anthophyllite series. Amer. Mineral. *33*, 263–323 (1948)
47. Warren, B.E.: The structure of tremolite, $H_2Ca_2Mg_5(SiO_3)_8$. Z. Krist. *72*, 42–57 (1930)
48. Whittaker, E.J.W.: The structure of Bolivian crocidolite. Acta Cryst. *2*, 312–317 (1949)
49. Whittaker, E.J.W.: Mineralogy, chemistry and crystallography of amphibole asbestos. In: Ledoux, R.L. (ed.): Short course in mineralogical techniques of asbestos determination, 1–34. Quebec, Mineralogical Society of Canada (1979)
50. Zussman, J.: The crystal structure of actinolite. Acta Cryst. *8*, 301–308 (1955)
51. Hutchison, J.L. et al.: High resolution electron microscopy and diffraction studies of fibrous amphiboles. Acta Cryst. A *31*, 794–801 (1976)
52. Jefferson, D.A. et al.: Multiple-chain and other unusual faults in amphiboles. Contr. Miner. Petrol. *66*, 1–4 (1978)
53. Chisholm, J.E.: Planar defects in fibrous amphiboles. J. Mater. Sci. *8*, 475–483 (1973)
54. Anhaeusser, C.R.: The nature of chrysotile occurrencesin Southern Africa: A review. Econ. Geol. *71*, 96–116 (1976)
55. Allen, C.C. et al.: The Jeffrey mine of Canadian Johns-Manville Company Limited. In: The geology of Canadian industrial mineral deposits. 6th Commonwealth Mining and metallurgical Congr. 27–36 (1957)

56. Avery, R.B. et al.: Selected annotated bibliography of asbestos resources in the United States and Canada. Bull. 1019-L, 817–865 U.S. Geological Survey (1958)
57. Bain, G.W.: Chrysotile asbestos. II. Chrysotile solutions. Econ. Geol. *27*, 281–297 (1932)
58. Barnes, I. et al.: Geochemical evidence of present day serpentinization. Science *156*, 830–832 (1967)
59. Bennington, K.O.: Role of shearing stress and pressure in differentiation as illustrated by some mineral reactions in the system MgO-SiO_2-H_2O. Jour. Geol. *64*, 558–560 (1956)
60. Büttner, W., Saager, R.: The geology of the Msauli asbestos mine, Barberton Mountain Land, South Africa. Erzmetall *35*(3), 147–151 (1982)
61. Büttner, W.: Mineralogische, geologische und geochemische Untersuchungen zur Genese der Msauli-Asbest-Lagerstätte, Barberton Greenstone Belt, Südafrika. Inaug.-Dissert. Univ. Köln, 245 S. (1983)
62. Chidester, A.K. et al.: Petrology, structure and genesis of the asbestos-bearing ultramatic rocks of the Belvidere Mountain area in Vermont. U.S. Geol. Survey, Prof. Paper 1016, 95 pp. (1978)
63. Clifton, R.A.: Asbestos. U.S. Bur. Mines, MCP 6, 17 pp (1977)
64. Cooke, H.C.: Asbestos deposits of Thetford district, Quebec. Econ. Geol. *31*, 355–376 (1935)
65. Denis, B.T.: Asbestos occurrence in Southern Quebec. Quebec Bur. Mines Ann. Rept. pt.D, 147–193 (1930)
66. Evans, B.W. et al.: Stability of chrysotile and antigorite in the serpentine multisystem. Schweiz. Miner. Petrogr. Mitt. *56*, 79–93 (1976)
67. Frankel, J.J.: South African asbestos fibres. Mining Magazine (London) No. *2*, 73–83, 89 (1953), No. *3*, 142–149 (1953)
68. Gabrielse, H.: The genesis of chrysotile asbestos in the Cassiar asbestos deposit, Northern British Columbia. Econ. Geol. *55*, 327–337 (1960)
69. Glen, R.A., Butt, B.C.: Chrysotile asbestos at Woodsreef. Econ. Geol. *76*, 1153–1169 (1981)
70. Graham, R.P.D.: Original of massive serpentine and chrysotile-asbestos, Black Lake-Thetford Area, Quebec. Econ. Geol. *12*, 154–202 (1917)
71. Grubb, P.L.C.: Serpentinization and chrysotile formation in the Matheson ultrabasic belt, northern Ontario. Econ. Geol. *57*, 1228–1246 (1962)
72. Hess, H.H.: The problem of serpentinization and the origin of certain chrysotile asbestos, talc and soapstone deposits. Econ. Geol. *28*, 634–647 (1933)
73. Ichimura, T.: Asbestos deposits of the Toyoda mine, Hualien Province, Formosa. Sci. Pap. Coll. Gen. Educ., Univ. Tokyo *2*, 99–115 (1952)
74. Keith, St.B., Bain, G.W.: Chrysotile Asbestos: I. Chrysotile veins. Econ. Geol. *27*, 169–190 (1932)
75. Kula, J., Wiser, J.P.: Msault Asbestos Hill. World Mining *23*, 26–29 Sept. (1970)
76. Laubscher, D.H.: The occurrence and origin of chrysotile asbestos and associated rocks, Shabani, Southern Rhodesia. In: The Geology of some ore deposits of Southern Africa, Vol. 2, 593–624 (1964) The Geological Society of South Africa
77. Laurent, R.: Petrology of asbestos serpentinites of Southern Quebec. Third Intern. Conf. Phys. Chem. Asbestos Min., Univ. Laval, Quebec, 14 pp. (1975)
78. Laurent, L.: Petrology of the alpine-type serpentinites of Asbestos and Thetford mines, Quebec. Schweiz. Miner. Petrogr. Mitt. *55*, 431–455 (1975)
79. Merenkov, B.Ya.: The genesis of chrysotile asbestos. Trans. Acad. Sci. U.S.S.R. *22* (1958)
80. Miller, W.B.: Asbestos in Yugoslavia. Asbestos 34(2), 2–10, (3)2–10, (4)2–6 (1952)
81. Mumpton, F.A., Thompson, C.S.: Mineralogy and origin of the Coalinga asbestos deposit. Clays and Clay Minerals *23*, 131–143 (1975)
82. Olsen, E.J.: Metamorphic differentiation during serpentinization. Ph.D. Thesis, Univ. Chicago 1959
83. Poldervaart, A.: Chrysotile asbestos produced by dolerite intrusions in dolomite. Colonial Geol. Min. Resources *1*, 239 (1950)
84. Proud, J.S., Osborne, G.D.: Stress-environment in the genesis of chrysotile, with special reference to the occurrence at Woodsreef, near Barraba, New South Wales. Econ. Geol. *47*, 13–23 (1952)

85. Ridge, J.D.: "Munro" and "Thetford-Black Lake" in: Annotated Bibliographies of mineral deposits in the Western Hemisphere, 86–87 and 142–146 (1972)
86. Riordon, P.H.: The genesis of asbestos in ultrabasic rocks. – Econ. Geol. *50,* 67–81 (1955)
87. Rohrbacher, R.G.: Asbestos in the Allamoore talc district, Hudspeth and Culberson Counties, Texas. Univ. Texas Geol. Circ., 73-1 (1973)
88. Sanford, R.F.: Mineralogical and chemical effects of hydration reactions and applications to serpentinization. Amer. Miner. *66,* 290–297 (1981)
89. Smitheringale, W.V.: The mine of Cassiar Asbestos Corporation Limited, Cassiar, B.C. The geology of Canadian industrial mineral deposits, 6th Commonwealth Mining and Metallurgical Congr., 49–53 (1957)
90. Straw, D.J.: A world survey of the main chrysotile asbestos deposits. Canad. Mining and Metall. Bull. *48,* 610–630 (1955)
91. Van Biljon, W.J.: The chrysotile deposits of the Eastern Transvaal and Swaziland. In: Geology of some ore deposits in Southern Africa, Vol. 2, 625–669 (1964) Geolog. Soc. of South Africa
92. Wicks, F.J., Whittaker, E.J.W.: Serpentine textures and serpentinization. Canad. Mineral. *15,* 459–488 (1977)
93. Yang, J.C.: The growth of synthetic chrysotile fiber. Amer. Mineral. *46,* 748–742 (1961)
94. Conn, H.M.K.: Geophysics and asbestos exploration. In: Mining and groundwater geophysics. Economic geology report No. 26, 485–491 (1967) Geol. Survey of Canada
95. Conn, H.M.K., Mann, E.L.: Evaluation of asbestos deposits. SME Preprint No. 71-H-27, 1–9 (1971) AIME Annual meeting, New York
96. Cossette, M., Delvaux, P.: Technical evaluation of chrysotile asbestos ore bodies. In: Ledoux, R.L. (ed.): Short course in mineralogical techniques of asbestos determination, 79–110 (1979) Quebec: Mineralogical Assoc of Canada
97. Dean, A.W., Mann, E.L.: The evaluation of chrysotile asbestos deposits. In: Ore reserve estimation and grade control. Spec. Vol. 9, 281–286 (1968) Canad. Inst. Min. Metall.
98. Lamarche, R.Y., Wicks, F.J.: Where to look for new asbestos deposits. 3rd Internat. Conf. Physics and Chemistry of Asbestos Minerals, Univ. Laval (1975)
99. Low, J.H.: Magnetic prospecting methods in asbestos exploration. Trans. Canad. Inst. Mining and Metall. *54,* 383–395 (1951)
100. Cilliers, J.J. le R. et al.: Crocidolite from the Koegas-Westerberg area, South Africa. Econ. Geol. *56,* 1421–1437 (1961)
101. Cilliers, J.J. le R., Genis, J.H.: Crocidolite asbestos in the Cape Province. In: The Geology of some ore deposits in Southern Africa Vol. 2, 543–570 (1964)
102. Cilliers, J.J. le R.: Amosite at the Penge asbestos mine. In: The Geology of some ore deposits in Southern Africa, Vol. 2, 579–591 (1964)
103. duToit, A.L.: The origin of the amphibole asbestos deposits of South Africa. Trans. Geol. Soc. South Africa *48,* 161 206 (1946)
104. Genis, J.H.: The formation of crocidolite asbestos. In: The Geology of some ore deposits in Southern Africa, Vol. 2, 571–578 (1964). The Geological Society of South Africa
105. Keep, F.E.: Amphibole asbestos in the Union of South Africa. Trans. 7th Commonwealth Mining and Met. Congr. *1,* 99–120 (1961)
106. Kirkman, H.L.: Some notes on crocidolite and amosite occurrences in the Union. Trans. Geol. Soc. South Afr. *33,* 13–18 (1930)
107. Miles, K.R.: The blue asbestos-bearing banded iron formation of the Hammersley Range, Western Australia. Bull. No. 100, Pt. 1, 5–37 (1942) Geol. Survey of Western Australia
108. Palomaki, A., Halonen, O.: Paakkilan antofylliittiasbestihoulos. (The Paakkila anthophyllite asbestos quarry). Vuoriteollisuus – Bergshantingen *26,* 92–98 (1968)
109. Peacock, M.A.: The nature and origin of the amphibole asbestos in South Africa. Amer. Mineral. *13,* 241–288 (1928)
110. Ssobolev, N.D., Ssobolev, M.V.: Genetic types of amphibole asbestos. In: Prospecting and protection of mineral Resources. USSR (1961)
111. Badollet, M.S.: Asbestos, a mineral of unparalled properties. Trans. Canad. Inst. Mining Metall. *54,* 151–160 (1951)
112. Hodgson, A.A.: Chemistry and physics of asbestos. In: Michaels, L., Chissick, S.S. (ed.): Asbestos – Properties, applications and hazards, Vol. 1, 67–114, Wiley 1979

113. Langer, A., Kerr, P.F.: Chemical and physical characteristics of soft and harsh chrysotile. 1st Internat. Conf. Physics and Chemistry of Asbestos Minerals, Oxford, Paper 2.2 (1967)
114. Page, N.J., Coleman, R.G.: Serpentine-mineral analysis and physical properties. U.S. Geol. Survey, Prof. Paper 575-B, 103–107 (1968)
115. Assuncao, J., Corn, M.: The effects of milling on diameters and lengths of fibrous glass and chrysotile asbestos fibers. Amer. Ind. Hyg. Ass. J. *36*, 811–819 (1975)
116. Bryans, R.G., Lincoln, B.: The role of the interfibrillar bond in the deformation and fracture of chrysotile asbestos. 3rd Internat. Conf. Physics and Chemistry of Asbestos Minerals, Quebec, Paper 6.24 (1975)
117. Burke, W.A., Esmen, N.: The inertial behavior of fibers. Amer. Ind. Hyg. ass. J. *39*, 400–405 (1978)
118. Holmes, S.: The definition of an asbestos fibre. Rochdale, Lanc., Engl. Asbestos Research Council
119. Maser, M. et al.: Chrysotile morphology. Amer. Mineral. *45*, 680–688 (1960)
120. Mueller, P.K. et al.: Asbestos fiber atlas. EPA 650/2-75-036 (1975) Washington, D.C.: U.S. Environmental Protection Agency
121. Naumann, A.W., Dresher, W.H.: The morphology of chrysotile asbestos as inferred from nitrogen adsorption data. Amer. Mineral. *51*, 711–725 (1966)
122. Stöber, W. et al.: The aerodynamic diameter of latex aggregates and asbestos fibers. Staub – Reinhalt. Luft. *30* (7), 277–285 (1970)
123. Choi, I., Smith, R.W.: Kinetic study of asbestos fibers in water. J. Colloid Interface Sci. *40*, 253–262 (1972)
124. Gaze, R.: Physical and molecular structure of asbestos. Ann. N.Y. Acad. Sci. *132*, 23–30 (1965)
125. Gibbs, G.W., Hwang, C.W.: Physical parameters of airborne asbestos. 3rd Internat. Conf. Physics and Chemistry of Asbestos Minerals, Quebec, Paper 8.37 (1975)
126. Gibbs, G.W., Hwang, C.W.: Physical parameters of airborne asbestos fibers in various work environments – preliminary findings. Amer. Ind. Hyg. Ass. J. *36*, 459–466 (1975)
127. Huggins, C.W., Shell, H.R.: Density of bulk chrysotile and massive serpentine. Amer. Mineral. *50*, 1058–1067 (1965)
128. Pundsack, F.L.: The properties of asbestos. II. The density and structure of chrysotile. J. Phys. Chem. *60*, 361–364 (1956)
129. Pundsack, F.L.: The pore structure of chrysotile asbestos. J. Phys. Chem. *65*, 30–33 (1961)
130. Robens, E. et al.: Die Porenstruktur von Filter- und Dichtungsmaterial aus Chrysotilasbest. GIT Fachz. Lab. *27* (4), 256–260 (1983)
131. Young, G.J., Healey, F.H.: The physical structure of asbestos. J. Phys. Chem. *58*, 881–886 (1954)
132. Badollet, M.S., Streib, W.C.: The heat treatment of chrysotile asbestos fibres. Trans. Canad. Inst. Mining Metall. *58*, 33–37 (1955)
133. Epprecht, E., Brandenberger, E.: Die Entwässerung von Chrysotil und Antigorit. Schweiz. Miner. Petrogr. Mitt. *26*, 229–256 (1946)
134. Hey, M.H., Bannister, F.A.: A note on the thermal decomposition of chrysotile. Miner. Mag. *28*, 333 (1948)
135. Hodgson, A.A. et al.: The thermal decomposition of crocidolite from koegas, South Africa. Mineral. Mag. *35*, 5–30 (1965)
136. Patterson, J.H., O'Connor, D.J.: Chemical studies of amphibole asbestos. I. Structural changes of heat-treated crocidolite, amosite and tremolite from IR absorption studies. Austr. J. Chem. *19*, 1155–1164 (1966)
137. Wondratschek, H.: Über die Vorgänge bei der Entwässerung des Chrysotils. Tercera Reunión Internat. sobre reactividad de los sólidos. Madrid 1956
138. Chamberlain, J.A.: Heazlewoodite and awaruite in serpentinites of the Eastern Townships, Quebec. Canad. Miner. *8*, 519–522 (1966)
139. Clark, S.G., Holt, P.F.: Studies on the chemical properties of chrysotile in relation to asbestosis. Ann. Occup. Hyg. *3*, 22–29 (1961)
140. Cooke, H.C.: The composition of asbestos and other fibres of Thetford district. Roy. Soc. Canada Tr., 3rd Ser., 29, sec. 4, 7–19 (1935)

141. Cralley, L.J. et al.: Characterization and solubility of metals associated with asbestos fibers. Amer. Ind. Hyg. Ass. J. *29*, 569–573 (1968)
142. Gibbs, G.W.: The organic geochemistry of chrysotile asbestos from the Eastern Townships, Quebec. Geochim. Cosmochim. Acta *35*, 485–502 (1971)
143. Hahn-Weinheimer, P., Hirner, A.: Major and trace elements in Canadian asbestos ore bodies. Analytical results and statistical interpretation. 3rd Internat. Conf. Physics and Chemistry of Asbestos Minerals, Quebec, Paper 6.25 (1975)
144. Harington, J.S.: Chemical studies of asbestos. Ann. N.Y. Acad. Sci. *131*, 31–47 (1965)
145. Harley, N.H. et al.: Radioactivity in asbestos. Environm. Intern. *1*, 161–165 (1978)
146. Martinez, E., Comer, J.J.: The concentration and study of the interstitial material in chrysotile asbestos. Amer. Mineral. *49*, 153–157 (1964)
147. Monkman, L.J.: Mineralogical and chemical contaminants in graded milled chrysotile. 3rd Internat. Conf. Physics and Chemistry of Asbestos Minerals, Quebec, Paper 3.13 (1975)
148. Reimschussel, G.P.: The association of trace metals with chrysotile asbestos. 3rd Intern. Conf. Physics and Chemistry of Asbestos Minerals, Quebec, Paper 3.12 (1975)
149. Atkinson, R.J.: Chrysotile asbestos: colloidal silica surfaces in acidified suspensions. J. Colloid Interface Sci. *42*, 624–628 (1973)
150. Atkinson, A.W., Rickards, A.L.: Acid decomposition of highly opened chrysotile. 2nd Internat. Conf. Physics and Chemistry of Asbestos Minerals, Paper 3.1 (1971) Louvain
151. Barbeau, C.: Evaluation of chrysotile asbestos by chemical methods. In: Ledoux, R.L. (ed.): Short course in mineralogical techniques of asbestos determination, 179–217 (1979). Quebec, Mineral. Ass. Canada
152. Barbeau, C. et al.: Reactivity of magnesium silicates in acid solutions: means of distinguishing between antigorite and chrysotile. 3rd Internat. Conf. Physics and Chemistry of Asbestos Minerals, Quebec, Paper 5.19 (1975)
153. Bleiman, C., Mercier, J.P.: Attaque acide et chloration de l'asbeste chrysotile. Bull. Soc. Chim. *3–4*, 529–534 (1975)
154. Donnet, J.B., Cosme, P.J.: Nouveaux résultats sur l'attaque protonique des macrofibres de chrysotile. 3rd Internat. Conf. Physics and Chemistry of Asbestos Minerals. Québec, Paper 4.18 (1975)
155. Faust, G.T., Nagy, B.S.: Solution studies of chrysotile, lizardite and antigorite. U.S. Geol. Survey, Prof. Paper 384-B, 93–105 (1967)
156. Harris, A.M., Grimshaw, R.W.: The leaching of ground asbestos. 3rd Internat. Conf. Physics and Chemistry of Asbestos Minerals, Quebec, Paper 4.17 (1975)
157. Hirner, A.: Beeinflussung der physikalischen und chemischen Eigenschaften von technisch aufbereitetem Chrysotilasbest durch hydrothermale Behandlung. Dissert. Techn. Univ. München, 88 pp. (1976)
158. Holt, P.F., Clark, S.G.: Dissolution of chrysotile asbestos in water, acid and alkali. Nature *185*, 327 (1960)
159. Hostetler, P.B., Christ, C.L.: Studies in the system MgO-SiO_2-CO_2-H_2O. 1: The activity product of chrysotile. Geochim. Cosmochim. Acta *32*, 485–493 (1900)
160. Monkman, L.J.: Some aspects of the reaction of chrysotile with inorganic and organic acids. 2nd Internat. Conf. Physics and Chemistry of Asbestos Minerals, Louvain, Paper 3 : 2 (1971)
161. Morgan, A. et al.: Solubility of chrysotile asbestos and associated trace metals in N hydrochloric acid at 25 °C. 2nd Internat. Conf. Physics and Chemistry of Asbestos Minerals, Louvain, Paper 2.8 (1971)
162. Noll, W. et al.: Adsorptionsvermögen und spezifische Oberfläche von Silicaten mit röhrenförmig gebauten Primärkristallen. Kolloid-Z. *157*, 1–11 (1958)
163. Otouma, T., Take, S.: Effect of anionic surface active agent on chrysotile. 3rd Internat. Conf. Physics and Chemistry of Asbestos Minerals, Quebec, Paper 5.21 (1975)
164. Page, N.J.: Chemical differences among the serpentine "polymorphs". Amer. Mineral. *53*, 201–215 (1968)
165. Papirer, M. et al.: Modifications physico-chimiques du chrysotile par attaque chimique ménagés: I. En milieu aqueux. Bull. Soc. Chim. *5–6*, 651–653 (1976)
166. Pundsack, F.L., Reimschussel, G.P.: The properties of asbestos. III. Basicity of chrysotile suspensions. J. Phys. Chem. *60*, 1218–1222 (1956)

167. Raghavendra, R.V. et al.: Acid resistance characteristics of amphibole asbestos. Indian. J. Techn. *16,* 317–322 (1978)
168. Weeks, T.J., Leineweber, J.P.: Adsorption of organics by chrysotile asbestos. 1st Internat. Conf. Physics and Chemistry of Asbestos Minerals, Oxford, Paper 2.7 (1967)
169. Whittaker, E.J.W., Wicks, F.J.: Chemical differences among the serpentine "polymorphs": A discussion. Amer. Mineral. *55,* 1025–1047 (1970)
170. Chowdhury, S.: Surface chemical studies of asbestos minerals. Ph.D. thesis, University of London 1973
171. Chowdhury, S., Kitchener, J.A.: The zeta-potentials of natural and synthetic chrysotiles. Intern. J. Mineral. Process, 16 pp. (1975)
172. Gracheva, O.I., Epinateva, V.I.: Electrokinetic properties of chrysotile from various USSR deposits. Chem. Abstr. *76,* 117 090 (1972)
173. Martinez, E., Zucker, G.L.: Asbestos minerals studied by zeta-potential measurements. J. Phys. Chem. *64,* 924–926 (1960)
174. Naumann, A.W., Dresher, W.H.: Colloidal suspensions of chrysotile asbestos: Surface charge enhancement. Jour. Colloid Interface Sci. *27,* 133–140 (1968)
175. Ney, P.: Zeta-Potentiale und Flotierbarkeit von Mineralen. Vol. 6 Applied Mineralogy, 180–182. Wien/New York: Springer 1973
176. Pundsack, F.L.: The properties of asbestos. I. The colloidal and surface chemistry of chrysotile. J. Phys. Chem. *59,* 892–895 (1955)
177. Anderson, H.V., Clark, G.L.: Application of X-rays in the classification of fibrous silicate minerals commonly termed asbestos. Industr. Enging. Chem. No. *10,* 924–933 (1929)
178. Campbell, W.J.: Identification of selected silicate minerals and their asbestiform varieties. Workshop on Asbestos, Gaithersburg, Mld., July 18–20 (1977). Nat. Bur. Stand. Spec. Publ. *506,* 201–220 (1978)
179. Campbell, W.J. et al.: Selected silicate minerals and their asbestiform varieties: mineralogical definitions and identification – characterization. U.S. Bur. Mines, Inf. Circ. 8751, 51 pp. (1977)
180. Champness, P.E. et al.: The identification of asbestos. J. Microscopy *108,* 231–249 (1976)
181. Chen, J.-Y.T.: Infrared studies of the effects of acid, base, heat and pressure on asbestos and structurally related substances. J. Assoc. Off. Anal. Chem. *60,* 1266–1276 (1977)
182. Coates, J.P.: The infrared analysis of quartz and asbestos. Beaconsfield, Bucks.: Perkin-Elmer Ltd. 1977
183. Cressey, B.A., Zussman, J.: Electron microscopic studies of serpentinites. Canad. Mineral. *14,* 307–313 (1976)
184. Fankuchen, I., Schneider, M.: Low angle X-ray scattering from chrysotiles. Jour. Amer. Chem. Soc. *66,* 500–501 (1944)
185. Lee, R.L. et al.: Important considerations in the identification and counting of mineral fragments. Nat. Bur. Stand. Spec. Publ. 506 (1978)
186. Matthes, S.: Mikroskopie der technisch nutzbaren Asbeste. In: Freund, H. (Hrsg.): Handbuch der Mikroskopie in der Technik, Bd. IV, Tl. 1, 783–796. Frankfurt/Main: Umschau-Verlag 1955
187. Whittaker, E.J.W., Zussman, J.: The serpentine minerals. In: Gard, J.A. (ed.): Electron optical investigation of clays, 159–191 (1971). London: The Mineralogical Society
188. Fish, R. et al.: Canada's asbestos mines: achieving environmental control with prosperity. Canad. Min. J. *98,* (11) 8–50 (1977)
189. Hoffmann, H.: Die Asbestgewinnung auf der Jeffrey-Mine. Gummi und Asbest *7,* 60–68 (1954)
190. Pryor, E.J.: Mineral Processing, 3rd edit., Chapter 23: Selected ore treatments: asbestos, 706–707. London: Elsevier 1965
191. Quebec Asbestos Mining Assoc.: Chrysotile asbestos test manual, 3rd edit., 1st revision (1978). Québec, PQ Canada GIR 2H2, 580 Grande Allée East
192. Anon.: The Quebec asbestos mining industry. At the crossroads. Industrial Minerals, No. 128, 51–55, May (1978)
193. Anon.: Greek asbestos breaks ground. Industrial Minerals, No. 129, 47–48, June (1978)
194. Anon.: How industrial minerals experts view asbestos, fluorspar, lithium. World Mining *29,* 44–45, Dec. (1976)

195. Anon.: Cement products the key to asbestos industry growth. Industrial Minerals, No. 28, 9–17, Jan. (1970)
196. Anon.: Asbestos – production losses prolong the shortage. Industrial Minerals No. 93, 19–33, June (1975)
197. Anon.: Woodsreef asbestos looks to the future. Industrial Miner. No. 113, 31–35 (1977)
198. Badollet, M.S.: Asbestos reinforcements. Modern Plastics Encyclopedia (1960)
199. Badollet, M.S., Ximenez, M.R.: The role of asbestos in plastics. CIM Trans. LIX, 283–288 (1956)
200. Berger, H.: Verarbeitung von Asbesten mit Kunststoffen und Kautschuken. Stuttgart: Gentner 1961
201. Bundesanstalt für Geowissenschaften und Rohstoffe: Untersuchungen über Angebot und Nachfrage mineralischer Rohstoffe. XV: Asbest. 215 S. Stuttgart: E. Schweizerbart 1982
202. Clarke, G.: Asbestos – a versatile mineral under siege. Industrial Minerals No. 174, 19–37 (1982)
203. Dickson, T.: Asbestos – some hope in a still cloudy future. Industrial Minerals No. 170, 69–71 (1981)
204. Fagan, D.M. (ed.): Asbestos. Annual International Review 1978, Pt. 1 – The economy and other factors. Pt. 2 – Asbestos mining industry. Pennsylvania: Stover Publishing Company
205. Foster, W.H. et al.: Asbestos Hill – Nordenham story. Canad. Min. Metall. Bull. *69,* Nr. 775, 43–98 (1976)
206. Harben, P.: Symposium Mondiale sur l'amiante / World Symposium on Asbestos, May 1982, Montreal. Industrial Minerals No. 178, 41–43, July (1982)
207. Harben, P.: What's new after asbestos? Industrial Minerals No. 156, 51–59, Sept. (1980)
208. Hodgson, A.A.: Asbestos – a world resource. Pt. 2: Amphibole Asbestos. Industrial Minerals No. 97, 37–41, Oct. (1975)
209. Hollingsworth, B.L.: New fibre-filled thermoplastics. Pt. 1: The future for asbestos. Composites, Sept. (1969)
210. Kuntze, R.A.: Test methods for the determination of the reinforcing value of chrysotile asbestos. 3rd Internat. Conf. Physics and Chemistry of Asbestos Minerals, Quebec, Paper 7.31 (1975)
211. Lincoln, B.: Asbestos – a world resource. Pt. 1: Chrysotile. Industrial Minerals No. 97, 31–37, Oct. (1975)
212. Rice, S.J.: California asbestos industry. Calif. Div. Mines & Geol. Inform. Serv. *16*(9), 1–7 (1963)
213. Rowbotham, P.I. (ed.): World asbestos industry. Industrial Minerals No. 28, 17–29 (1970)
214. Schäfer, H.: Baustoffe für die Wärmedämmung im Industrieofenbau. 151 pp. Moers: I. Szimarowsky und U. Agst 1980
215. Speil, S., Leineweber, J.P.: Asbestos minerals in modern technology. Environm. Res., 2, 166–208 (1969)
216. Winer, A.: Mineral wool insulation from asbestos tailings. Canad. Min. Metall. Bull. *67,* No. 752, 97–104 (1974)
217. Baris, Y.I.: Pleural mesothelioma and asbestos pleurisies due to environmental asbestos exposure in Turkey: An analysis of 120 cases. Hacettepe Bulletin of medicine surgery *8,* 166–185 (1975)
218. Baris, Y.I. et al.: An outbrak of pleural mesothelioma in the village of Karain, Ürgüp – Anatolia. Kasner, the Turkish journal of cancer *5,* No. 2, 1–14 (1975)
219. Pooley, F.D.: Report of the examination of Turkish samples. Dept. of Mineral Exploitation, University College Cardiff, Jan. (1978) 9 pp.
220. Fowler, D.P.: Disposals and emission of asbestos in the United States. (1977). Menlo Park, Calif.: SRI International
221. Gilson, J.C.: Problems and perspectives: The changing hazards of exposure to asbestos. Ann. N.Y. Acad. Sci. *132,* 696–702 (1965)
222. Lohrer, W.: Asbestemission aus Produkten. Gefahr für die Umwelt. U – das technische umweltmagazin, 26–31, August (1978)
223. Newhouse, M.L.: Asbestos in the workplace and the community. Ann. Occup. Hyg. *16,* 97–107 (1973)

224. Nicholson, W.J., Pundsack, F.L.: Asbestos in the environment. In: Biological effects of asbestos. IARC Scient. Publ. No. 8, 126–130 (1973) Lyon
225. Reimschussel, G.R.: Asbestos in the environment. (Kramer, J.R. et al. ed.). Rept. Intern. Joint Comm., Great Lakes Res. Advis. Board, McMaster Univ. (1974)
226. Robock, K.: Gewinnung, Vorkommen und Verwendung von Asbest und Situationen der Staubgefährdung in den Betrieben der Vergangenheit. Sem. f. Weiterbildg. d. Ärztekammer Westfalen-Lippe, Bochum, 5. 3. 1980
227. Robock, K.: The asbestos problem today. Presentat. to the IXth MEDICHEM, Aswan/ Egypt, 21.–23. 9. 1981
228. Dement, J.M. et al.: Discussion paper: asbestos fiber exposures in a hard rock gold mine. Ann. N.Y. Acad. Sci. *271,* 345–352 (1976)
229. Gibbs, G.W., Lachance, M.: Dust exposure in the chrysotile asbestos mines and mills of Quebec. Arch. Environ. Health *24,* 189–197 (1972)
230. Gillam, J.D. et al.: Mortality among hard rock gold miners exposed to an asbestiform mineral. Ann. N.Y. Acad. Sci. *271,* 336–344 (1976)
231. McDonald, J.C., Liddell, F.D.K.: Dust exposure and mortality in chrysotile mining 1910–1975. Brit. Journ. Industr. Medic. *37,* 11–24 (1980)
232. Rohl, A.N. et al.: Environmental asbestos pollution related to use of quarried serpentine rock. Science (New York) *196,* 1319–1322 (1977)
233. Anon.: Asbestos fallout. Engng. News Rec. *193* (13) 41 (1974)
234. Laamanen, A. et al.: Observations on atmospheric air pollution caused by asbestos. Ann. N.Y. Acad. Sci. *132,* 240–254 (1965)
235. Merley, G.E.: Asbestos in the air. Lancet *I,* 1075 (1976)
236. Murchio, J.C. et al.: Asbestos fibres in ambient air at California. Berkeley, Calif.: California Air Resources Board (1973)
237. Nicholson, W.J. et al.: Asbestos contamination of air in public buildings. Mt. Sinai School of Medicine, EPA-450/3-76-001 (1973)
238. Owens, J.S.: Suspended impurities in the air. Proc. Roy. Soc., Ser. A, *101,* 18–37 (1922)
239. Rickards, A.L., Badami, D.V.: Chrysotile asbestos in urban air. Nature (London) *234,* 93–94 (1971)
240. Selikoff, I.J. et al.: Asbestos air pollution. Arch. Environ. Health *25,* 1–12 (1972)
241. Thompson, R.J., Morgan, G.B.: Determination of asbestos in ambient air. Presented at "Identification and measurement Pollutants Symposium", Ottawa, Ont., Canada, June 14–17 (1971)
242. Thomson, J.G. et al.: Asbestos as a modern urban hazard. South Afric. Med. *37,* 77 (1963)
243. Vigliani, E.C. et al.: Presence and identification of fibres in the atmosphere of Milan. Medna Lav. *67,* 551–567 (1976)
244. Conseil Superieur d'Hygiène Publique de France: Amiante et alimentation. Rapport d'enquètes effectués en octobre 1976 par l'inspection de la repression des fraudes et du controle de la qualité chez les productués et negociants en vins. Séance du 18 Octobre (1976)
245. Cunningham, H.M., Pontefract, R.D.: Asbestos fibres in beverages and drinking water. Nature (London) *232,* 332–333 (1971)
246. Hallenbeck, W.H. et al.: Is chrysotile asbestos released from asbestos-cement pipe into drinking water? J. Amer. Water Works Ass. *70,* 97–102 (1978)
247. Kuschner, M. et al.: Problem of asbestos in water. Does the use of asbestos cement pipe for potable water constitute a health hazard? J. Amer. Water Works Ass. *66,* 1–22 (1974)
248. Stewart, E.M.: Asbestos in the Water supplies of ten regional cities. EPA Publ. No. 560/ 6-76-017 (1976). Washington, D.C., Environmental Protection Agency
249. Wehman, H.J., Plantholt, B.A.: Asbestos fibrils in beverages. Bull. Envir. Contam. Toxicol. *11,* 267–272 (1974)
250. Bryon, J.C. et al.: A dust survey, carried out in buildings incorporating asbestos-based materials in their construction. Ann. Occup. Hyg. *12,* 144–145 (1969)
251. Kühnen, G.: Baumusterprüfung von Geräten zur Bearbeitung asbesthaltiger Produkte im Hinblick auf die Schadstoffemission. Tagungsber. Asbest, 6. u. 7. 3. 1980, Bonn-Bad Godesberg, Schriftenr. d. Hauptverb. d. gewerbl. Berufsgenossensch. e. V., 91–92 (1980)

252. Löffler, W.: Besondere Probleme und Maßnahmen in der Bauwirtschaft. Tagungsber. Asbest, 6. u. 7. 3. 1980, Bonn-Bad Godesberg, Schriftenr. d. Hauptverb. d. gewerbl. Berufsgenossensch. e. V., 90 (1980)
253. Meyer, A.: Faserbeton. Zement-Taschenbuch 1979/1980, 453–477. Wiesbaden: Bauverlag GmbH
254. Sawyer, R.N.: Asbestos exposure in a Yale Building. Envir. Res. *13,* 146–169 (1977)
255. Schütz, A., Heidermanns, G.: Gesundheitsgefährliche Stäube im Hoch- und Tiefbau. Staub-Reinhalt.-Luft. 37(7) 273 (1977)
256. Wehmeyer, H.P.: Umweltbelastung durch Asbest: Gegenüberstellung von Emissionen und Immissionen bei der Asbestzementherstellung. Tagungsber. Asbest, 6. u. 7. 3. 1980, Bonn-Bad Godesberg, Schriftenr. d. Hauptverb. d. gewerbl. Berufsgenossensch. e. V., 103–104 (1980)
257. Castleman, B.C. et al.: The hazards of asbestos for brake mechanics. Publ. Health Rept. *90,* 254–256 (1975)
258. Förster, H.: Gesundheitsgefahren durch Stäube asbesthaltiger Bremsbeläge in Kraftfahrzeug-Werkstätten. Tagungsber. Asbest, 6. u. 7. 3. 1980, Bonn-Bad Godesberg, Schriftenr. d. Hauptverb. d. gewerbl. Berufsgenossensch. e. V., 88–89 (1980)
259. Harwood, C.F.: Asbestos air pollution from the wear of brake linings. IITRI, Chicago, April (1972)
260. Heidermanns, G. et al.: Untersuchungen über die Gefährdung durch Stäube asbesthaltiger Reibbeläge. STF-Report 1, 1–46 (1975) Bonn: Staubforschungsinst. d. gewerbl. Berufsgenossensch. e. V.
261. Jacko, M.G. et al.: How much asbestos do vehicles emit? Auto Eng. *81,* 38–40 (1973)
262. Lynch, J.R.: Brake lining decomposition products. J. Air Pollut. Contr. Ass. *18,* 824–826 (1968)
263. Anon.: Proposed wildlife withdrawal – asbestos. Mineral. Inform. Serv., Calif. Div. Mines *10,* 6 (1957)
264. Anon.: Asbestos hazards seen in patching, sprackling, and repair of brakes. Chem. Mark. Rep., Aug. 18 (1975)
265. Anon.: Unusual application of chrysotile asbestos: The elimination of used oil emulsions. Clays and Clay Minerals *29,* 69–70 (1981)
266. Gibbs, G.W.: Fibre release from asbestos garments. Ann. Occup. Hyg. *18,* 143–149 (1975)
267. Harries, P.G.: Asbestos dust concentrations in ship repairing: a practical approach to improving asbestos hygiene in naval stockyards. Ann. Occup. Hyg. *14,* 241 (1971)
268. Harwood, C.F. et al.: Asbestos emissions from baghouse controlled sources. Amer. Ind. Hyg. Ass. J. *36,* 595–602 (1975)
269. Kauschitz, R.: Wieviel Asbeststaub wird beim Tragen asbesthaltiger Hitzeschutzkleidung frei? Ergebnisse von Messungen. Tagungsber. Asbest, 6. u. 7. 3. 1980, Bonn-Bad Godesberg, Schriftenr. d. Hauptverb. d. gewerbl. Berufsgenossensch. e. V., 101–102 (1980)
270. Lumley, K.P.S. et al.: Buildings insulated with sprayed asbestos: a potential hazard. Ann. Occup. Hyg. *14,* 255–257 (1971)
271. Margolin, S.V., Igwe, B.U.N.: Economic analysis of effluent guide lines: The textiles, friction, and sealing materials segment of the asbestos manufacturing industry. EPA Publ. EPA 230/2-74/030, A.D. Little Inc. & U.S. Environmental Protection Agency (1975)
272. Murphy, R.L. et al.: Floor tile installations as a source of asbestos exposure. Amer. Rev. Resp. Dis. *104,* 576–580 (1971)
273. Nicholson, W.J. et al.: Occupational and community asbestos exposure from wallboard finishing compounds. Bull. N.Y. Acad. Med. *51,* 1180–1181 (1975)
274. Riediger, G.: Untersuchungen zum Freiwerden von faserigem Staub beim Verarbeiten und beim Einsatz von Asbestgeweben für Hitzeschutzzwecke unter besonderer Berücksichtigung der Wirksamkeit des Imprägnierens zur Staubunterdrückung. Moderne Unfallverhütung H. 21, 120–126 (1979)
275. Rohl, A.N. et al.: Exposure to asbestos in the use of consumer sprackling, patching, and taping compounds. Science (New York) *189,* 551–553 (1975)
276. Cralley, L.J. et al.: Fibrous and mineral content of cosmetic talcum products. Amer. Ind. Hyg. Ass. J. *29,* 350–354 (1968)

277. Merliss, J.R.: Talc and asbestos contaminant of rice. J. Amer. Med. Ass. *216*, 2144 (1971)
278. National Bureau of Standards: A report on the fiber content of 80 industrial talc samples obtained from, and using the procedures of the Occupational Safety and Health Administration, Analytical Chemistry Div., Inst. Mat. Res., Natl. Bur. Stand., May (1977) Washington, D.C.
279. Pooley, F.D., Rowlands, N.: Chemical and physical properties of British talc powders. In: Walton, W.H. (ed.): Inhaled Particles IV, Pt. 2, 639–646. Oxford: Pergamon Press 1976
280. Rohl, A.N. et al.: Consumer talcums and powders: mineral and chemical characterization. J. Toxic. Environ. Health *2*, 255–284 (1976)
281. Shelz, J.P.: The detection of chrysotile asbestos at low levels in talc by differential thermal analysis. Thermochim. Acta *8*, 197–204 (1974)
282. Cunningham, H.M., Pontefract, R.D.: Symposium on industrial chemicals as food contaminants. J. Ass. Off. Analyt. Chem. *56*, 976–981 (1973)
283. Eisenberg, W.V.: Inorganic particle content of food and drugs. Envir. Health Perspect. *9*, 183–191 (1974)
284. Nicholson, W.J. et al.: Asbestos contamination of parenteral drugs. Science (New York) *177*, 171–173 (1972)
285. Wolff, A.H., Oehme, F.W.: Carcinogenic chemicals in food as an environmental issue. J. Amer. Vet. Med. Ass. *164*, 623–629 (1974)
286. Council on Dental Therapeutics: Council on dental materials and devices: Hazards of asbestos in dentistry. J. Amer. Dent. Assoc. *92*, 777–778 (1976)
287. Infante, P.F., Lemen, R.A.: Asbestos in dentistry. J. Amer. Dent. Ass. *93*, 221–222 (1976)
288. Fuller, W.H.: Movement of selected metals, asbestos, and cyanide in soil: applications to waste disposal problems. U.S. EPA Municipal Environmental Research Laboratory, EPA-600/2-77-020, April (1977) Cincinnati, Ohio
289. Harwood, C.F. et al.: Study of the effect of asbestos waste piles on ambient air. In: Proceedings of the Symposium on Fugitive Emission Measurement and Control. EPA-Publ. 600/2-76-246, 183–202 (1976)
290. Doll, R., Peto R.: The causes of cancer: Quantitative estimates of avoidable risks of cancer in the United States today. J. Natl. Cancer Inst. *66* (6) (1981)
291. Elmes, P.C.: Current information on the health risk of asbestos. Roy. Soc. Health J. *96*, 248–252 (1976)
292. Gilson, J.C.: Asbestos cancer as an example of the problem of comparative risks. INSERM *52*, 107–116 (1976)
293. Habs, M., Klein, R.: Stellungnahme zum möglichen Krebsrisiko durch Luftverunreinigungen mit Asbest. Institut für Toxikologie und Chemotherapie, Deutsches Krebsforschungszentrum, Heidelberg (1980)
294. International Agency for Research on Cancer: Evaluation of the carcinogenic risk of chemicals to man, Vol. 14 (1977): Asbestos. Lyon
295. Newhouse, M.L.: The medical risks of exposure to asbestos. The Practitioner *199*, 285 (1967)
296. Selikoff, I.J., Stokinger: Cancer risk to allergy patients from syringes with asbestos-wound plunger. Questions and answers. J. Amer. Med. Ass. *225*, 423 (1973)
297. Selikoff, I.J. et al.: Cancer risk of insulation workers in the United States. In: Biological effects of asbestos, IARC Scient. Publ. No. 8, 209–216 (1973) Lyon
298. Selikoff, I.J.: Cancer risk of asbestos exposure. In: Origins of Human Cancer. Cold Spring Harbor Laboratory, 1765–1784 (1977)
299. Selikoff, I.J., Hammond, E.C.: Multiple risk factors in environmental cancer. Persons at high risk of cancer, 467–483 (1975)
300. Woitowitz, H.-J.: Aktuelle Aspekte der Risikobeurteilung kanzerogener Arbeitsstoffe. Verh. Dtsch. Ges. Arbeitsmedizin e. V., 19. Jahrestagung in Münster, 2.–5. 5. 1979. Stuttgart: A. W. Gentner
301. Zielhuis, R.L.: Public health risks of exposure to asbestos. Pergamon Press 1977
302. Beck, E.G.: Biologische Wirkung von faserförmigen Stäuben. Arbeitsmedizin, Sozialmedizin, Präventivmedizin *10*, 178 (1975)
303. Berkley, C. et al.: The detection and localization of mineral fibres in tissue. Ann. N.Y. Acad. Sci. *132*, 48–63 (1965)

304. Britton, D.C.: Exposure to asbestos dust. Lancet *II*, 175 (1976)
305. Celikoglu, Z. et al.: Contamination aerienne par les fibres d'amiante dans les regions rurales en Turquie. Archieves l'union médicale Balkanique *13*, 45 (1975)
306. Coope, B. (ed.): Asbestiform minerals in Tucson. Industrial Minerals No. 147, 57–59, Dec. (1979)
307. Cralley, L.J. et al.: Source and identification of respirable fibers. Am. Ind. Hyg. Ass. J. *29*, 129–135 (1968)
308. Elmes, P.C.: Fibrous minerals and health. J. Geol. Soc. London *137*, 525–535 (1980)
309. Gilson, J.C.: Environmental mineralogy: medicine and mineralogy. Phil. Trans. Roy. Soc. London A *286*, 585–592 (1977)
310. Gross, P. et al.: The pulmonary response to fibrous dusts of diverse compositions. Amer. Ind. Hyg. Ass. J. *31*, 125–131 (1970)
311. Gross, P. et al.: Ingested mineral fibers, do they penetrate tissue or cause cancer? Arch. Environ. Health *29*, 341–347 (1974)
312. Harington, J.S. et al.: Mineral fibers: Chemical, physicochemical and biological properties. Adv. Pharmac. Chemother. *12*, 291–402 (1975)
313. Harris, R.L. et al.: The influence of fibre shape in lung deposition – mathematical estimates. Inhaled Particles IV, Pt. 1, 75–89
314. Harris, R.L., Fraser, D.A.: A model for the deposition of fibers in the human respiratory system. Amer. Ind. Hyg. Ass. J. *37*, 73–89 (1976)
315. Langer, A.M., Pooley, F.D.: Identification of single asbestos fibres in human tissue. In: Biological effects of asbestos, IARC Scient. Publ. No. *8*, 119–125 (1973). Lyon
316. Langer, A.M. et al.: Chrysotile asbestos in the lungs of persons in New York City. Arch. Environ. Health *14*, 559–563 (1967)
317. Merewether, E.R.A., Price, C.W.: Report on the effects of asbestos dust on the lungs and dust suppression in the asbestos industry. HMSO, London, 34–206 (1930)
318. Morgan, A. et al.: Significance of fiber length on the clearance of asbestos fibers from the lung. Brit. J. Ind. Med. *35*, 146–153 (1978)
319. Pott, F.: Some aspects on the dosimetry of the carcinogenic potency of asbestos and other fibrous dusts. Staub-Reinhalt. Luft *38*, 486–490 (1978)
320. Pott, F. et al.: Die tumorerzeugende Wirkung faserförmiger Stäube. CEC, WHO, EPA Internat. Sympos., Paris, 24.–28. 6. 1974. Vol. II, 715 (1974)
321. Pott, F.: Significance and size of inhalable fibres and their carcinogenic effect. Hefte Unfallheilkunde *126*, 593–594 (1975)
322. Pott, F.: Krebserzeugende faserige Feinstäube. Arbeitsmed., Sozialmed., Präventivmed. *12*, 172–176 (1977)
323. Pylev, L.N.: Carcinogenic action of commercial serpentine asbestos. Vopr. Onkol. *20* (10), 87–88 (1974)
324. Robock, K., Klosterkotter, W.: Untersuchungen über die Zytotoxizität von Asbest-Stäuben. Staub-Reinhalt. Luft *33*, 279–282 (1973)
325. Schmidt, K.G.: Asbestsorten, ihre Untersuchung mit optischen Mitteln und ihre krankmachende Wirkung. Staub *20*, 173–204 (1960)
326. Selikoff, I.J.: Air pollution and asbestos carcinogenesis: investigation of possible synergism. IARC Scient. Publ. No. *16*, 247–253 (1977)
327. Stanton, M.F., Layard, M.: The carcinogenicity of fibrous minerals. In: Proceed. Workshop on Asbestos : Definitions and Measurement
328. Stanton, M.F., Layard, M.: The carcinogenicity of fibrous minerals. In: Proceed. Workshop on Asbestos, Gaithersburg, Maryld., 18.–20. 7. 1977, 143–151. Washington, D.C., N.B.S.
329. Stanton, M.F.: Fiber carcinogenesis: Is asbestos the only hazard? J. Nat. Cancer Inst. *52*, 633–634 (1974)
330. U.S. Department of Labor, Mine Safety and Health Administration: Asbestiform and/or fibrous minerals in mines, mills and quarries. Informat. Report IR 1111 (1980)
331. Timbrell, V.: The inhalation of fibrous dusts. Ann. N.Y. Acad. Sci. *132*, 255–273 (1965)
332. Timbrell, V.: Possible biological importance of fibre diameters of South African amphiboles. Nature *232*, 55–56 (1971)
333. Timbrell, V.: Physical factors as etiological mechanisms. IARC Sci. Publ. No. *8*, 295–303 (1973)

334. Timbrell, V.: Alignment of respirable asbestos fibers by magnetic fields. Ann. Occup. Hyg. *18*, 299–311 (1975)
335. Timbrell, V. et al.: Characteristics of respirable asbestos fibres. In: Pneumoconiosis: Proceed. Internat. Conf., Johannesburg, 120–125. Cape Town: Oxford Univ. Press 1969
336. VDI-Bericht 475: Faserige Stäube – Messung, Wirkung, Abhilfe. Hrsg.: Verein Dtsch. Ingen. Düsseldorf: VDI-Verlag 1983
337. Woitowitz, H.-J., Rödelsperger, K.: Asbeststaub als Ursache bösartiger Tumoren. Tagungsber. Asbest, 6. u. 7. 3. 1980, Bonn-Bad Godesberg, Schriftenr. Hauptverb. d. gewerbl. Berufsgenossensch. e. V., 16–23 (1980)
338. Bignon, J. et al.: Long term pulmonary clearance of fibrous particles in man. In: Réactions Bronchopulmonaires aux Pollutants Atmospheriques, C. Voisin (ed.). Paris: INSERM 1974
339. Blount, M. et al.: The protein coating of asbestos bodies. Biochem. Journ. *101*, 204–207 (1966)
340. Brain, J.D. et al.: Pulmonary distribution of particles given by intratracheal instillation or by aerosol inhalation. Environ. Res. *11*, 13–33 (1976)
341. Burell, R.: Immunological reflections on asbestos. Envir. Health Perspect. *9*, 297–298 (1974)
342. Cralley, L.J: Electromotive phenomenon in metal and mineral particulate exposures: relevance to exposures to asbestos and occurrence of cancer. Am. Ind. Hyg. Ass. J. *32*, 653–661 (1971)
343. Davis, J.M.G. et al.: The mineral dust load of human lungs. Bronchopneumologie *26*, 103–113 (1976)
344. Merchant, J.A. et al.: The HL-A system in asbestos workers. Brit. Med. J *1*, 189–191 (1975)
345. Morgan, A. et al.: Studies of the solubility of constituents of chrysotile asbestos in vivo using radioactive tracer techniques. Environ. Res. *4*, 558–570 (1971)
346. Morgan, A., Holmes, A.: Neutron activation techniques in investigations of the composition and biological effects of asbestos. In: Pneumoconiosis: Proceedings of the Internat. Conf., Johannesburg, 52–56. Cape Town: Oxford Univ. Press 1969
347. Pernis, B. et al.: Rheumatoid factor in serum of individuals exposed to asbestos. Ann. N.Y. Acad. Sci. *132*, 112–120 (1965)
348. Turner Warwick M., Parkes, W.R.: Circulating rheumatoid and antinuclear factors in asbestos workers. Brit. Med. J. *III*, 492–495 (1970)
349. Westlake, G.E. et al.: Penetration of colonic mucosa by asbestos particles. Lab. Invest. *14*, 2029 (1965)
350. Blejer, H.P., Arlon, R.: Talc: a possible occupational and environmental carcinogen. J. Occup. Med *15*, 92–97 (1973)
351. Cralley, L.J. et al.: Exposure to metals in the manufacture of asbestos textile products. Am. Ind. Hyg. Ass. J. *28*, 452–461 (1967)
352. Cummins, B.T., Gibbs, G.W.: Contaminating organic material in asbestos. Brit. J. Cancer *23*, 358–362 (1969)
353. Dixon, J.R. et al.: The role of trace metals in chemical carcinogenesis: asbestos cancers. Cancer Research *30*, 1068–1074 (1970)
354. Harington, J.S., Roe, F.J.C.: Studies of carcinogenesis of asbestos fibres and their natural oils. Ann. N.Y. Acad. Sci. *132*, 439–450 (1965)
355. Harington, J.S.: Natural occurrence of oils containing 3.4-benzpyrene and related substances in asbestos. Nature *193*, 43–45 (1973)
356. Harington, G.S.: Chemical factors (including trace elements) as aetiological mechanisms. In: Biological effect of asbestos, IARC Scient. Publ. No. 8, 304.311 (1973). Lyon
357. Constantinidis, K.: Asbestos exposure – its related disorders. Brit. J. Clin. Pract. *31*, 89–101 (1977)
358. Dobbertin, S.: Gesundheitsgefahren durch Luftverunreinigungen mit Asbest. Staub-Reinhalt.Luft *37*, 65 (1977)
359. Gilson, J.C.: Asbestos cancer: Past and future hazards. Proc. Roy. Soc. Med. *66*, 395 (1973)
360. Graham, J., Graham, R.: Ovarian cancer and asbestos. Environ. Res. *1*, 115–128 (1967)

361. Guidotti, T.L. et al.: Asbestos and carcinoma of the larynx. JAMA *228,* 1571 (1974)
362. Hughes, D.T.: Lung disease related to asbestos. Med. Sci. Law *14,* 147–151 (1974)
363. International Union against Cancer: Working group on asbestos and cancer. Arch. Environ. Health *11,* 221–229 (1965)
364. Jones, J.S.: Pathological and environmental aspects of asbestos associated diseases. Med. Sci. Law *14,* 152–158 (1974)
365. Ledwan, K.: Asbesterkrankungen aus der Sicht der Betroffenen. Tagungsber. Asbest, 6. u. 7. 3. 1980, Bonn-Bad Godesberg, Schriftenr. d. Hauptverb. d. gewerbl. Berufsgenossensch. e. V., 56–58 (1980)
366. Parkes, W.R.: Asbestos related disorders. Brit. J. Dis. Chest *67,* 261–300 (1973)
267. Parkes, W.R.: Occupational lung disorders, 270–357. London: Butterworths 1974
368. Rodriguez-Roisin, R. et al.: Asbestos exposure and small airways disease. Scand. J. Respir. Dis. *57,* 318 (1976)
369. Selikoff, I.J. et al.: Asbestos exposure and neoplasia. JAMA *188,* 22–26 (1964)
370. Skidmore, W.J.: In symposium on diseases associated with asbestos dust exposure. Int. Pathol., 9–17 (1973)
371. Stell, P.M., McGill, T.: Asbestos and laryngeal cancer. Lancet *II,* 416–417 (1973)
372. Timbrell, V.: Inhalation and biological effects of asbestos. In: Mercer, T.T. et al. (ed.): Assessment of Airborne Particles. Proceed. 3rd Rochester Internat. Conf. on Environm. Toxicity, 429–455 (1970)
373. Von Chossy, R.: Pneumoconiosis due to silica and asbestos dust. ZFA (Stuttgart) *52,* 1667–1670 (1976)
374. Wagner, J.C.: The sequelae of exposure to asbestos dust. Ann. N.Y. Acad. Sci. *132,* 691–695 (1965)
375. Webster, I.: The pathology of asbestos. In: Rogan, M. (ed.): Medicine in the mining industries. London: William Heinemann Med. Books Ltd. 1972
376. Weill, H. et al.: Differences in lung effects resulting from chrysotile and crocidolite exposure. Inhaled Particles IV, Pt. 2, 799–813 (1977)
377. Woitowitz, H.-J.: Problematik der Einwirkung von krebserzeugenden Arbeitsstoffen: Einführung am Beispiel von asbesthaltigem Staub. Schriftenr. d. Hauptverb. d. gewerbl. Berufsgenossensch. e. V., 47–55. Stuttgart: A. W. Gentner 1979
378. Wright, G.D.: Asbestos and health. Amer. Rev. Resp. Dis. *2,* 467–479 (1969)
379. Bellote, J.A. et al.: Asbestosis in West Virginia and Eastern Ohio. W. Va. Med. J. *72,* 341–344 (1976)
380. Burilkov, T., Michailova, L.: Asbestos content of soil and endemic asbestosis. Environ. Res. *3,* 443 (1970)
381. Boutin, C. et al.: Benign asbestosic pleurities (àpropos of 3 cases). Poumon Coeur *31,* 111–118 (1975)
382. Britton, M.G. et al.: A survey of patients diagnosed as having asbestosis. Med. Sci. Law *16,* 279–284 (1976)
383. Britton, M.G. et al.: Seral pulmonary function tests in patients with asbestosis. Thorax *32,* 45–52 (1977)
384. Buchanan, W.D.: Asbestosis and primary intrathoracic neoplasm. Ann. N.Y. Acad. Sci. *132,* 507–518 (1965)
385. Cooke, W.E.: Fibrosis of the lungs due to inhalation of asbestos dust. Brit. Med. J. *2,* 147 (1924)
386. Cooke, W.E.: Pulmonary asbestosis. Brit. Med. J. *4,* 1024–1025 (1927)
387. Davis, J.M.G.: Electron microscope studies of asbestosis in man and animals. Ann. N.Y. Acad. Sci. *132,* 98–111 (1965)
388. Enticknap, J.P., Smither, W.J.: Peritoneal tumours in asbestosis. Brit. J. Ind. Med. *21,* 20 (1964)
389. Evans, C.C. et al.: Frequency of HLA antigen in asbestos workers with and without pulmonary fibrosis. Brit. Med. J. *I,* 603–605 (1977)
390. Gloyne, S.R.: The morbid anatomy and histology of asbestosis. Tubercle *14,* 445–451, 493–497, 550–558 (1932–1933)
391. Gloyne, S.R.: Two cases of squamous carcinoma of the lung occurring in asbestosis. Tubercle *17,* 5 (1935)

392. Gloyne, S.R.: Silicosis and Asbestosis (ed. A. J. Lanza). London: Oxford University Press 1938
393. Gloyne, S.R.: Pneumoconiosis. A histological survey of necropsy in 1205 cases. Lancet *I*, 810 (1951)
394. Hajdukowiecz, Z. et al.: Pulmonary asbestosis and lung neoplasms. Patol. Pol. *26*, 551–556 (1975)
395. Heard, B.E., Williams, R.: The pathology of asbestosis with reference to lung function. Thorax *16*, 264–281 (1961)
396. Hourihane, D.O'B., McCaughey, W.T.E.: Pathological aspects of asbestosis. Postgrad. Med. J. *42*, 613–622 (1966)
397. Kagan, E. et al.: Immunological studies of patients with asbestosis: I. Studies of cell-mediated immunity. Clin. Exp. Immunol. *28*, 261–267 (1977)
398. Kagan, E. et al.: Immunological studies of patients with asbestosis: II. Studies of circulating lymphoid cell numbers and humoral immunity. Clin. Exp. Immunol. *28*, 268–275 (1977)
399. Keal, E.E.: Asbestosis and abdominal neoplasms. Lancet *II*, 1211–1216 (1960)
400. Kiviluoto, R.: Asbestosis: aspects of its radiological features. In: Peneumoconiosis: Proceedings of the Internat. Conf., Johannesburg, 253–255. Cape Town: Oxford Univ. Press 1969
401. McVittie, J.C.: Asbestosis in Great Britain. Ann. N.Y. Acad. Sci. *132*, 128–138 (1965)
402. Merewether, E.R.A.: The occurrence of pulmonary fibrosis and other pulmonary affections in asbestos workers. J. Ind. Hyg. Toxicol. *12*, 198–222, 239–257 (1930)
403. Richards, R.J., Morris, T.G.: Collagen and mucopolysaccharide production in growing lung fibroblasts exposed to chrysotile asbestos. Life Sciences *12*, Pt. II, 441–451 (1973)
404. Shevchenko, A.M. et al.: On the mechanism of lessening the fibrogenic activity of dust processed with calcium hydroxide. Gigiena Truda i Professionalniye Zabol. *21* (8), 31–34 (1977)
405. Smither, W.J.: Asbestos and asbestosis. Ann. Occup. Hyg. *13*, 3–5 (1970)
406. Smither, W.J.: Secular changes in asbestosis in an asbestos factory. Ann. N.Y. Acad. Sci. *132*, 166–181 (1965)
407. Smither, W.J., Lewinsohn, H.C.: Asbestosis in textile manufacturing. In: Biological effects of asbestos, IARC Scient. Publ. No. *8*, 169–174 (1973). Lyon
408. Toivanen, A.: Pulmonary asbestosis and autoimmunity. Brit. Med. J. *1*, 691–692 (1976)
409. Trunevski, M. et al.: Non-occupational asbestosis. God. Zb. Med. Fak. Skopje *20*, 451–459 (1974)
410. Webster, I.: The Pathogenesis of asbestosis. In: Pneumoconiosis: Proceed. Internat. Conf., Johannesburg, 117–119. Cape Town: Oxford Univ. Press 1969
411. Wilson, M.R. et al.: Activation of the alternative complement pathway and generation of chemotactic factors by asbestos. J. Allergy Clin. Innumol. *60*, 218–222 (1977)
412. Woitowitz, H.J. et al.: Asbestosis and asbestos related tumours, assessment of disablement. Prax. Pneumol. *31*, 153–159 (1977)
413. Wood, W.B., Gloyne, S.R.: Pulmonary asbestosis. Lancet *II*, 445–448 (1930)
414. Wood, W.B., Gloyne, S.R.: Pulmonary asbestosis – a review of one hundred cases. Lancet *II*, 1382–1385 (1934)
415. Wyers, H.: Asbestosis. Postgrad. Med. J. *25*, 631–638 (1949)
416. Zolov, C. et al.: Pleural asbestosis in agricultural workers. Envir. Res. *1*, 287 (1967)
417. Burilkov, T., Babadjov, L.: Endemic occurrence of bilateral pleura calcification. Prax. Pneumonol. *24*, 7 (1970)
418. Gibbs, G.W.: The aetiology of pleural calcification: A study of Quebec chrysotile asbestos miners and millers. Arch. Envir. Health *31* (1979)
419. Hourihane, D.O'B. et al.: Hyaline and calcified pleural plaques as an index of exposure to asbestos. Brit. Med. J. *I*, 1069–1074 (1966)
420. Jones, J.S.P., Sheers, G.: Pleural plaques. In: Biological effects of asbestos, IARC Scient. Publ. No. *8*, 243–248 (1973). Lyon
421. Kiviluoto, R.: Pleural calcification as a roentgenologic sign of non-occupational anthophyllite asbestosis. Acta Radiol. *194*, 1–67 (1960) suppl.
422. Kiviluoto, R.: Pleural plaques and asbestosis: further observations on endemic and other non-occupational asbestos. Ann. N.Y. Acad. Sci. *132*, 235 (1965)

423. Klemperer, P., Rabin, C.B.: Primary neoplasms of the pleura. Arch. Pathol. *11*, 385 (1931)
424. Le Bouffant, L. et al.: Structure and function of pleural plaques. In: Biological effects of asbestos, IARC Scient. Publ. No. *8*, 249–257 (1973). Lyon
425. Mattson, S.B. et al.: Pleural asbestoses. Lakartidningen *72*, 3802–3804 (1975)
426. Mattson, S.B.: Editorial. Pleural plaques and asbestos. Lakartidningen *73*, 496–497 (1975)
427. Mattson, S., Ringqvist, T.: Pleural plaques and exposure to asbestos. Scand. J. Resp. Dis., Suppl. 75 (1970)
428. Meurman, L.: Asbestos bodies and pleural plaques in a finnish series of autopsy cases. Acta Path., Microbiol. Scand. *181*, Suppl. (1966)
429. Meurman, L.: Pleural fibrocalcific plaques and asbestos exposure. Environ. Res. *1*, 30–46 (1968)
430. Moigneteau, C. et al.: Asbestosic pleural calcifications and the associated pathology (32 cases). Poumen Coeur *32*, 101–106 (1977)
431. Navratil, M., Dobias, J.: Development of pleural hyalinosis in long term studies of persons exposed to asbestos dust. Environ. Res. *6*, 455 (1973)
432. Navratil, M.: Pleural calcifications due to asbestos exposure compared with relevant findings in the non-exposed population. In: Inhaled Particles III, Vol. 2, 695–701. London: Unwin Broth. 1971
433. Raunio, V.: Occurrence of unusual pleural calcification in Finland. Ann. Med. Intern. Fenn., Suppl. 47 (1966)
434. Selikoff, I.J.: The occurrence of pleural calcification among asbestos insulation workers. Ann. N.Y. Acad. Sci. *132*, 351–367 (1965)
435. Thomson, J.G.: The pathogenesis of pleural plaques. In: Pneumoconiosis: Proceed. Internat. Conf., Johannesburg, 138–141. Cape Town: Oxford Univ. Press 1969
436. Yazicioglu, S.: Pleural calcification associated with exposure to chrysotile asbestos in Sourth East Turkey. Chest *70*, 43–47 (1976)
437. Anspach, M. et al.: Ein Beitrag zur Ätiologie des diffusen malignen Pleura-Mesothelioms. – Arch. Gewerbepath. Gewerbehyg. *21*, 392–407 (1965)
438. Ashcroft, T., Heppleston, A.G.: Mesothelioma and asbestos on Tynside – a pathological and social study. In: Pneumoconiosis: Proceedings of the Internat. Conf., Johannesburg, 177–179. Cape Town: Oxford Univ. Press 1969
439. Beck, B., Irmscher, G.: Extrathorakle Mesotheliome durch Inhalation von Asbeststaub. Z. Erkrank. Atm.-Org. *152*, 282–293 (1979)
440. Bianchi, C. et al.: Diffuse mesothelioma of the peritoneum and exposure to asbestos (reflection on 2 cases). Pathologica *68*, 975–976 (1976)
441. Boersma, A. et al.: Diffuse mesothelioma: biochemical stages in the diagnosis, detection and measurement of hyaluronic acid in the pleural fluid. In: Biological Effects of Asbestos, IARC Pub. No. *8*, 65–67 (1973). Lyon
44.2 Butler, E.B.B., Berry, A.V.: Diffuse mesotheliomas: diagnostic criteria using exfoliative cytology. In: Biological effects of asbestos, IARC Publ. No. *8*, 68–73 (1973). Lyon
443. Champion, P.: Two cases of malignant mesothelioma after exposure to asbestos. Amer. Rev. Resp. Dis. *103*, 821 (1971)
444. Churg, J. et al.: Histological characteristics of mesothelioma associated with asbestos. Ann. N.Y. Acad. Sci. *132*, 614–622 (1965)
445. Davis, J.M.G.: Ultrastructure of human mesothelioma. J. Natl. Cancer Inst. *52*, 1715 (1974)
446. Elmes, P.C.: Therapeutic openings in the treatment of mesothelioma. In: Biological effects of asbestos, IARC Publ. No. *8*, 277–280 (1973). Lyon
447. Elmes, P.C.: The natural history of diffuse mesothelioma. In: Biological effects of asbestos, IARC Publ. No. *8*, 267–272 (1973). Lyon
448. Elmes, P.C., Simpson, M.J.C.: The clinical aspects of mesothelioma. Q.J. Med., New Ser. *45* (179) 427–449 (1976)
449. Embleton, M.J. et al.: Assessment of cell-mediated immunity to malignant mesothelioma by microcytotoxicity tests. Int. J. Cancer *17*, 597–601 (1976)
450. Glyn Owen, W.: Mesothelial tumours and exposure to asbestos dust. Ann. N.Y. Acad. Sci. *132*, 674–679 (1965)

451. Godwin, M.C.: Diffuse mesotheliomas. Cancer *10,* 28 (1957)
452. Greenberg, M.: Mesothelioma register. Brit. J. Ind. Med. *31,* 91 (1974)
453. Haider, M. et al.: Mesothelioma cases and asbestos exposure in Austria. Hefte Unfallheilkunde *126,* 547–549 (1975)
454. Hourihane, D.O'B.: The pathology of mesotheliomata and an analysis of their association with asbestos exposure. Thorax *19,* 268–278 (1964)
455. Hourihane, D.O'B.: A biopsy series of mesotheliomata and attempts to identify asbestos within some of the tumours. Ann. N.Y. Acad. Sci. *132,* 647–673 (1965)
456. Hasan, F.M. et al.: The significance of asbestos exposure in the diagnosis of mesothelioma, a 28 year experience from a major urban hospital. Ann. Rev. Resp. Dis. *115,* 761–768 (1977)
457. Heller, R.M. et al.: The radiological manifestations of malignant pleural mesothelioma. Amer. J. Roentgenol. *108,* 53 (1970)
458. Klempman, S.: The exfoliate cytology of diffuse pleural mesothelioma. Cancer *15,* 691–704 (1962)
459. Leiben, J., Pistawka, H.: Mesothelioma and asbestos exposure. Arch. Envir. Health *14,* 559–563 (1967)
460. Milne, J.E.: 32 cases of mesothelioma in Victoria. Australia: A retrospective survey related to occupational asbestos exposure. Brit. J. Med. *33,* 115–122 (1976)
461. Naylor, B.: The exfoliative cytology of diffuse malignant mesothelioma. J. Path. Bact. *86,* 293–298 (1963)
462. Newhouse, M.L., Thompson, H.: Mesothelioma of pleura and peritoneum following exposure to asbestos in the London area. Brit. J. Ind. Med. *22,* 261 (1965)
463. Otto, H.: Mesothelioma and asbestos exposure. Hefte Unfallheilkunde *126,* 555–559 (1975)
464. Stanton, M.F., Wrench, C.: Mechanisms of mesothelioma induction with asbestos and fibrous glass. J. Nat. Cancer Inst. *48,* 797–821 (1972)
465. Vollhaber, H.H.: Mesotheliom aus der Sicht einer Thoraxklinik. Tagungsber. Asbest, 6. u. 7. 3. 1980, Bonn-Bad Godesberg. Schriftenr. d. Hauptverb. d. gewerbl. Berufsgenossensch. e. V., 51–53 (1980)
466. Taryle, D.A. et al.: Pleural mesothelioma: An analysis of 18 cases and review of the literature. Medicine *55,* 153 ff. (1976)
467. Wagner, J.C. et al.: Diffuse pleural mesothelioma and asbestos exposure in North Western Cape Province. Brit. J. Ind. Med. *13,* 250(1960)
468. Wagner, J.C. et al.: Diffuse pleural mesothelioma. Brit. J. Ind. Med. *17,* 260–271 (1960)
469. Wagner, J.C. et al.: Histochemical demonstration of hyaluronic acid in pleural mesotheliomas. J. Path. Bact. *84,* 73–78 (1962)
470. Webster, I.: Mesotheliomatous tumours in South Africa: pathology and experimental pathology. Ann. N.Y. Acad. Sci. *132,* 623–646 (1965)
471. Whitwell, F., Rawcliffe, R.M.: Diffuse malignant pleural mesothelioma and asbestos exposure. Thorax *26,* 6–22 (1971)
472. Whitwell, F. et al.: Relationships between occupations and asbestos fibre content in patients with pleural mesothelioma by cancer and other diseases. Thorax *32,* 377–386 (1977)
473. Winslow, D.J., Taylor, H.B.: Malignant peritoneal mesotheliomas: a clinico-pathological analysis of 12 cases. Cancer *13,* 127 (1960)
474. Kannerstein, M., Churg, J.: Pathology of carcinoma of the lung associated with asbestos exposure. Cancer *30,* 14–21 (1972)
475. Leicher, F.: Primärer Deckzellentumor des Bauchfelles bei Asbestose. Arch. Gewerbehyg. *13,* 382 (1954)
476. Levy, B.S. et al.: Investigating possible effects of asbestos in city water: surveillance of gastrointestinal cancer incidence in Duluth, Minnesota. Am. J. Epidemiol. *103,* 362–368 (1976)
477. Libshitz, H.I. et al.: Asbestos and carcinoma of the larynx. JAMA *228,* 1571 (1974)
478. Lynch, K.M., Smith, W.A.: Carcinoma of the lung in asbestos-silicosis. Amer. J. Cancer *24,* 56–64 (1935)
479. Lynch, K.M., Smith, W.A.: Pulmonary asbestosis III. Carcinoma of lung in asbestos workers. Am. J. Cancer *24,* 56 (1935)

480. Martischnig, K.M. et al.: Unsuspected exposure to asbestos and bronchogenic carcinoma. Brit. Med. J. *1*, 746–749 (1977)
481. McCaughey, W.T.E.: Primary tumours of the pleura. J. Path. Bact. *76*, 517 (1958)
482. Meyer, K., Chaffee, F.: Hyaluronic acid in the pleural fluid associated with a malignant tumour involving the pleura and peritoneum. J. Biol. Chem. *133*, 83–91 (1940)
483. Selikoff, I.J.: Environmental cancer associated with inorganic microparticulate air pollution. In: Clinical Implications of Air Pollution Research (A. J. Finkel, W. C. Duel, eds.), 49–66 (1974). Acton, Mass., Publishing Sciences Groups
484. Webster, I.: Malignancy in relation to crocidolite and amosite. In: Biological effects of asbestos, IARC Scient. Publ. No. *8*, 195–198 (1973)
485. Wedler, H.-W.: Über den Lungenkrebs bei Asbestose. Dtsch. Arch. klin. Med. *191*, 189 (1943)
486. Whitwell, F. et al.: A study of the histological types of lung cancer in workers suffering from asbestosis in the United Kingdom. Brit. J. Ind. Med. *31*, 298–303 (1974)
487. Berry, G. et al.: Combined effect of asbestos exposure and smoking on mortality from lung cancer in factory workers. Lancet *II*, 476–479 (1972)
488. Hammond, E.C., Selikoff,I.J.: Relation of cigarette smoking to risk of death of asbestos-associated disease among insulation workers in the United States. In: Biological effects of asbestos, IARC Scient. Publ. No. *8*, 312–317 (1973). Lyon
489. Selikoff, I.J. et al.: Asbestos exposure, smoking and neoplasia. J. Amer. Med. Ass. *204*, 106 (1968)
490. Shettigara, P.T., Morgan, R.W.: Asbestos, smoking and laryngeal carcinoma. Arch. Environ. Health *30*, 517–518 (1975)
491. Weiss, W.: Cigarette smoking, asbestos and pulmonary fibrosis. Amer. Rev. Resp. Dis. *104*, 223–227 (1971)
492. Attia, O.M. et al.: Sputum picture in workers at an Egyptian asbestos cement pipe factory. J. Egypt. Med. Ass. *58*, 227–233 (1975)
493. Bader, M.E. et al.: Pulmonary function and radiographic changes in 598 workers with varying duration of exposure to asbestos. Mt. Sinai J. Med. *37*, 492 (1970)
494. Becklake, M.R. et al.: Lung function in relation to chest radiographic changes in Quebec asbestos workers. In: Pneumoconiosis: Proceed. Internat. Conf., Johannesburg 1969. Cape Town: Oxford Univ. Press 1970
495. Berge, T., Gröntoft, O.: Cytologic diagnosis of malignant pleural mesothelioma. Acta Cytol. *9*, 207–212 (1965)
496. Cauna, D. et al.: Asbestos bodies in human lungs at autopsy. J. Am. Med. Ass. *192*, 371 (1965)
497. Cohen, D.: Ferromagnetic contamination in the lungs and other organs of the human body. Science *180*, 745–748 (1973)
498. Doniach, I. et al.: Prevalence of asbestos bodies on a necropsy series in East London. Brit. J. Ind. Med. *32*, 16–34 (1975)
499. Fletcher, D.E., Edge, J.R.: The early radiological changes in pulmonary and pleural asbestosis. Clin. Radiol. *21*, 355–365 (1970)
500. Fondimare, A., Desbordes, J.: Asbestos bodies and fibers in lung tissues. Environ. Health Perspect. *9*, 147 (1974)
501. Fukui, T. et al.: Biopsy study of asbestosis found 18 years after 1 year exposure. Jap. J. Thorac. Dis. *14*, 17–20 (1976)
502. Gold, C.: A simple method of detecting asbestos in tissues. J. Clin. Pathol. *20*, 674 (1967)
503. Gough, J.: Differential diagnosis in the pathology of asbestosis. Ann. N.Y. Acad. Sci. *132*, 368–372 (1965)
504. Gloyne, S.R.: The asbestos body. Lancet *I*, 1351–1355 (1932)
505. Greenberg, S.D. et al.: Sputum cytopathological findings in former asbestos workers. Tex. Med. *72*, 39–43 (1976)
506. Gross, P. et al.: Asbestos bodies, their non-specifity. Amer. Ind. Hyg. Ass. J. *28*, 541–544 (1967)
507. Hinson, K.F.W. et al.: Criteria for the diagnosis and grading of asbestos. In: Biological effects of Asbestos, IARC Scient. Publ. No. *8*, 54–57 (1973). Lyon
508. Kane, P.B. et al.: Diagnosis of asbestosis by transbronchial biopsy. A method to facilitate demonstration of ferruginous bodies. Amer. Rev. Resp. Dis. *115*, 689–694 (1977)

509. Kannerstein, M. et al.: Histochemical studies in the diagnosis of mesothelioma. In: Biological effects of asbestos, IARC Scient. Publ. No. *8*, 62–64 (1973). Lyon
510. Leathart, G.L.: Pulmonary function tests in asbestos workers. Trans. Soc. Occup. Med. *18*, 49–55 (1968)
511. Liddell, D. et al.: Radiological changes over 20 years in relation to chrysotile exposure in Quebec. Inhaled Particles IV, Pt. 2, 799–813 (1975)
512. Matsuda, M.: Asbestos bodies. 1. Detection of asbestos bodies in the lung at autopsy. Jap. J. Thorac. Dis. *13*, 40–44 (1975)
513. McCaughey, W.T.E.: Criteria for diagnosis of diffuse mesothelial tumours. Ann. N.Y. Acad. Sci. *132*, 603–613 (1965)
514. McAughey, W.T.E., Oldham, P.D.: Diffuse mesotheliomas; observer variation in histological diagnosis. In: Biological effects of asbestos, IARC Scient. Publ. No. *8*, 58–61 (1973). Lyon
515. McPherson, P. Davidson, J.K.: Correlation between lung asbestos count at necroscopy and radiological appearances. Brit. Med. J. *I*, 355–357 (1969)
516. Naylor, B.: The role of the cytology laboratory in the diagnosis of diffuse malignant mesothelioma. In: Proceedings of Internat. Conf. "Biologische Wirkungen des Asbestes", Dresden, 288–295 (1968)
517. Pooley, F.D.: Asbestos bodies, their formation, composition and character. Envir. Res. *5*, 363–379 (1972)
518. Raithel, H.J.: Computertomographische Nachweismöglichkeiten asbeststaubinduzierter Lungen- und Pleuraveränderungen. Tagungsber. Asbest, 6. u. 7. 3. 1980, Bonn-Bad Godesberg, Schriftenr. d. Hauptverb. d. gewerbl. Berufsgenossensch. e. V., 60–62 (1980)
519. Roberts, H.: Asbestos bodies in lungs at necropsy. J. Clin. Pathol. *20*, 570–573 (1967)
520. Rossiter, C.E. et al.: Radiographic changes in chrysotile asbestos mine and mill workers of Quebec. Arch. Envir. Health *24*, 388–400 (1972)
521. Shishido, S. et al.: Asbestos pollution of the lung. 1. Incidence of asbestos bodies found in the lung and sputum. Jap. J. Thorac. Dis. *14*, 728–735 (1976)
522. Simson, F.W., Strachan, A.S.: Asbestosis bodies in the sputum; a study of specimens from fifty workers in an asbestos mill. J. Path. Bact. *34*, 1 (1931)
523. Weyer, R.V.: Radiological aspects of asbestosis. J. Belge Radiol. *58*, 347–361 (1975)
524. Barnes, R.: Asbestos spraying and occupational and environmental hazard. Med. J. Austral. *2*, 599–602 (1976)
525. Berkovitch, I.: Hazards of asbestos in construction practice. A review of the U.K. sources of information and advice. London: Construction Industries Research and Information Association (1976)
526. Bonser, G.M. et al.: Occupational cancer of the urinary tract in dye-stuffs operatives, and of the lung in asbestos textile workers and iron-ore miners. Am. J. Clin. Pathol. *25*, 126 (1955)
527. Gibbs, G.W.: Qualitative aspects of dust exposure in the Quebec asbestos mining and milling industry. In: Walton, W.H. (ed.): Inhaled Particles III, 783–799 (1970). Unwin Broth.
528. Hammond, E.C. et al.: Neoplasia among insulation workers in the United States with special reference to intraabdominal neoplasia. Ann. N.Y. Acad. Sci. *132*, 519–525 (1965)
529. Harries, P.G.: Experience with asbestos disease and its control in Great Britain's naval dokkyards. Environ. Res. *11*, 261–267 (1976)
530. Harries, P.G.: Asbestos hazards in Naval dockyards. Ann. Occup. Hyg. *11*, 135–145 (1968)
531. Jacob, G., Anspach, M.: Pulmonary neoplasia among Dresden asbestos workers. Ann. N.Y. Acad. Sci. *132*, 536–548 (1965)
532. Jones, J.S.P. et al.: Factory populations exposed to crocidolite asbestos – a continuing survey. In: Environmental Pollution and Carcinogenic Risks, INSERM Symposia series Vol. 52, IARC Scient. Publ. No. *13*, 117–120 (1976). Lyon
533. Keen, R.C., Mumford, C.J.: A preliminary study of hazards to toxic waste disposal operators on ten landfill sites in Britain. Ann. Occup. Hyg. *18*, 213–228 (1975)
534. Kelly, R.T.: Asbestos health hazards in perspective. Constructional use. Roy. Soc. Health J. *96*, 246–248 (1976)
535. Kinsey, J.S. et al.: A preliminary survey of the hazards to operators in the disposal of asbestos waste. Ann. Occup. Hyg. *20*, 85–89 (1977)

536. Kogan, F.M. et al.: Working conditions of woman track maintenance workers in asbestos mines. Gig. Sanit. *9*, 19–23 (1977)
537. Maugh, T.H.: Workplace carcinogens: A new look of industry. Science *197*, No. 4310, 1268 (1977)
538. Mayer, P.: Zusammenführung medizinischer Daten (Befunde) und meßtechnischer Daten bei asbeststaubexponierten Personen. Tagungsber. Asbest, 6. u. 7. 3. 1980, Bonn-Bad Godesberg, Schriftenr. d. Hauptverb. d. gewerbl. Berufsgenossensch. e. V., 58–59 (1980)
539. McDonald, J.C.: Cancer in chrysotile mines and mills. In: Biological effects of asbestos, IARC Scient. Publ. No. *8*, 189–193 (1973). Lyon
540. McDonald, J.C. et al.: The health of chrysotile asbestos mine and mill workers. Arch. Environ. Health *28*, 61–68 (1974)
541. Mihajlov, P. et al.: Occupational lesions associated with the production and processing of asbestos and mica. Hihiena i Zdraveopazvane (Bulgarien) *11*, No. 2, 145–150 (1968)
542. Muir, D.C.F.: Health hazards of thermal insulation products. Ann. Occup. Hyg. *19*, 139–145 (1976)
543. Newhouse, M.L.: Cancer among workers in the asbestos textile industry. In: Biological effects of asbestos, IARC scient. Publ. No. *8*, 203–208 (1973). Lyon
544. Seaton, T.: Regional lung function in asbestos workers. Thorax *32*, 40–44 (1977)
545. Smither, W.J.: Asbestos in the workplace and the community. Environ. Health Perspect. *9*, 327–329 (1974)
546. Valentin, H., Otto, H.: Kriterien zur Anerkennung bösartiger Neubildungen als Berufskrankheit. Die Berufsgenossenschaft *30*, 151–156 Nr. 4 (1976)
547. Versen, P.: Berufsbedingte Krebserkrankungen und ihre versicherungsrechtliche Beurteilung. Die Berufsgenossenschaft *30*, (4), 290–299 (1978)
548. Versen, P.: Berufsbedingte Erkrankungen durch Asbest im Bereich der gewerblichen Berufsgenossenschaften. Tagungsber. Asbest, 6. u. 7. 3. 1980, Bonn-Bad Godesberg. Schriftenr. d. Hauptverb. d. gewerbl. Berufsgenossenschaften e. V., 24–38 (1980)
549. Weill, H. et al.: Lung function consequences of dust exposure in asbestos cement manufacturing plants. Arch. Envir. Health *30*, 88–97 (1975)
550. Weill, H. et al.: Influence of dose and fibre type on respiratory malignancy in asbestos cement manufacturing. Amer. Rev. Resp. Dis. *120*, 345–353 (1979)
551. Becklake, M.R.: Asbestos related diseases of the lung and other organs: their epidemiology and implication for clinical practice. Amer. Rev. Resp. Dis. *114*, 187–227 (1976)
552. Dalquen, R. et al.: Pleuraplaques, Asbestose und Asbestexposition, eine epidemiologische Studie aus dem Hamburger Raum. Pneumologie *143*, 23 (1970)
553. Enterline, P.: Pitfalls in epidemiological research. An examination of the asbestos literature. J. Occup. Med. *18*, 150–156 (1976)
554. Gilson, J.C.: Proceedings: Biological effects of asbestos, unanswered questions posed by epidemiological studies. Clin. Sci. Mol. Med. *47* (3) (1974) 11 p.
555. Gilson, J.C.: Process in epidemiology. In: Biological effects of asbestos, IARC Scient. Publ. No. *8*, 5–10 (1973). Lyon
556. Hain, E.: Current results of epidemiological studies on the asbestos problem in Northern Germany. Hefte Unfallheilk *126*, 536–538 (1975)
557. Hany, A.: Asbestos problems in Switzerland. Hefte Unfallheilk *126*, 542–545 (1975)
558. McDonald, J.C. et al.: Epidemiology of primary mesothelial tumors in Canada. Cancer *26*, 914–919 (1970)
559. McDonald, J.C., McDonald, A.D.: Epidemiology of mesothelioma from estimated incidence. Prev. Med. *6*, 426–446 (1977)
560. McDonald, A.D. et al.: Epidemiology of primary malignant mesothelial tumours in Canada. In: Pneumoconiosis: Proceedings of the Internat. Conf., Johannesburg, 197–200. Cape Town: Oxford Univ. Press 1969
561. McEwen, J. et al.: Mesothelioma in Scotland. Brit. Med. J. *4*, 575 (1970)
562. McNulty, J.C.: Asbestos exposure in Australia. In: Pneumoconiosis: Proceedings of the Internat. Conf., Johannesburg, 201–203. Cape Town: Oxford Univ. Press 1969
563. Newhouse, M.L., Thompson, H.: Epidemiology of mesothelial tumors in the London area. Ann. N.Y. Acad. Sci. *132*, 579 (1965)

564. Nicholson, W.J. et al.: Epidemiological evidence on asbestos. N.B.S. Spec. Publ. *506*, 71–84 (1978). Proceedings of the workshop: Definitions and measurement methods
565. Nurminen, M.: The epidemiologic relationship betwen pleural mesothelioma and asbestos exposure. Scand. J. Work. Envir. Hlth. *1*, 128 (1975)
566. Otto, H.: Zur Epidemiologie des berufsbedingten Mesothelioms in der Bundesrepublik Deutschland. Tagungsber. Asbest, 6. u. 7.3. 1980, Bonn-Bad Godesberg. Schriftenr. d. Hauptverb. d. gewerbl. Berufsgenossensch. e.V., 49–50 (1980)
567. Plantedydt, H.T.: Mesothelioma and asbestos in the Netherlands. Hefte Unfallheilk. *126*, 549–555 (1975)
568. Rubino, G.F. et al.: Epidemiology of pleural mesothelioma in north-western Italy (Piedmont). Brit. J. Ind. Med. *29*, 436 (1972)
569. Sarraci, R.: Asbestos and lung cancer: an analysis of the epidemiological evidence on the asbestos smoking reaction. Int. J. Cancer *20*, 323–331 (1977)
570. Selikoff, I.J.: Asbestkrankheiten in den Vereinigten Staaten von 1918–1975. Vortrag auf der IBM-Weltkonferenz Oslo, 16.–19.8. 1976
571. Selikoff, I.J.: Epidemiological investigations of asbestos workers in the United States. Hefte Unfallheilk. *126*, 512–520 (1975)
572. Selikoff, I.J., Hammond, E.C.: Asbestos bodies in the New York City population in two periods of time. In: Pneumoconiosis: Proceed. Inter. Conf. Johannesburg, 180. Cape Town: Oxford Univ. Press 1969
573. Wagner, J.C. et al.: Epidemiology of asbestos cancers. Brit. Med. Bull. *27*, 71–76 (1971)
574. Webster, I.: Asbestos exposure in South Africa. In: Pneumoconiosis: Proceed. Internat. Conf., Johannesburg, 209–212. Cape Town: Oxford Univ. Press 1969
575. Woitowitz, H.J.: Arbeitsmedizinisch-epidemiologische Untersuchung zu der unmittelbaren Gesundheitsgefahr durch Asbest. Arbeit und Gesundheit, Heft 86 (1972)
576. Doll, R.: Mortality from lung cancer in asbestos workers. Brit. J. Ind. Med. *12*, 81–86 (1955)
577. Elmes, P.C., Simpson, M.J.C.: Insulation workers in Belfast mortality 1940–1966. Brit. J. Ind. Med. *28*, 226–236 (1971)
578. Elmes, P.C., Wade, O.L.: Relationship between exposure to asbestos and pleural malignancy in Belfast. Ann. N.Y. Acad. Sci. *132*, 549–557 (1965)
579. Enterline, P.E.: Mortality among asbestos products workers in the United States. Ann. N.Y. Acad. Sci. *132*, 156–165 (1965)
580. Enterline, P.E., Marsh, G.M.: Final report on part one of mortality among man-made mineral fiber workers in the United States, submitted to the Medical and Scientific Committee of the Thermal Insulation Manufacturers Association (1979)
581. Enterline, P.H. et al.: Mortality in relation to occupational exposure in the asbestos industry. J. Occup. Med. *14*, 897–903 (1972)
582. Fears, T.R.: Cancer mortality and asbestos deposits. Am. J. Epidemiol. *104*, 523–526 (1976)
583. Fletcher, D.E.: A mortality study of shipyard workers with pleural plaques. Brit. J. Ind. Med. *29*, 142–145 (1972)
584. Knox, J.F. et al.: Mortality from lung cancer and other causes among workers in an asbestos textile factory. Brit. J. Ind. Med. *25*, 293–303 (1968)
585. Konstanty, R.: Dunkelziffer asbestbedingter Erkrankungen. Tagungsber. Asbest. 6. u. 7.3. 1980, Bonn-Bad Godesberg. Schriftenr. f. Hauptverb. d. gewerbl. Berufsgenossensch. e.V., 53–55 (1980)
586. Konzen, J.L.: Comments on mortality patterns among fibrous production workers. U.S. Dept. Commerce, Natl. Techn. Inform. Service No. PB-257-784 (1976)
587. Lacquet, L.M. et al.: Roentgenographic lung changes, asbestosis and mortality in a Belgian asbestos cement factory. IARC Scient. Publ. (1979). Lyon
588. Mason, I.J. et al.: Asbestos-like fibers in Duluth water supply: relation to cancer mortality. JAMA *228*, 1019 (1974)
589. McDonald, J.C. et al.: Mortality in the chrysotile asbestos mines and mills of Quebec. Arch. Environ. Health *22*, 677–686 (1971)
590. McDonald, J.C. et al.: Chrysotile fibre concentration and lung cancer mortality: A preliminary report. Symposium on Biological effects of Mineral Fibres, Lyon (1979)

591. Meurman, L.O. et al.: Mortality and morbidity of employees of anthophyllite asbestos mines in Finland. In: Biological effects of asbestos, IARC Scient. Publ. No. *8*, 199–202 (1973). Lyon
592. Newhouse, M.L.: A study of the mortality of workers in an asbestos factory. Brit. J. Ind. Med. *26*, 294–301 (1969)
593. Newhouse, M.L., Berry, G.: Predictions of mortality from mesothelioma tumours in asbestos factory workers. Brit. J. Ind. Med. *33*, 147–151 (1976)
594. Newhouse, M.L. et al.: A study of the mortality of female asbestos workers. Brit. J. Ind. Med. *29*, 134–141 (1972)
595. Selikoff, I.J. et al.: Mortality experiences of asbestos insulation workers. In: Pneumoconiosis: Proceedings of the Internat. Conf., Johannesburg, 1969, 180–186. Cape Twon: Oxford Univ. Press
596. Wigle, D.T.: Cancer mortality in relation to asbestos in municipal water supplies. Arch. Environ. Health *32*, 185 (1977)
597. Allison, A.C.: Experimental methods-cell tissue culture: effects of asbestos particles on macrophages, mesothelial cells and fibroblasts. In: Biological effects of asbestos, IARC Scient. Publ. No. *8*, 89–93 (1973). Lyon: Internat. Agency for Res. on Cancer
598. Beck, E.G.: Reactions of macrophages cultivated in vitro towards particulate and fibrous dusts. Proc. Internat. Sympos. Industr. Toxicol., Lucknow, Nov. 4–7 (1975)
599. Bolton, R.E. et al.: The short term effect of chronic asbestos injection in rats. Ann. Occup. Hyg. *19*, 121–123 (1976)
600. Brown, R.C. et al.: In vitro biological effects of glass fibres. J. Envir. Pathol. & Toxicol. *2*, 1369–1383 (1979)
601. Davis, J.M.G. et al.: "Ferrugineous bodies" in guinea pigs. Arch. Pathol. *89*, 364 (1970)
602. Engelbrecht, F.M. et al.: Mesothelial reaction of asbestos and other irritants after intraperitoneal injection. South Afr. Med. J. *49*, 87–90 (1975)
603. Holt, P.F. et al.: Experimental asbestosis with four types of fibres: importance of small particles. Ann. N.Y. Acad. Sci. *132*, 87–97 (1965)
604. Holt, P.F. et al.: The early effects of chrysotile asbestos dust on the rat lung. J. Pathol. and Bacteriol. *87*, 15–23 (1964)
605. Kanazawa, K. et al.: Migration of asbestos fibres from subcutaneous injection sites in mice. Brit. J. Cancer *24*, 96–106 (1970)
606. Kuschner, M., Wright, G.W.: The influence of varying lengths of glass and asbestos fibers on tissue response in guinea pigs. Inhaled Particles IV, Vol. 2, 455–474 (1977)
607. Peacock, P.R., Peacock, A.: Asbestos-induced tumors in white leghorn fowls. Ann. N.Y. Acad. Sci. *132*, 501–503 (1965)
608. Pelnar, P.V. (ed.): Fibres for biological experiments. Conference of the Institute of Occupational and Environmental Health (I.O.E.H.), Montreal, October 29–30, 1973. Transcript, 89 (1974)
609. Pott, F. et al.: Ergebnisse aus Tierversuchen zur kanzerogenen Wirkung faserförmiger Stäube und ihre Bedeutung im Hinblick auf die Tumorentstehung beim Menschen. Zbl. Bakt. Hyg., I. Abt., Orig. ber. *162*, 467–505 (1976)
610. Pott, F. et al.: Tumorigenic effects of fibrous dusts in experimental animals. Envir. Health Persp. *9*, 313–315 (1974)
611. Pott, F.: Animal experiments on biological effects of mineral fibres – a review. Sympos. on biol. effects of mineral fibres, Lyon, 25.–27.9. 1979
612. Pylev, L.N. et al.: Mechanism of the induction of asbestos mesotheliomas in the pleura of rats. Vopr. Onkol. *22*, 63–68 (1976)
613. Pylev, L.N., Shabad, L.M.: Some results of experimental studies in asbestos carcinogenesis. In: Biological effects of asbestos, IARC Scient. Publ. No. *8*, 99–105 (1973). Lyon
614. Richards, R.J. et al.: Light microscope studies on the effect of chrysotile asbestos and fiber glass on the morphology and reticulin formation of cultured lung fibroblasts. Environ. Res. *11*, 112–121 (1976)
615. Smith, W.E. et al.: Tests for carcinogenicity of asbestos. Ann. N.Y. Acad. Sci. *132*, 456–488 (1965)
616. Stanton, M.F. et al.: Experimental pulmonary carcinogenesis with asbestos. Amer. Ind. Hyg. Ass. J. *30*, 236–244 (1969)

617. Tetley, T.D. et al.: Chrysotile induced asbestosis: changes in the free cell population, pulmonary surfactant and whole lung tissue of rats. Brit. J. Exp. Pathol. *5*, 505–514 (1976)
618. Vorwald, A.J. et al.: Experimental studies of asbestosis. Arch. Ind. Hyg. *3*, 1–43 (1951)
619. Wagner, J.C.: Asbestosis in experimental animals. Arch. Industr. Hyg. *20*, 1–12 (1963)
620. Wagner, J.C. et al.: Mesotheliomata in rats after inoculation with asbestos and other materials. Brit. J. Cancer *28*, 173–185 (1973)
621. Beckett, S.T.: The generation and evaluation of UICC asbestos clouds in animal exposure chambres. Ann. Occup. Hyg. *18*, 187–198 (1975)
622. Timbrell, V. et al.: Exposure chambers for inhalation experiments with standard reference samples of asbestos of the International Union against Cancer (UICC). J. Aerosol Sci. *1*, 215–223 (1970)
623. Berkley, C. et al.: The detection and localization of mineral fibres in tissue. Ann. N.Y. Acad. Sci. *132*, 48–63 (1965)
624. Churg, A. et al.: A simple method for preparing ferruginous bodies for electron microscope examination. Am. J. Clin. Pathol. *68*, 513–517 (1977)
625. Davies, R. et al.: Identification to toxic mineral dusts using mammalian cells. Trans. Roy. Soc. Edinburgh: Earth Sci. *71*, 181–184 (1980)
626. Gleit, C.E., Holland, W.D.: Use of electrically excited oxygen for the low temperature decomposition of organic substances. Analyt. Chem. *34*, 1454–1457 (1962)
627. Gold, C.: A simple method of detecting asbestos in tissues. J. Clin. Pathol. *20*, 674 (1967)
628. Morgan, A. et al.: Studies of the solubility of constituents of chrysotile asbestos in vivo using radioactive tracer techniques. Envir. Res. *4*, 558–570 (1971)
629. Pooley, F.D.: Proceedings: The recognition of various types of asbestos as minerals, and in tissues. Clin. Sci. Mol. Med. *47* (3), 11P–12P (1974)
630. Pooley, F.D.: An examination of the fibrous mineral content of asbestos lung tissue from the Canadian chrysotile mining industry. Envir. Res. *12*, 281–298 (1976)
631. Pooley, F.D. et al.: The detection of asbestos in tissues. In: Pneumoconiosis: Proceed. Internat. Conf. Johannesburg 1969, 108–116. Cape Town: Oxford University Press 1970
632. Talvitie, N.A., Brewer, L.W.: Separation and analysis of dust in lung tissue. Amer. Ind. Hyg. Ass. J. *23*, 58–61 (1962)
633. Carton, B., Kauffer, E.: The metrology of asbestos. Atmosph. Environm. *14*, 1181–1186 (1980)
634. Gibbs, G.W.: Physical parameters of airborne asbestos fibres in various work environments – preliminary findings. Amer. Ind. Hyg. Ass. J. *36*, 459–466 (1975)
635. Walkenhorst, W., Coenen, W.: Zusammenhänge zwischen den äußeren Abmessungen von Aggregaten und ihrem aerodynamischen Durchmesser. Reinhalt. Luft *37*, 106–109 (1977)
636. Walkenhorst, W.: Modellversuche zur Bestimmung des dynamischen Formfaktors nichtisometrischer Teilchen. Staub-Reinhalt.Luft *36*, 149–155 (1976)
637. Timbrell, V.: Alignment of amphibole asbestos fibres by magnetic fields. Microscope *20*, 365–368 (1972)
638. Timbrell, V.: Alignment of respirable fibres by magnetic fields. Ann. Occup. Hyg. *18*, 299–312 (1975)
639. Asbestos International Association: Referenz-Methode zur Bestimmung der Asbestfaserkonzentration im Schwebestaub am Arbeitsplatz durch Lichtmikroskopie. (Membran-Filter-Methode). 1. 1. 1982., 60 pp.
640. Beckett, S.T. et al.: A comparison of airborne asbestos fibre counting with and without an eyepiece graticule. Ann. Occup. Hyg. *19*, 69–76 (1976)
641. Bobeth, W., Müller, U.: Zur Identifizierung anorganischer Faserstoffe. Faserforsch. u. Textiltechn. *7*, 497–504 (1956)
642. Bobeth, W.: Bestimmungsschlüssel für anorganische Faserstoffe. Silikattechn. *12* (12), 543 (1961)
643. Brown, K.M., McCrone, W.C.: Dispersion staining: Part I. Theory, method and apparatus. Microscope *13*, 311–322 (1963)
644. Brown, K.M. et al.: Dispersion staining: Part II. The systematic application to the identification of transparent substances. Microscope *14*, 39–54 (1963)
645. Champness, P.E. et al.: The identification of asbestos. J. Microscopy *108*, 231–249 (1976)

646. Chao, A.T.C.: The application of quantitative interference microscopy to mineralogic and petrologic investigations. Amer. Mineral. *61*, 212–228 (1976)
647. Cherkasov, J.A.: Applications of "focal screening" to measurement indices of refraction by immersion method. Internat. Geol. Rev. *2*, 218–235 (1960)
648. Dodgson, J.: Use of interference microscopy for the mineralogical analysis of samples of airborne dust obtained with the thermal precipitator. Nature (London) *199*, 245–247 (1963)
649. Goni, J. et al.: Emploi de la microscopie en contraste de phase pour l'identification des particules fibreuses présentes dans la poumon humain. Bull. Soc. Franç. Minér. Cristallogr. *98*, 294–298 (1975)
650. Heidermanns, G.: Asbestos content determination by optical, chemical, radiographic and IR spectrographic analysis procedures. Staub (engl.) *33*, 67–72 (1973)
651. Heidermanns, G. et al.: Asbestbestimmung in industriellen Feinstäuben und in Lungenstäuben. Staub *36*, 107–111 (1976)
652. Heidermanns, G. et al.: Asbestos determination in industrial microdusts and in pulmonary dusts. Hefte Unfallheilk. *126*, 617–623 (1975); 574–584 (1975)
653. Heidermanns, G.: Asbestgehaltsbestimmung durch optische, chemische, röntgenographische und infrarotspektrographische Analysenverfahren. Staub-Reinhaltg.-Luft *2*, 66–70 (1973)
654. Julian, Y., McCrone, W.C.: Identification of asbestos fibres by microscopical dispersion staining. Microscope *18*, 1–10 (1970)
655. McCrone, W.C.: Detection and identification of asbestos by mycroscopical dispersion staining. Envir. Health Perspect. *9*, 57–61 (1974)
656. McCrone, W.C.: Identification of asbestos by polarized light microscopy. N.B.S. Spec. Publ. *506*, 235–247 (1978)
657. Schmidt, K.G.: Asbestsorten, ihre Untersuchung mit optischen Mitteln und ihre krankmachende Wirkung. Staub *20*(6), 173–180 (1960)
658. Schmidt, K.G., Heidermanns, G.: Untersuchung von Staubproben mit dem Phasenkontrastmikroskop, insbesondere bei Verwendung von Membranfiltern. Staub *19* (12), 413–416 (1959)
659. Walton, W.H., Beckett, S.T.: A microscope eyepiece graticule for the evaluation of fibrous dust. Ann. Occup. Hyg. *20*, 19–23 (1977)
660. Abraham, J.L., DeNee, P.B.: Biomedical applications of backscattered electron imaging – one year's experience with SEM histochemistry. Scanning Electron Microscopy (Part 1), ITT Res. Inst., Chicago, Ill., 251–255 (1974)
661. Ashcroft, T., Heppelston, A.G.: The optical and electron microscopic determination of pulmonary asbestos fibre concentration and its relation to human pathological reaction. J. Clin. Pathol. *26*, 224–234 (1973)
662. Beaman, D.R., Walker, H.J.: Difficulties encountered in the identification of asbestos fibres by analytical transmission electron microscopy. Sympos. on Electron Microscopy of Microfibers, 98–105 (1976). Washington: U.S. Gov. Printing Office
663. Beckett, S.T.: The evaluation of airborne asbestos fibres using a scanning electron microscope. Ann. Occup. Hyg. *16*, 405–408 (1973)
664. Boles, M.O. et al.: Identification of asbestos by X-ray diffraction electron microscope, microprobe analysis and by optical dispersion staining methods. Acta Cryst., Pt. S 4 (1978)
665. Chatfield, E.J. et al.: Comparison of optical and transmission electron microscope fiber counts on membrane filters. Ontario Research Foundation Report, May 24 (1978)
666. Chatfield, E.J.: Identification and measurement of asbestos fibers by electron microscopy. Asbestos, 6–13, March (1978)
667. Chatfield, E.J., Dillon, M.J.: Some aspects of specimen preparation and limitations of precision in particulate analysis by SEM and TEM. Scanning Electron Microscopy, Vol. I, 476–493 (1978)
668. Clark, R.L., Ruud, C.O.: Transmission electron microscopy standards for asbestos. Micron *5*, 83–88 (1974)
669. Davis, J.G.M.: Electron microscope studies of asbestosis in man and animals. Ann. N.Y. Acad. Sci. *132*, 98–111 (1965)
670. Hayashi, H. et al.: Semiquantitative chemical analysis of asbestos fibres and clay minerals with an analytical electron microscope. Clays and Clay Minerals *26*, 181–188 (1978)

671. Huggins, C.W.: Electron micrographs of some unusual inorganic fibers. Rep. Inv. 6020 (1962) Pittsburg, Pa.: U.S. Bur. Mines
672. Kalmus, H.: Preparation of aerosols for electron microscopy. J. Appl. Phys. *25*, 87 (1954)
673. Langer, A.M. et al.: Electron microscopical investigation of asbestos fibres. Envir. Hlth. Perspect. *9*, 63–80 (1974)
674. Lee, R.J.: Computerized SAED and the electron optical identification of particulates. Symposium on the Electron Microscopy of Microfibers, F.D.A., 60 (1976)
675. Miller, J.L.: Identification of selected silicate minerals and their asbestiform varieties by electron optical and x-ray techniques. Bull. Electron optics EM 113(2), 1–11 (1979)
676. Millette, J.R., McFarren, E.F.: EDS of waterborne asbestos fibers in TEM, SEM and STEM. Scanning Electron Microscopy, Part III, 451 (1976) Chicago, Ill., IITRI
677. Millette, J.R.: Sizing of particulates for environmental health studies. Scanning Electron Microscopy, 253–258 (1978). Chicago, Ill., SEM Inc.
678. Ortiz, L.W., Isom, B.L.: Transfer technique for electron microscopy of membrane filter samples. Amer. Ind. Hyg. Ass. J. *35*, 423–425 (1974)
679. Pattnaik, A., Meakin, J.D.: Development of scanning electron microscopy of airborne asbestos concentrations. Report EPA 650/2-73-016 (1973). Springfield, Va. 22161: Natl. Techn. Inform. Service
680. Pooley, F.D.: Electron microscope characteristics of inhaled chrysotile asbestos fibre. Brit. J. Ind. Med. *29*, 146–153 (1972)
681. Pooley, F.D.: The identification of asbestos dust with an electron microscope microprobe analyzer. Ann. Occup. Hyg. *18*, 181–186 (1975)
682. Pooley, F.D.: Methods for assessing asbestos fibres and asbestos bodies in tissue by electron microscopy. In: Biological effects of asbestos, IARC Scient. Publ. No. *8*, 50–53 (1973). Lyon
683. Rickards, A.L.: Estimation of submicrogram quantities of chrysotile asbestos by electron microscopy. Analyt. Chem. *45*, 809–811 (1973)
684. Rubin, I.B., Maggiore, C.J.: Elemental analysis of asbestos fibres by means of electron probe technique. Envir. Health Perspect. *9*, 81–94 (1974)
685. Ruud, C.O. et al.: Selected area electron diffraction and energy dispersive X-ray analysis for the identification of asbestos fibres, a comparison. Micron *7*, 115–132 (1976)
686. Ruud, C.O.: An overview of electron microscopy methods. N.B.S. Spec. Publ. 506, Proceed. of the workshop on asbestos: Definitions and measurement methods, 221–232 (1978)
687. Samudra, A.V. et al.: Electron microscope measurement of airborne asbestos concentrations. A provisional methodology manual. Report 600/2-77-178, August 1977. Research Triangle Park, North Carolina 27711: U.S. Environmental Protection Agency
688. Seshan, K.: Explanation for the insensivity to tilts of the electron diffraction patterns of amphibole asbestos fibres. Envir. Res. *14*, 46–58 (1977)
689. Skikne, M.I. et al.: Electron diffraction patterns of UICC asbestos samples. Envir. Res. *4*, 141–145 (1971)
690. Spurny, K.R. et al.: The sampling and electron microscopy of asbestos aerosol in ambient air by means of Nuclepore filters. J. Air Pollut. Control Ass. *26*, 496–498 (1976)
691. Taylor, J.C.: Presentation of the Magiscan. 2nd Colloqium on dust measuring technique and strategy, Washington (1978)
692. Crable, J.V.: Quantitative determination of chrysotile, amosite and crocidolite by X-ray diffraction. Amer. Ind. Hyg. Ass. J. *27*, 293–298
693. Crable, J.V., Knott, M.S.: Application of X-ray diffraction to the determination of chrysotile in bulk or settled dust samples. Amer. Ind. Hyg. Ass. J. *27*, 383–387 (1966)
694. Goodhead, K., Martindale, K.W.: The determination of amosite and chrysotile in airborne dusts by an X-ray diffraction method. Analyst (London) *94*, 985–988 (1969)
695. Hayashi, H.: Quantitative determination of airborne asbestos dust in occupational environment by X-ray diffraction using conventional and rotating anode X-ray tube. Ind. Hlth. *11*, 225–236 (1973)
696. Nenadic, C.M., Crable, J.V.: Application of X-ray diffraction to analytical problems of occupational health. Amer. Ind. Hyg. Ass. J. *32*, 529–538 (1971)
697. Ocella, E., Maddalon G.: X-ray diffraction characteristics of some types of asbestos in relation to different techniques of comminution. Medna. Lav. *54*, 628–636 (1963)

698. Omoto, M. et al.: Application of X-ray microanalyzer to analysis of asbestosis. Jap. J. Hyg. *30*, 1 (1975)
699. Plowman, C.: Effect of heat on the identification of asbestos in X-ray diffraction. Nature (London) *244*, 280 (1973)
700. Rickards, A.L.: Estimation of trace amounts of chrysotile asbestos by X-ray diffraction. Analyt. Chem. *44*, 1872–1873 (1972)
701. Rohl, A.N., Langer, A.M.: Identification and quantification of asbestos in talc. Envir. Health Perspect. *9*, 95–109 (1974)
702. Talvitie, N.A.: X-ray diffraction analysis of industrial dust. Amer. Ind. Hyg. Ass. J. *23*, 214–221 (1962)
703. Taylor, M.: Methods for the quantitative determination of asbestos and quartz in bulk samples using x-ray diffraction. Analyst *103*, 1009–1012 (1978)
704. Whittaker, E.J.W., Zussman, J.: The characterization of serpentine minerals by X-ray diffraction. Mineral. Mag. *31*, 107–126 (1956)
705. Bagioni, R.P.: Separation of chrysotile asbestos from minerals that interfere with its infrared analysis. Envir. Sci. Technol. *9*, 262–263
706. Beckett, S.T. et al.: The use of infrared spectrophotometry for the estimation of small quantities of single varieties of UICC asbestos. Ann. Occup. Hyg. *18*, 313–320 (1975)
707. Coates, J.P.: The infrared analysis of quartz and asbestos. Perkin-Elmer Infrared Bull. 141
708. Flick, K.: Beitrag zur Untersuchung von silikogenen und asbesthaltigen Stäuben mit Hilfe der Infrarot-Spektralphotometrie. Arbeitsschutz*7*, 161–167 (1969)
709. Gadsen, J.A. et al.: Determination of chrysotile in airborne asbestos dusts by an infrared spectrometric technique. Atm. Environm. *4*, 667–670 (1970)
710. Luce, R.W.: Identification of serpentine varieties by infrared absorption. U.S. Geol. Survey, Prof. Paper 750-B, 199–201 (1972)
711. Morgan, A., Holmes, A.: Neutron activation techniques in investigations of the composition and biological effects of asbestos. In: Pneumoconiosis: Proceed. Internat. Conf., Johannesburg 1969, 52–56. Cape Town: Oxford University Press 1970
712. Shelz, J.P.: The detection of chrysotile asbestos at low levels in talc by differential thermal analysis. Thermochim. Acta *8*, 197–204 (1974)
713. Taylor, T.G. et al.: Infrared spectra for mineral identification. Amer. Ind. Hyg. Ass. J. *31*, 100–108 (1970)
714. Williams, A.D., Hoddinott, S.G.M.: Automated quantitative analysis with the IR data station, tremolite in talc. Perkin-Elmer Infrared Bull. 84 (1981) 5 pp.
715. Addingley, C.G.: Asbestos dust and its measurements. Ann. Occ. Hyg. *9*, 73–82 (1966)
716. Advisory Committee on Asbestos: Asbestos measurements and monitoring of asbestos in air (2nd report). Health and Safety Commission (U.K.), 3–28 (1978)
717. AIHA-ACGIH: Recommended procedures for sampling and counting asbestos fibres. Procedures for the evaluation of occupational exposures to airborne asbestos. Amer. Ind. Hyg. Ass. J. *36*, 83–90 (1975)
718. Almich, B. et al.: A theoretical and laboratory evaluation of a portable direct-reading particulate mass concentration instrument. HEW Publication, NIOSH, 76–114 (1975)
719. Anderson, C.H., Lomg, J.M.: Preliminary interim procedure for fibrous asbestos. U.S. Environm. Protection Ageny, Analyt. Branch, Environm. Res. Laboratory, July 3th 1976. Athens, Georgia 30605
720. Archambault, G.: Etude des séparateurs de poussière dans les moulins d'amiante. Rapport interne AMAQ, 99 p., Avril 1978
721. Asbestosis Research Council: a) Technical Note 1 (1971): The measurement of airborne asbestos dust by the membrane filter method (Rev.); b) Technical Note 2 (1971): Dust sampling procedures for use with the asbestos regulations. Rochdale, Lancs.
722. Australian Department of Health: Membrane filter method for estimating airborne asbestos dust. Canberra 1976
723. Avol, E.L., Clark, W.E.: An experimental evaluation of the Philips PWG 790 and the GCA model APM Beta attenuation mass monitors. Technical Report submitted to EPA by Rockwell International, Febr. 1976, 36 pp.
724. Ayer, H.E. et al.: A comparison of impinger and membrane filter techniques for evaluating air samples in asbestos plants. Ann. N.Y. Acad. Sci. *132*, 274–275 (1965)

725. Bartosiewicz, L.: Improved techniques of identification and determination of airborne asbestos. Amer. Ind. Hyg. Ass. J. *34*, 252–259 (1973)
726. Bayer, St.G., Zumwalde, R.D.: Sampling and evaluating airborne asbestos dust. Divis. of Training, NIOSH, U.S. Dept. of Health, Education and Welfare (1973). Cincinnati, Ohio
727. Deaman, D.R., File, D.M.: Quantiative determination of asbestos fiber concentrations. Analyt. Chem. *48*, 101–110 (1976)
728. Beckett, S.T., Attfield, M.D.: Inter-laboratory comparisons of the counting of asbestos fibres on membrane filters. Am. Occup. Hyg. *17*, 85–96 (1974)
729. Bignon, J. et al.: Measurement of asbestos retention in the human respiratory system related to health effects. N.B.S. Spec. Publ. 506, Proceed. of the Workshop on Asbestos: Definitions and Measurement Methods, 95–118 (1978)
730. Bresling, J.A., Stein, R.L.: Efficiency of dust sampling inlets in calm air. Amer. Ind. Hyg. Ass. J. *36*, 576–583 (1975)
731. Breuer, H., Robock, K.: Das Tyndallometer TM digital zur unmittelbaren Bestimmung der Feinstaubkonzentration in Ergänzung zu Langzeitwerten gravimetrischer Staubmeßgeräte. Silikosebericht Nordrhein-Westfalen, Bd. 10, 77–87 (1975)
732. Chatfield, E.J.: Measurement of asbestos fibers in the workplace and in the general environment. In: Ledoux, R.L. (ed.): Short course in mineralogical techniqes of asbestos determination, 165–196 (1979). Quebec
733. Chatfield, E.J.: Quantitative analysis of asbestos minerals in air and water. Proceed. 32nd Annual Proc. EMSA, 528 (1974). St. Louis, Miss.
734. Chatfield, E.J.: Asbestos background levels in three filter media used for environmental monitoring. 33rd Annual Proc. EMSA (1975). Las Vegas, Nev.
735. Coenen, W.: Ein neues Meßverfahren zur Beurteilung fibrogener Stäube am Arbeitsplatz. Staub-Reinhaltg. Luft *33*, (3), 99–103 (1973)
736. Coenen, W.: Feinstaubmessung mit dem VC 25. Neuere Untersuchungen und praktische Erfahrungen. Staub-Reinhaltg. Luft *35*, 452–458 (1975)
737. Cooper, D.W. et al.: Fiber counting: a source of error corrected. Amer. Ind. Hyg. Ass. J. *39*, 362 (1978)
738. Cralley, L.J. et al.: Source and identification of respirable fibers. Amer. Ind. Hyg. Ass. J. *29*, 129–135 (1968)
739. Dagbert, M.: Etude de correlation des mésures d'empoussièrages dans l'industrie de l'amiante. Comité d'étude sur la salubrité dans l'industrie de l'amiante, documents, 1–114 (1976). Montreal, Quebec, Canada
740. Du Toit, R.S.: A review of early results of comparative tests with the konimeter and thermal precipitator in asbestos mines. Ann. Occup. Hyg. *20*, 279–281 (1977)
741. Tu Toit, R.S.J.: Methods used in the measurement of asbestos-dust exposure. Proceed. Asbestos Symposium Johannesburg, S.A., Oct. 3–7 (1978) 19–36. Natl. Inst. for Metallurgy
742. Du Toit, R.S., Gilfillan, T.C.: Simultaneous airborne dust samples with konimeter, thermal precipitator, and dosimeter in asbestos mines. Ann. Occup. Hyg. *20*, 333–344 (1977)
743. Edwards, G.H., Lynch, J.R.: The method used by the U.S. Public Health Service for enumeration of asbestos dust on membrane filters. Ann. Occup. Hyg. *11*, 1–6 (1968)
744. GCA Corp., Technology Div.: Respirable dust monitor, Model RDM 101 for short term measurements. Brochure. (1976). Bedford, Mass.
745. GCA Corp., Technology Div.: Operational and performance characteristic of GCA Model APM Ambient mass monitor. July (1976)
746. GCA Corp.: The instruction manual. Aerosol Mass Monitor Model APM. Oct. (1977)
747. GCA Corp., Precision Scient. Group: Fibrous aerosol monitor, model FAM for both short and extended time monitoring of airborne fibers. Brochure. 11-77-CP-2M (1977)
748. Gentry, J. et al.: Measurements of collection efficiency of amosite fibers. In: Aerosole in Naturwissenschaft, Medizin und Technik, 26.–28. 9. 1978, Wien, 400–405
749. Gentry, J.W., Spurný, K.R.: Measurements of collection and efficiency of Nuclepore filters for asbestos fiber. Journ. Colloid Interface Sci. *65*, 174–180 (1978)
750. Gibbs, G.W.: Comments to methods used in the measurement of asbestos dust exposures by R. S. J. Du Toit. Proceed. Asbestos Symposium Johannesburg (1978)
751. Gibbs, G.W.: Techniques of asbestos determination – research perspective. In: Ledoux, R.L. (ed.): Short course in mineralogical techniques of asbestos determination, 253–278 (1979). Quebec, Canada

752. Gibbs, G.W. et al.: A summary of asbestos fibre counting experience in seven countries. Ann. Occup. Hyg. *20,* 321–332 (1977)
753. Gravatt, C.C. et al.: Proceedings of a workshop on asbestos: Definitions and measurement methods. Nat. Bur. Stand. Spec. Publ. 506, 496 pp. (1978) Washington, D.C.: U.S. Dept. of Commerce
754. Harness, I.: Airborne asbestos dust evaluation. Ann. Occup. Hyg. *16,* 397–404 (1973)
755. Hauptverband der gewerblichen Berufsgenossenschaften: Regeln zur Messung und Beurteilung gesundheitsgefährlicher mineralischer Stäube. Ausg.: 4. (1977). Bestell-Nr.: ZH 1/561, 23 S.
756. Heffelfinger, R.E. et al.: Development of a rapid survey method of sampling and analysis for asbestos in ambient air. Columbus, Ohio: Battelle Laboratories, Contract No. CPA 22-69-110 (1972)
757. Heidermanns, G.: Determination of asbestos fine dust according to mass concentration. In: Ledoux, R.L. (ed.): Short course in mineralogical techniques of asbestos determination, 165–196 (1979). Québec, Canada
758. Heidermanns, G.: Methoden zur Identifikation und quantitativen Analyse von Asbest bei Anwendung der Technischen Richtkonzentrationen für Asbest. STF-Report Nr. 2/78
759. Heidermanns, G.: Methoden zur Identifikation und quantitativen Analyse von Asbest bei Anwendung der Technischen Richtkonzentrationen für Asbest. STF Report Nr. 2/78
760. Holmes, S.: Development in dust sampling and counting techniques in the asbestos industry. Ann. N.Y. Acad. Sci. *132,* 288–297 (1965)
761. Holt, P.F., Young, D.K.: Asbestos fibres in the air of towns. Atmos. Envir. *7,* 481–483 (1973)
762. Horai, Z. et al.: Correlation between asbestos dust concentration and frequency of radiographic detection of pulmonary asbestosis. Japan. J. Thorac. Dis. *13,* 33–39 (1975)
763. Keenan, R.G., Lynch, J.R.: Techniques for the detection, identification and analysis of fibres. Amer. Ind. Hyg. Ass. J. *31,* 587–597 (1970)
764. Knight, G.: Overlap problems in counting fibres. Amer. Ind. Hyg. Ass. J. *36,* 113–114 (1975)
765. Knight, G.: Gravimetric sampling in asbestos mines and mills. QAMA technical report Nov. 1977
766. Knight, G.: Asbestos dust – review and research. QAMA technical report 1978
767. Leitz, Wetzlar: Tyndallometer TM digital. Optical fine dust measuring instrument. Instructions 650-10
768. Lilienfeld, P.: Beta-absorption-impactor aerosol mass monitor. Amer. Ind. Hyg. Ass. J. *31,* 727–729 (1970)
769. Lilienfeld, P.: Design and operation of dust measuring instrumentation based on the beta-radiation method. Internat. Tagg. "Stäube und Gase am Arbeitsplatz", 18.–20. 6. 1975, 79–86 (1975) Bonn-Bad Godesberg
770. Lippman, M., Harris, W.G.: Size selective samplers for estimating respirable dust concentration. Health Physics *8,* 155 (1962)
Lippman, M. et al.: "Respirable" dust sampling. Amer. Ind. Hyg. Ass. J. *31,* 138 (1970)
771. Losman, D.: Detection and analysis of asbestos in environmental samples collected on membrane filters: a multiple choice procedure integrating the LM, the SEM and the EMP. Bull. Environ. Contam. Toxicol. *23,* 170–178 (1979)
772. Lynch, J.R. et al.: The interrelationships of selected asbestos exposure indices. Amer. Ind. Hyg. Ass. J. *31,* 598–604 (1970)
773. Mark, D.: Problems associated with the use of membrane filters for dust sampling when compositional analysis is required. Ann. Occup. Hyg. *17,* 35–40 (1974)
774. Marple, V.A., Rubow, K.L.: An evaluation of the GCA respirable dust monitor. Amer. Ind. Hyg. Ass. J. *39* (1978)
775. Middleton, A.P.: On the occurrence of fibres of calcium sulphate resembling amphibole asbestos in samples taken for the evaluation of airborne asbestos. Ann. Occup. Hyg. *21,* 91–93 (1978)
776. Nicholson, P.D. et al.: Asbestos contamination of the air in public buildings. EPA Contract No. 68-02-1346, Washington, D.C.: U.S. Govt. Printing Office (1975)
777. Ortiz, L.W. et al.: Calibration standards for counting asbestos. Amer. Ind. Hyg. Ass. J. *36,* 104–112 (1975)

778. Quilliam, J.H.: The value of the konimeter for dust control purposes in routine mine sampling. J. Mine Vent. Soc. South Afr. *9*, 115–116 (1976)
779. Rajhans, G.S., Bragg, G.M.: A statistical analysis of asbestos fibre counting in the laboratory and industrial environment. Amer. Ind. Hyg. Ass. J. *36*, 909–915 (1975)
780. Robock, K.: Referenzmethode zur Messung von Asbestfaser-Konzentrationen. Asbestveranstalt. d. Hauptverb. d. gewerbl. Berufsgenossensch. e. V., Bonn-Bad Godesberg, 6. u. 7. 3. 1980, 7 S
781. Robock, K.: Besonderheiten der Emissionsmessung bei Fasern. VDI-Kolloquium „Faserige Stäube", Straßburg, 4.–8. 10. 1982
782. Robock, K.: Vergleich verschiedener Meßverfahren im Hinblick auf die Verhältnisse am Arbeitsplatz. Symposium des Bundesgesundheitsamtes „Zur Beurteilung der Krebsgefahren durch Asbest", Berlin, 17.–19. 2. 1982
783. Rüttner, J.R.: Asbestos problems from the morphological and dust-analytical viewpoint. Hefte Unfallheilk. *126*, 559–565 (1975)
784. Spurný, K. et al.: Identifizierung und krebserzeugende Wirkung von faserhaltigem Aktinolith aus einem Steinbruch. In: 7. Conf. Aerosols in Science, Medicine and Technology, 3.–5. 10. 1979, Düsseldorf, 138–143
85. Spurný, K.R. et al.: Size-selective preparation of inorganic fibers for biological experiments. Amer. Ind. Hyg. Ass. J. *40*, 20–38 (1979)
786. Spurný, K.R. et al.: Sampling and analysis of fibrous aerosol particles. In: Shaw, D.T.: Fundamentals of aerosol science. Wiley 1972
787. Spurný, K.R., Stober, W. 1972: Asbestos measurements in ambient air. Clean Air *9*, 38–41 (1975)
788. Teichert, U.: Equipment and strategy of dust measurements within the German A/C-industry. Internat. Colloquium on dust measuring technique and strategy, Warmensteinach (1977)
789. Thaer, A.: Instruments for dust measurement and dust analysis. LEITZ-Journal, 312–315 (1976) Wetzlar
790. Thaer, G. et al.: Untersuchungen über den Asbestanteil im Staub der Außenluft. Bericht des Battelle-Inst., Frankfurt/M. (1973)
791. Timbrell, V. et al.: UICC standard reference samples of asbestos. Intern. J. Cancer *3*, 406–408 (1968)
792. Trudeau, M.: Methods for the evaluation of asbestos dust concentration at the workplace. In: Ledoux, R.L. (ed.): Short course in mineralogical techniques of asbestos determination, 213–252 (1979) Quebec, Mineral. Assoc. of Canada
793. Walton, W.H. et al.: An international comparison of counts of airborne asbestos fibres sampled on membrane filters. Ann. Occup. Hyg. *19*, 215–224 (1976)
794. Walton, W.H.: Problems in the standardization of asbestos dust measurements. Internat. Colloq. on dust measuring techniques and strategy, Warmensteinach, 38–52 (1977)
795. Winters, J.W.: A simple small light-weight personal dust sampling unit for full shift determination of asbestos dust exposure. Ann. Occup. Hyg. *19*, 77–80 (1976)
796. Wood, J.D.: Review of personal sampling pumps. Ann. Occup. Hyg. *20*, 3–17 (1977)
797. American Water Works Association: A study of the problems of asbestos in water. Amer. Water Works Ass. J. *66* (9) Pt. 2, 1–22 (1974)
798. Anon.: Gefährdung des Trinkwassers durch Asbestzement. Ergebnisse einer Fachtagung an der ETH. Zürich: Neue Züricher Ztg., 5. 5. 1982, S. 23
799. Berry, E.E.: Thermal analysis of various chrysotiles using evolved water analysis techniques. 2nd Internat. Conf. on the Physics and Chemistry of Asbestos Minerals, Paper 2:7 (1971) Louvain, Belgium
800. Biles, B., Emerson, T.R.: Examination of fibres in beer. Nature *219*, 93 (1968)
801. Bonner, W.P. et al.: Identification of asbestos-type materials in suspended solids. Res. Rept. 61 (1977). Knoxville, Tenn.: University of Tennessee, Water Resource Research Center
802. Chatfield, E.J.: A new technique for preparation of beverage samples for asbestos measurement by electron microscopy. 9th Internat. Cong. on Electron Microscopy, Toronto, Vol. II, 102–103 (1978). Toronto M5S 1A1, 150 College Street, University of Toronto, Microscopical Society of Canada

803. Chatfield, E.J. et al.: Preparation of water samples for asbestos fiber counting by electron microscopy. EPA Report EPA-600/4-78-011, 1–118 Jan. (1978)
804. Chatfield, E.J., Glass, R.W.: Analysis of water samples for asbestos: sample storage and technique development studies. Symposium on Electron Microscopy of Microfibers, F.D.A., 123–137, Aug. (1976). Washington, D.C., U.S. Govt. Printing Office
805. Chopra, K.S.: Interlaboratory measurements of amphibole and chrysotile fiber concentration in water. Journ. Testing and Evaluat. *6,* No. 4, 241–247 (1978)
806. Cook, P.M. et al.: X-ray diffraction and electron beam analysis of asbestiform minerals in Lake Superior waters. Proceed. Internat. Conf. Environm. Sensing and Assessment, IEEE, New York, 2, 34-1-1 (1975)
807. Cook, P.M. et al.: Asbestiform amphibole minerals: detection and measurement of high concentrations in municipal water supplies. Science, N.Y. *185,* 853–855 (1974)
808. Cook, P.M.: Semiquantitative determination of asbestiform amphibole mineral concentrations in western Lake Superior water samples. Adv. X-ray Anal. *18* (1975)
809. Cooper, R.C., Murchio, J.C.: Preliminary studies of asbestiform fibers in domestic water supplies. University of California, Berkeley, AMRL-TR-125, Paper No. 5 (1974)
810. Melton, C.W.: Progress in development of rapid extraction methods for chrysotile asbestos in H_2O. Symposium on Electron Microscopy of Microfibers, F.D.A., 166–169 (1976)
811. Millette, J.R.: Analyzing for asbestos in drinking water – News Environ. Res., Cincinnati, Jan. 16 (1976). EPA, Washington, D.C.
812. Mudroch, O., Kramer, J.R.: Enumeration and identification of asbestos fibres in water. 32nd Annual EMSA Proceedings, 526 (1974)
813. Nicholson, W.J.: Analysis of amphibole asbestiform fibers in municipal water supplies. Envir. Health Perspect. *9,* 165–172 (1974)
814. Auribault, M.: Sur l'hygiène et la sécurité des ouvriers dans les filatures et tissages d'amiante. Bull. de l'inspect. du travail *14,* 120–132 (1906)
815. Baumann, E.R.: Diatomite filters for asbestiform fibre removal from water. In: Proceed. of the American Water Works Assoc., 95th Annual Conf. (1975), Minneapolis
816. Beierl, L.: Zentrale Erfassungsstelle asbeststaubgefährdeter Arbeitnehmer. Asbest-Veranst. d. Hauptverb. d. gewerbl. Berufsgenossensch. e. V., 6. u. 7. 3. 1980, Bonn-Bad Godesberg, 39–41
817. Berufsgenossenschaft der Chemischen Industrie: Unfallverhütungsvorschrift „Schutz gegen gesundheitsgefährlichen mineralischen Staub“ (UVV Staub) (1973)
818. BMI: Asbestanteil in der Außenluft. Umwelt, Information des Bundesministers des Inneren zur Umweltplanung und zum Umweltschutz Nr. 60, S. 14 (1977)
819. Chissick, S.S.: Commercially available products and services. In: Chapter 1: Introduction to Asbestos, Properties, Applications and Hazards, Michaels, L., Chissick, S.S. (eds.) Vol. 1, 14–42. Chichester: J. Wiley & Sons 1979
820. Crowder, J.V., Wood, G.H.: Control techniques for asbestos air pollutants. EPA Publ. No. A:–117 (1972)
821. DGB: DGB schlägt 17-Punkte-Programm gegen Asbestkrebs in der Arbeitswelt vor. Düsseldorf 12. 2. 1981: Bundespressestelle des Deutschen Gewerkschaftsbundes
822. DFG: DFG-Senatskommission zur Prüfung gesundheitsschädlicher Arbeitsstoffe. Mit. XV. Maximale Arbeitsplatzkonzentrationen 1979, Technische Richtkonzentrationen, 40 S., Boppard: Boldt-Verlag 1979
823. DFG: Maximale Arbeitsplatzkonzentrationen 1981, Mitt. XVII der Senatskommission zur Prüfung gesundheitsschädlicher Arbeitsstoffe
824. EG: Richtlinie des Rates vom 27. 7. 1976 zur Angleichung der Rechts- und Verwaltungsvorschriften für Beschränkungen des Inverkehrbringens und der Verwendung gewisser gefährlicher Stoffe und Zubereitungen. ABL Nr. 262 v. 27. 9. 1976, S. 201
825. EG: Entwurf einer Entschließung des Rates der EG über Leitlinien für ein Aktionsprogramm zur Bekämpfung der Gefahren, die sich durch Asbestverunreinigung für die menschliche Gesundheit ergeben. DokNr. V/F/1316/77d (1977)
826. EG: Entwurf einer Entschließung des Rates der EG über ein Aktionsprogramm für Sicherheit und Gesundheit am Arbeitsplatz. ABL Nr. C 9 vom 11. 1. 1978, S. 2
827. EG: Richtlinie des Rates vom 26. 6. 1967 zur Angleichung der Rechts- und Verwaltungsvorschriften für die Einstufung, Verpackung und Kennzeichnung gefährlicher Stoffe. ABL Nr. 196 v. 16. 8. 1967, S. 1

828. Förster, H.: Gesundheitsgefahren durch Stäube asbesthaltiger Bremsbeläge in Kraftfahrzeug-Werkstätten. Asbest-Veranst. d. Hauptverb. d. gewerbl. Berufsgenossensch. e.V., Bad-Godesberg, 6. u. 7.3. 1980, 88–89
829. Hain, E.: Arbeitsmedizinische Vorsorgeuntersuchungen bei Einwirkung durch Asbeststaub: Erfahrungen und Perspektiven. ibid., 42–48
830. Health and Safety Executive: Asbestos: Precautions in industry. 2nd edit. (Health and Safety at work, No. 44) (1975) London: H.M. Stationary Office
831. Hoffmann, E.: Industrielle Verwendung von Asbest: Erfassung der Betriebe, der Tätigkeiten und der Versicherten – Schutzmaßnahmen. Asbest-Veranst. d. Hauptverb. d. gewerbl. Berufsgenossensch. e.V., Bonn-Bad Godesberg, 6. u. 7.3. 1980, 80–87
832. Jones, J.S.P. et al.: Factory population exposed to crocidolite asbestos – a continuing survey. IARC Sci. Publ. No. *13*, 117.120 (1976)
833. Kesting, A.: 20 Jahre Entwicklung technischer und persönlicher Schutzmaßnahmen bei der Verarbeitung von Asbest. Asbest-Veranst. d. Hauptverb. d. gewerbl. Berufsgenossensch. e.V., Bonn-Bad Godesberg, 6. u. 7.3. 1980, 72–79
834. Kleinfeld, M. et al.: A study of workers exposed to asbestiform minerals in commercial talc manufacture. Envir. Res. *6*, 132–143 (1973)
835. Knobloch, S.: Zweiter Nachtrag zur Unfallverhütungsvorschrift „Schutz gegen gesundheitsgefährlichen mineralischen Staub" (VBG 119) – Vorschläge für Verwendungsbeschränkungen von Asbest. Asbest-Veranst. d. Hauptverb. d. gewerbl. Berufsgenossensch. e.V., Bonn-Bad Godesberg, 6. u. 7.3. 1980, 108–111
836. Kommission der Europäischen Gemeinschaften: Merkblatt zu der Berufskrankheitenliste der EG. Köln: Verlag d. Bundesanzeigers 1972
837. Landt, E.: Physikalische Betrachtungen zum Faserfilter. Gesundheits-Ing., 139–145 (1956)
838. Lane, R.E. et al.: Hygiene standards for chrysotile asbestos dust. Ann. Occup. Hyg. *11*, 47–69 (1968)
839. Lawrence, J. et al.: Asbestos: Its removal from potable water. Canad. Res., Dev., Nov./Dec. (1974) 29–30
840. Lawrence, J. et al.: Removal of asbestos fibres from potable water by coagulation and filtration. Water Res. *9*, 397–400 (1975)
841. Levine, R.J.: How the industrial physician can reduce mortality from asbestos-related diseases. J. Occup. Med. *20*, 464–468 (1978)
842. Lewinsohn, H.C.: The medical surveillance of asbestos workers. Roy. Soc. Health J. *92* (2), 69–77 (1972)
843. McDonald, J.C.: Problems in the determination of safety standards for asbestos exposed workers. Hefte Unfallheilk. *126*, 603–607 (1975)
844. Mürmann, H.: Möglichkeiten der Entstaubung am Arbeitsplatz. Umwelt *1*, 21–24 (1980)
845. N.N.: Schutz gegen gesundheitsgefährlichen mineralischen Staub. VBG 119 v. 1.4. 1973 in der Fassung vom 1.10. 1981. Köln: Carl Heymanns Verlag KG
846. N.N.: Spezifische Einwirkungsdefinitionen. (ZH 1/600). Entwurf: 10.1979. Köln: Carl Heymanns Verlag KG
847. N.N.: Berufsgenossenschaftliche Grundsätze für arbeitsmedizinische Vorsorgeuntersuchungen. Gefährdung durch gesundheitsgefährlichen mineralischen Staub. Fassung v. Mai 1974. Stuttgart: A.W. Gentner-V.
848. N.N.: Regeln zur Messung und Beurteilung gesundheitsgefährlicher mineralischer Stäube. (ZH 1/1561) Ausgabe: 4.1977 Köln: Carl Heymanns Verlag KG
849. Peto, J.: The hygiene standard for chrysotile asbestos. Lancet *I*, 848–489 (1978)
850. Roach, S.A.: Hygiene standards for chrysotile asbestos dust. Ann. Occup. Hyg. *11*, 47–60 (1968)
851. Ochs, H.-J.: Technik filternder Abscheider. Umwelt *1*, 39–43 (1980)
852. Robock, K.: Monitoring the workplace: Problems and progress state of the art and international efforts for standardization of methods. World Sympos. on Asbestos, Montreal/Can., 25.–27.5. 1982
853. Pott, F.: Some aspects on the dosimetry of the carcinogenic potency of asbestos and other fibrous dusts. Staub-Reinhalt.Luft *38* (12), 486–490 (1978)
854. Schütz, A., Heidermann, G.: Gesundheitsgefährliche Stäube im Hoch- und Tiefbau. Staub-Reinhalt.Luft *37* (7), 273 (1977)

855. Schütz, A., Woitowitz, H.J.: Technische Richtwerte für die zulässige Arbeitsplatzkonzentration von Chrysotil-Asbest. Staub-Reinhalt.Luft *33* (12), 469–474 (1973)
856. Schütz, A., Coenen, W.: Grenzwerte für Asbest, Kriterien zur Feststellung der Einwirkung, meßtechnische Überwachung der Betriebe. Asbest-Veranst. d. Hauptverb. d. gewerbl. Berufsgenossensch. e. V., Bonn-Bad Godesberg, 6. u. 7. 3. 1980, 66–71
857. Schütz, A. et al.: Staub am Arbeitsplatz, Arbeitssicherheit. In: Handbuch für Unternehmensleitung, Betriebsrat und Führungskräfte. Freiburg: Rudolf Haufe Verlag 1975
858. Skidmore, J.W., Jones, J.S.P.: Monitoring an asbestos spray process. Ann. Occup. Hyg. *18*, 151–156 (1975)
859. Smith, M.K., White, R.W.: Composition and method for inhibiting asbestos fiber dust. U.S. Pat. 3,928,060, Dec. 23 (1975)
860. VDI-Richtlinie 2262: Staubbekämpfung am Arbeitsplatz
861. Wirtschaftsverband Asbestzement e. V.: Hinweise für die Bearbeitung von Asbestzementprodukten. (1977) Neuss
862. Wirtschaftsverband Asbestzement e. V.: Arbeiten Sie umweltfreundlich. Nutzen Sie die neuen Bearbeitungsgeräte für Asbestzement (1980)
863. Zentralstelle für Unfallverhütung: Richtlinie Nr. 138 (1966): Schutzmaßnahmen gegen Staubgefahr in asbestverarbeitenden Betrieben.
864. Wagg, R.M.: Safety measures when handling asbestos. Roy. Soc. Health J. *96*, 252–255 (1976)
865. Hain, E. et al.: Asbest: Gesundheitsschäden, Grenzwerte, Prävention. Staub-Reinhaltg-Luft *33*, 51–57 (1973)
866. Heide, H. et al.: Untersuchung und Bewertung von Asbestemissionen bei der Bearbeitung von Asbestzement und asbesthaltigen Fußbodenbelägen. Berichte 1/80 des Umweltbundesamtes Berlin, 100 pp., Berlin: Erich Schmidt Verlag 1980
867. Kogan, F.M.: Maximum permissible exposure level (MPEL) of asbestos powders. Med. Lavoro *68* (2), 142–148 (1977)
868. McDonald, J.C.: Problems in the determination of safety standards for asbestos exposed workers. Hefte Unfallheilk. *126*, 603–607 (1975)
869. Peto, J.: The hygiene standard for chrysotile asbestos. Lancet *I*, 484–489 (1978)
870. Schütz, A., Coenen, W.: Grenzwerte für Asbest, Kriterien zur Feststellung der Einwirkung, Meßtechnische Überwachung der Betriebe. Veranstalt. d. Hauptverb. d. gewerbl. Berufsgenossensch. e. V., Bonn-Bad Godesberg, 6. u. 7. 3. 1980, 66–71
871. Umweltbundesamt: Berichte 7/80: Umweltbelastung durch Asbest und andere faserige Feinstäube. 400 pp., Berlin: Erich Schmidt Verlag 1980
872. Umweltbundesamt: Berichte 4/78: Analyse der Asbestindustrie. Erarbeitet vom Battelle-Institut e. V. Frankfurt/M. 92 pp. Berlin: Erich Schmidt Verlag 1978
873. Wagner, J.C.: Disputes on the safety of asbestos. New Scientist *7*, 609 (1974)
874. Webster, I.: Control of asbestos-exposed workers in South Africa. Hefte Unfallheilk. *126*, 614–617 (1975)
875. Anon.: Mineral fibres – A general review. Industrial Minerals, No. 169, 19–31, Oct. (1978)
876. Blumberg, H., Hillermeier, K.H.: Aramid als Asbestsubstitution. Tagungsbericht Asbest, 6. u. 7. 3. 1980, Bonn-Bad Godesberg, Schriftenreihe d. Hauptverb. d. gewerbl. Berufsgenossensch. e. V., 149–150 (1980)
877. Bornemann, P.: Industrielle Anwendung von Ersatzstoffen für Asbest – Erfahrungen – Probleme – Perspektiven. Tagungsbericht Asbest, 6. u. 7. 3. 1980, Bonn-Bad Godesberg, Schriftenreihe des Hauptverb. d. gewerbl. Berufsgenossensch. e. V., 125–136 (1980)
878. Clark, F.M.: Insulating materials for design and engineering practice. New York/London: Wiley 1962
879. deGroot-Böhlhoff, H.: Bremsen bald ohne Asbeststaub. Suche nach Ersatz für den Risiko-Werkstoff zeigt erste Erfolge. Die Welt, Nr. 79, IV, 3. 4. 1982
880. deGroot-Böhlhoff, H.: Oft ist der Asbest schon austauschbar. Die Welt, Nr. 157, IV, 10. 7. 1982
881. Green, A.K., Pye, A.M.: The search for asbestos substitutes. Asbestos *58* (11) (1977)
882. Heide, H. et al.: Untersuchungen von produktspezifischen Anforderungen an Asbest zur Erarbeitung von Vorschlägen für umweltfreundliche Substitutionsprodukte. (1978). Frankfurt/Main: Bericht des Battelle Institutes für das Umweltbundesamt

883. Jentzsch, L.: Grenzen der Substitution von Asbest im Hochdruckdichtungsmaterial. Tagungsber. Asbest, 6. u. 7.3. 1980, Bonn-Bad Godesberg, Schriftenreihe d. Hauptverb. d. gewerbl. Berufsgenossensch. e.V., 142–143 (1980)
884. Katz, H.S., Milewski, J.V. (Hrsg.): Handbook of fillers and reinforcements for plastics. New York: Van Nostrand Reinhold 1978
885. Klingholz, R.: Technology and production of man-made mineral fibres. Ann. Occup. Hyg. *20,* 153–159 (1977)
886. Lippmann, H.D.: Siltemp-Silikat-Hitzegewebe. Tagungsber. Asbest, 6. u. 7.3. 1980, Bonn-Bad Godesberg, Schriftenr. d. Hauptverb. d. gewerbl. Berufsgenossensch. e.V., 148 (1980)
887. Loewenstein, K.L.: The manufacturing technology of continuous glass fibres. Amsterdam: Elsevier 1973
888. Mohr, J.G., Rowe, W.P.: Fiber glass. New York: Van Nostrand Reinhold 1978
889. Lohrer, W., Poeschel, E.: Ersatzstoffe für Asbest – Einsatzmöglichkeiten – Gesundheitliche Bedeutung. Tagungsber. Asbest, 6. u. 7.3. 1980, Bonn-Bad Godesberg, Schriftenr. d. Hauptverb. d. gewerbl. Berufsgenossensch. e.V., 112–124 (1980)
890. Lubin, G. (ed.): Handbook of fiberglass and advanced plastics composites. New York: Van Nostrand Reinhold 1969
891. Merkel, H.: Grenzen der Substitution von Asbest in Stopfbuchspackungen. Tagungsber. Asbest, 6. u. 7.3. 1980, Bonn-Bad Godesberg, Schriftenr. d. Hauptverb. d. gewerbl. Berufsgenossensch. e.V., 144–146 (1980)
892. Nixdorf, J.: Anorganische Faserwerkstoffe. Ber. Dtsch. Keram. Ges. *45* (4), 141–148 (1968)
893. Ortlepp, W.: Industrielle Möglichkeiten und Ergebnisse des Asbestersatzes in Asbestzement-Produkten. Forsch. Ber. BMFT, März 1979
894. Pilkington, Ltd.: Cem-FIL: Alkali-resistant glassfibre for reinforced cement. information Cem-FIL 1978
895. Pye, A.H.: A review of asbestos substitute materials in industrial applications. Journ. Hazard. Mat. *3,* 125–147 (1979)
896. Ritchie, K.M.: Mineral wool – rock, slag and glass wool. – In: Industrial Minerals and Rocks, 3rd edit., No. 30, 595–604 (1960). New York: Amer. Inst. Mining Metall., Petrol. Engin., Inc.
897. Rühmann, H.: Der GFK-Markt in der Bundesrepublik Deutschland und Europa – Ist-Zustand und Prognosemöglichkeiten. 18. öffentliche Jahrestag. der Arbeitsgemeinschaft Verstärkte Kunststoffe e.V., Freudenstadt, 5.–7.10. 1982, 3.1–3.5
898. Schmidt, K.A.F.: Textilglas für die Kunststoffverstärkung. 2. Aufl. Speyer: Zechner und Hüthig 1972
899. Schütz, A.: Ersatzstoffe für Asbest-Gefährdung durch Asbestschutzkleidung. Sichere Arbeit *3,* 10–13 (1979)
900. Selden, P.H.: Glasfaserverstärkte Kunststoffe. Berlin: Springer-Verlag 1967
901. Smith, H.V.: History, processes and operations in the manufacturing and uses of fibrous glass – one company's experience. – In: Occupational Exposure to Fibrous Glass – Proceedings of a symposium, HEW Publ. No. (NIOSH) 76–151, 19–26 (1976)
902. von Falkai, B. (Hrsg.): Synthesefasern. Grundlagen, Technologie, Verarbeitung und Anwendung. Weinheim: Verlag Chemie 1981
903. Anon.: A report on health aspects of fibrous glass. Medical Ser., Bull. No. 13–68, 1–7 (1968). Pittsburgh: Industrial Hygiene Foundation of America
904. Bayliss, D.L. et al.: Mortality patterns among fibrous glass production workers. Ann. N.Y. Acad. Sci. *271,* 324–335 (1976)
905. Botham, S.H., Holt, P.F.: The development of glass fibre bodies in the lungs of guinea pigs. Jour. Pathol. *103,* 149–156 (1971)
906. Champeix, J.: Fiberglass pathology and hygiene of factories. Arch. Malad. Prof. *6,* 91–94 (1944)
907. Cirla, P.: Occupational pathology from spun glass. Med. Lavoro *39,* 152–157 (1948)
908. Dement, J.M.: Environmental aspects of fibrous glass production and utilization. Environm. Res. *9,* 295–312 (1979)
909. Esmen, N. et al.: Summary of measurements of employee exposure to airborne dust and fiber in sixteen facilities producing man-made mineral fibers. Amer. Ind. Hyg. Ass. J. *40,* 108–177 (1979)

910. Esmen, N. et al.: Estimation of employee exposure to total suspended particulate matter and airborne fibers in insulation installation operations. A report to the Thermal Insulation Manufacturers Association (1980)
911. Gross, P. et al.: The pulmonary reaction to high concentration of fibrous glass dust. Arch. Environ. Health *20*, 696–704 (1970)
912. Gross, P. et al.: Lungs of workers exposed to fiber glass. Arch. Environ. Health *23*, 67–76 (1971)
913. Harris, R.L.: Aerodynamic considerations: What is a respirable fiber of fibrous glass? In: Occupational exposure to fibrous glass. Proceedings of a symposium. HEW Publ. No. (NIOSH) 76–151, 51–56 (1976)
914. Hill, J.W.: Health aspects of man-made fibres – a review. Ann. Occup. Hyg. *20*, 161–173 (1977)
915. Hill, H.W. et al.: Glass fibers: Absence of pulmonary hazard in production workers. Brit. J. Ind. Med. *30*, 174–179 (1973)
916. Konzen, J.L.: Man-made vitreous fibres and health. Industrial Minerals No. 163, 61–71, April (1981)
917. Lippmann, M. et al.: Deposition of fibrous glass in the human respiratory tract. In: Occupational exposure to fibrous glass-Proceedings of a Symposium. HEW Publ. No. (NIOSH) 76–151, 57–61 (1976)
918. Mayer, P.: Neuere Erkenntnisse über Gefährdungen durch Glas- und Mineralfasern. Zbl. Arbeitsmed. *30*, 280–285 (1980)
919. Milby, T.H., Wolf, C.R.: Respiratory tract irritation from fibrous glass inhalation. Jour. Occup. Med. *11*, 409–410 (1969)
920. Morgan, R.W. et al.: Mortality study of Owens-Corning fiberglass production workers. Report prepared by SRI International for Owens-Corning Fiberglass Corporation (1980)
921. Mungo, A.: Pathology from processing glass woll stratified materials. Folia Medica *43*, 962–970 (1960)
922. Nasr, N.M. et al.: The prevalence of radiographic abnormalities in the chests of fiber glass workers. Jour. Occup. Med. *13*, 371–376 (1971)
923. Poeschel, E. et al.: Umweltrelevanz künstlicher Fasern als Substitut für Asbest. Berlin: Erich Schmidt Verlag
924. Poeschel, E. et al.: Umweltrelevanz künstlicher Fasern als Substitute für Asbest. Forsch.ber. 78-10408301 (1978) Bonn: Bundesminist. d. Inn.
925. Possick, P.A. et al.: Fibrous glass dermatitis. Amer. Ind. Hyg. Ass. J. *31*, 12–15 (1970)
926. Pott, F.: Krebsrisiko durch künstliche Mineralfasern? Tagungsber. Asbest, 6. u. 7. 3. 1980, Bonn-Bad Godesberg, Schriftenr. d. Hauptverb. d. gewerbl. Berufsgenossensch. e. V., 137–139 (1980)
927. Riediger, G.: Staubkonzentrationen bei der Herstellung sowie Be- und Verarbeitung von künstlichen Mineralfasern und mineralfaserhaltigen Produkten. Tagungsber. Asbest, 6. u. 7. 3. 1980, Bonn-Bad Godesberg, Schriftenr. d. Hauptverb. d. gewerbl. Berufsgenossensch. e. V., 140–142 (1980)
928. Riediger, G.: Künstliche Mineralfasern in der Atemluft – Eine Pilotstudie für den Arbeitsplatz. Staub-Reinhal.Luft *37* (4), 147–151 (1977)
929. Robinson, C.F. et al.: Mortality patterns of rock and slag mineral wool production workers, presented at the Natl. Inst. of Occup. Safety and Health Sympos., Rockville/Maryland (1979)
930. Roche, I.: The pulmonary hazards in the glass fiber industry. Arch. Malad. Profess. *7*, 27–28 (1946)
931. Siebert, W.J.: Fiberglass health hazard investigation. Industr. Med. *11*, 6–9 (1942)
932. Stanton, M.F. et al.: Carcinogenity of fibrous glass: pleural response in the rat in relation to fiber dimension. J. Nat. Cancer. Inst. *58*, 587–603 (1977)
933. Sulzberger, M.B., Baer, R.: The effects of fiberglass on animal and human skin. Industr. Med. *11*, 482–484 (1942)
934. Wagner, J.C. et al.: Studies of carcinogenic effects of fiber glass of different diameters following intrapleural inoculation in experimental animals. In: Occupational exposure to fibrous glass – Proceeding of a symposium. HEW Publ. No. (NIOSH) 76–151, 193–197 (1976)

Carbon Black

D. Rivin

Cabot Corporation, Concord Road
Billerica, MA 01821, USA

Summary

Carbon black is a high purity colloidal carbon produced in large quantities world-wide for myriad industrial and consumer applications. Commercial carbon black differs in important respects from soots and other environmental carbonaceous particles but it is useful as an idealized model for investigating the adsorption properties, atmospheric reactions, and to a lesser extent, the environmental effects of these materials.

The toxicology of carbon black and related particulate carbons is examined in depth, with emphasis on major routes of exposure (e.g. inhalation). The absence of significant acute health effects is noted and the potential for chronic health effects is considered, particularly in regard to the fate of carbon particles in vivo and the bioavailability of strongly adsorbed trace organic impurities on carbon black.

Human exposure to carbon black dust occurs mainly in the workplace. The results of pertinent industrial hygiene surveys, and clinical and epidemiologic studies of occupationally exposed workers are summarized and evaluated.

Carbon Black Process

Production and Applications

Carbon black was first produced commercially in China by burning purified animal or vegetable oil in porcelain pots. For two thousand years this lampblack process underwent only minor evolutionary change until the advent of the modern Carbon Black Industry in 1872. In that year, a small plant was built in Pennsylvania to produce channel black from natural gas. As the supply of byproduct gas from the oil fields in Pennsylvania diminished, the industry moved to new gas fields in West Virginia and then to Louisiana, Oklahoma, and Texas.

About 500,000 pounds of channel black and lampblack were produced in 1881, increasing to 3 million pounds per year (mainly channel black) by 1895 [1]. Carbon black was used primarily as a pigment in printing inks, paints, and lacquers until the early 1900's when its use as a reinforcing filler for rubber became important following the discovery in England of carbon black's ability to strengthen and toughen rubber.

The rubber industry soon became the major market for carbon black. Its requirements led to the development of more efficient, lower cost, high-volume furnace processes for the production of carbon black. In the first of these, a limited range of carbon blacks was obtained from natural gas. The gas furnace process was developed in the USA in 1922 and employed for about 40 years. A method for producing carbon black from heavy aromatic liquids was introduced in the USA in 1943. Today, the oil furnace process accounts for over 95% of world production.

Although no channel black has been produced in the USA since 1976, a channel-type black is produced from vaporized coal tar liquids by the impingement process in the Federal Republic of Germany. Current production of impingement black is only a few percent of the maximum USA production of channel black which was 307,000 mt in 1948 [2].

Small amounts of carbon black are produced by the thermal process and by the exothermic dissociation of acetylene. The 60 year old thermal process yields only large particle size carbon blacks by batch pyrolysis of gas or vaporized liq-

Table 1. Furnace black capacity[a], 1984 Est

	10^3 mt	Plant sites
Western Europe		
France	235	Ambes, Berre, Port Jerome
Holland	115	Rotterdam [2]
Italy	170	Ravenna, [2] S. Martino di Trecate
Germany (FR)	360	Dortmund, Hamburg, Hanau, Hannover, Kalscheuren
Great Britain	160	Bristol, Ellesmere Port
Spain	100	Ciervana, San Roque, Santander
Portugal	20	Sines
Sweden	35	Malmo
Switzerland	15	Aigle
North and Central America		
USA	1,515	24 plants in 9 states (ALA., ARK., CA., KAN., OK., OH., LA., W. VA., TX.)
Canada	120	Hamilton, Sarnia
Mexico	115	Altamira, Salamanca, Tampico
South America		
Argentina	55	Campana
Brazil	150	Cubatao, Piassaguera, Triunfo[b]
Colombia	10	Cartagena
Peru	25	Talara
Venezuela	50	Valencia
Africa		
Rep. South Africa	55	Port Elizabeth
Asia and Oceania		
Australia	90	Altona, Cronulla, Kurnell
South Korea	155	Pohang, Seoul
India	120	Baroda, Bellary[b], Calcutta, Durgapur, Ghaziabad, Mathura[b], Renukot[b], Thana, Karimugal
Indonesia	10	Banteng[b], Pangkalan Susu
Iran	15	Ahwaz
Japan	630	Ichihara, Ishinomaki, Iwaki, Kitakyushu [3], Niigata, Osaka, Shimonoseki, Tahara, Taketoyo, Yokkaichi [2], Yokoshiba
Malaysia	25	Port Dickson
Pakistan	10	Karachi
Phillipines	20	
Taiwan	20	Lin Yuan
Thailand	20	Ang-Thong, Bangkok[b]
Turkey	30	Yarimca/Izmit
Total	4,450	

[a] Based in part on information in [3]
[b] Planned or under construction

uids. Acetylene black is a conductive carbon black which was first produced by continuous decomposition in Canada in the 1930's.

Carbon black plants are distributed widely, with the majority in the USA and Western Europe. The estimated 1984 production capacity for furnace black outside the eastern bloc countries is listed by country and plant site in Table 1. Fur-

Table 2. 1984 Estimated production capacity of non-furnace carbon black[a], 10^3 mt

Country	Thermal black	Acetylene black	Lampblack
Canada	20		
France		12	
Germany (FR)	[b]	10	[b]
Great Britain	[b]		[b]
Japan	5–10	20	
USA	<30	<15	6

[a] [2, 4]
[b] Plant of unknown capacity

Table 3. Carbon black demand[a], 10^6 mt

	1983	1990
United States	1.2	1.4
Soviet Bloc/China	1.2	1.8
Rest of World	2.2	3.0
Total	4.6	6.2

[a] [5]

nace black is produced in at least seven communist countries (China, Czechoslovakia, Germany-Democratic Republic, Poland, Rumania, USSR, and Yugoslavia) but reliable production data are not available. The extent of communist production of non-furnace blàcks is also unknown. Estimated 1984 production capacities in the West are summarized for thermal black, acetylene black and lampblack in Table 2.

The average annual growth rate for carbon black manufacture in the United States decreased to 6% following an average annual rate of 15% in the first quarter of this century. Future growth is expected to be less than 6% per year. Substantial year to year variations occur in response to activity in the rubber industry and in the national economy. Growth rates in Western Europe and Japan exhibit a similar trend.

In recent years, the USA has supplied 40%–50% of the non-communist world requirements of carbon black. Expanding industrialization in Asia and elsewhere has decreased the role of USA production. This shift should continue as indicated by the comparison in Table 3 of projected carbon black demand in 1990 as compared with that in 1983.

The changing contribution of the three major manufacturing processes is illustrated in Fig. 1. Until 1950, most carbon black was produced by the channel process, but now the oil furnace process is dominant because of its greater flexibility, better economics, and lower pollution potential. Consumption of thermal black has remained relatively flat, with imports from Canada compensating for the decrease in USA production due to plant closures. Additional small quantities of

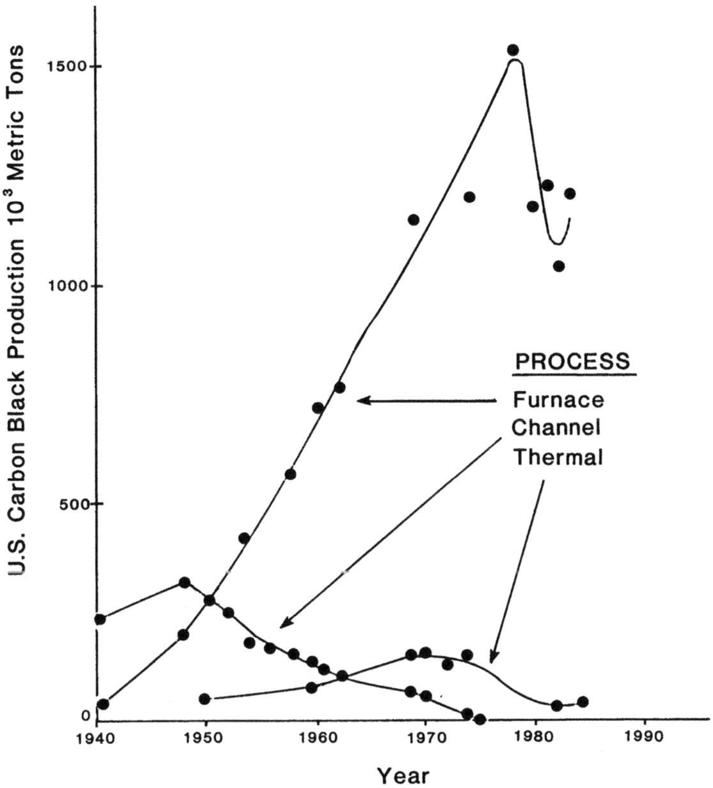

Fig. 1. Annual US production of carbon black by process, from 1940–1984

furnace black are imported to the USA and some special furnace grades for non-rubber use are exported but the total amounts to only percent of USA production.

Carbon black is an essential ingredient in thousands of industrial products, however, over 90% of the carbon black produced is used as a reinforcing filler in elastomers, mainly in the manufacture of rubber tires. Some other important applications are: pigment in inks, paints, plastics, and paper; conductive filler; radio-frequency insulator; dry cell batteries; magnetic tapes; UV stabilizer and antioxidant in plastics; and photocopy toners. A few of the grades used in inks, enamels, and toners are aftertreated with ozone, nitric acid or nitrogen oxides to yield an acidic, oxidized product. These grades account for $<1\%$ of total furnace black production. Table 4 lists the relative consumption by major application categories for furnace black.

Thermal black and lampblack are used mainly in rubber but also find use as tinting agents, low-reinforcing polymer extenders and in electrical components (e.g. resistors, brushes, electrodes). Acetylene black is used primarily for its high conductivity in plastics and batteries, etc. Impingement blacks are used in applications similar to those for medium to high surface area furnace blacks, particularly where surface acidity is useful. Some grades are sold in both pelleted and

Table 4. Carbon black consumption (1980–1982)[a], %

<table>
<tr><th>End user</th><th>World (Non-Comecon)</th><th>United States</th><th>W. Europe</th><th>Japan</th></tr>
<tr><td>Rubber, tires</td><td>62</td><td>62</td><td>60</td><td>73</td></tr>
<tr><td>Rubber, non-tire</td><td>29</td><td>30</td><td>30</td><td>22</td></tr>
<tr><td>Ink, paint, plastics</td><td>9</td><td>6</td><td rowspan="2">10</td><td rowspan="2">5</td></tr>
<tr><td>Other</td><td>1</td><td>2</td></tr>
</table>

[a] [6]

fluffy versions, with pellets the standard form for high shear processing such as rubber compounding whereas fluffy blacks are used mainly in lower shear liquid media.

Technology of Manufacture

Mechanism of Formation. In all carbon black processes except that for acetylene black, a liquid or gaseous hydrocarbon feedstock is pyrolyzed at 1200–1700 °C. The resultant molecular fragments polymerize in the vapor phase to polycyclic aromatic species which condense to form liquid nuclei. Small amounts of stable polycyclic aromatic hydrocarbons (PAH) are formed also as a minor by-product. Subsequent coalescence and carbon deposition yields spherical particles with diameters of 5–20 nm in the furnace and channel processes and up to 500 nm in the thermal process. Progressive dehydrogenation leads to an increase in viscosity and "stickiness" so that further collisions cause the particles to cohere and partially fuse but not to coalesce into spherical form. Continued dehydrogenation and carbon deposition yields carbon aggregates of characteristic morphology (Fig. 2) made up of fused particles having turbostratically oriented graphite-like carbon layers. When solid aggregates are present the temperature must be decreased to retard oxidation of the carbon in the presence of the high concentration of water vapor in the flue gas. This process appears to proceed via hydroxyl radical attack to produce porosity and loss of surface carbon [7].

Aggregates are the smallest dispersible units of carbon black, but in the dry state in the absence of sufficient shear forces, they are held together in large agglomerates by weak van der Waals forces.

Channel Process. The original process employed a sheet metal building containing thousands of natural gas flames quenched by overhead reciprocating iron channels. A limited air supply was admitted at the base of the building with combustion products vented to the atmosphere. Most of the carbon deposited on the channel and was scraped off and collected in hoppers. The product from many hot houses was converged to a central processing unit where coke and foreign materials were removed. Yields were very low, usually <5% of the theoretical carbon, with up to 20% of the carbon black lost as smoke through the vents.

In the German impingement modification, naphthalene or anthracene coal tar residue is vaporized in coke oven gas at 350–400 °C. The vapor is burned at gas

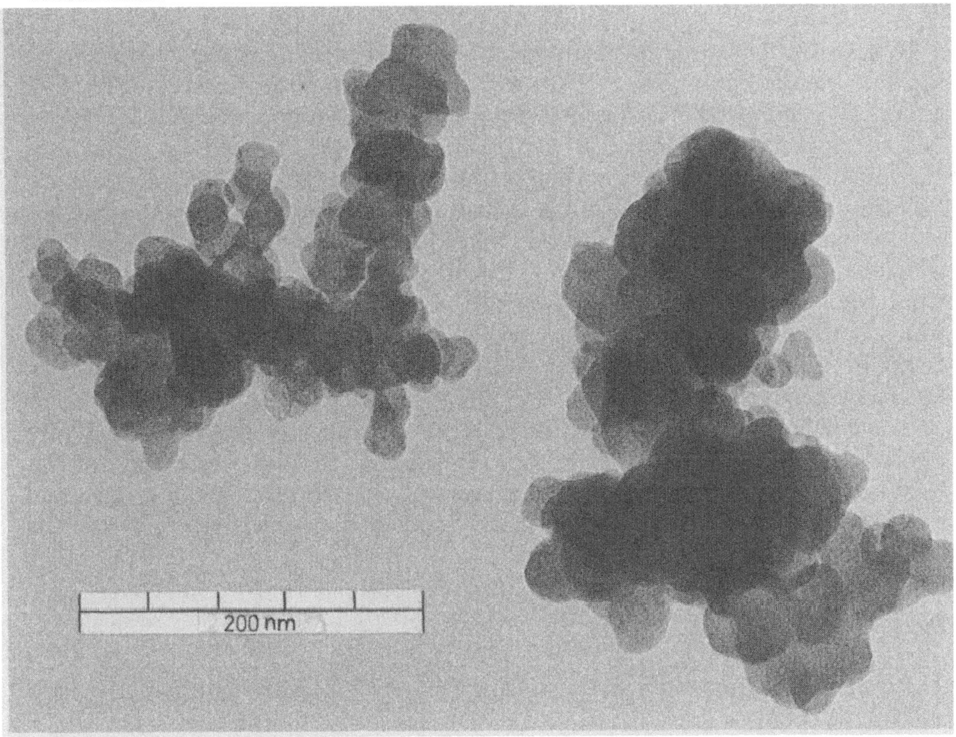

Fig. 2. Aggregates of an oil furnace carbon black

tips with limited air and collected on a cooled mild steel pipe. Yield is >60% of the theoretical carbon.

Acetylene Black Process. Flowing acetylene gas is decomposed exothermically to carbon and hydrogen at ~1,000 °C in a water cooled refractory lined metal reactor. Almost all input carbon is converted to product since no oxygen or other oxidant is present.

Thermal Process. Gas or vaporized oil is pyrolysed in a preheated refractory brick retort at 1,300–1,500 °C to produce carbon suspended in an off-gas composed of >85% hydrogen plus methane and heavier hydrocarbons. After cooling with a water spray, the carbon is removed by cyclone separators followed by bag filters or wet scrubbers. Two reactors are operated in tandem; one being heated by combustion of recycled off-gas in air while the other is producing carbon. Yields approach 80% of theoretical carbon when the feedstock is oil and 60% from gas.

Gas Furnace Process. Natural gas is injected into a gas-air flame at 1,400 °C in a refractory lined furnace with the combined pyrolysis and combustion products subsequently cooled to 200–300 °C by water sprays. In the original process, carbon black was collected using electrostatic precipitators in series with cyclone separators but process modifications led to the use of more efficient bag filters. Yields are 10%–30% of theoretical carbon; inversely related to the surface area of the carbon black product.

Oil Furnace Process. In this modern successor to the gas furnace process, a highly aromatic liquid feedstock derived from coal or petroleum is sprayed into a flame at 1,300–1,700 °C in a refractory lined or water-cooled reactor. The carbon laden gas is cooled to ~300 °C by water sprays then filtered through coated glass fiber or teflon fabric filter bags to remove and collect fluffy carbon black. Yields range from 25%–75% of the carbon in the oil feed depending on grade.

For many applications, the carbon black must be in a densified, pelleted form which requires additional processing. Usually, water and small amounts of pelletizing additives (e.g. molasses, lignosulfonates) are added to the fluffy carbon which is then passed through a pin pelletizer into an externally heated rotary dryer. Product densities range from 0.1 g/cc for certain fluffy blacks to 0.5 g/cc for the densest pelleted grades.

Most pelleted carbon black is transported in bulk via specially designed hopper cars (~100,000 lb.cap.) whereas, fluffy blacks are usually packaged in 50 lb. bags for rail or truck shipment. Large users of pelleted blacks employ enclosed pneumatic systems for rail car unloading and internal distribution within the workplace.

Process Emissions

The progression from channel to gas furnace to oil furnace technology has led to a dramatic decrease in particulate emissions. This is partly balanced by the emission of increased amounts of sulfur compounds arising from the use of oil feedstocks.

A schematic of the oil furnace process which emphasizes emission sources is shown in Fig. 3. The process generates very small amounts of liquid and solid wastes, mainly from maintenance. Wash water and rain run-off are sent to settling pounds after removal of oil and suspended solids. Solid wastes are discarded in approved landfills or incinerated.

Nitrogen, water, and carbon dioxide account for almost 90% of the gaseous emissions in the absence of recycle or combustion of process gas. These gases approach 98% of the emissions in plants which employ thermal incinerators or flares, or which maximize use of process gas for its heating value.

Particulate emissions are maintained at a very low level using high-efficiency bag filters (>99.9% retention) for both economic and esthetic reasons. Other stack emissions vary with the composition of the feedstock and the grade of carbon black being produced. Representative emission factors for a plant producing a reinforcing grade of carbon black from catalytic cracker decant oil, are presented in Table 5. It is assumed that the process off-gas is neither used as fuel or combusted although this is not true for many modern plants. The PAH and trace element categories in Table 5 each contain many individual components, however, pyrene and acenapthene represent >65% of the former, whereas alkali and alkali earth metals account for more than half the trace metal content.

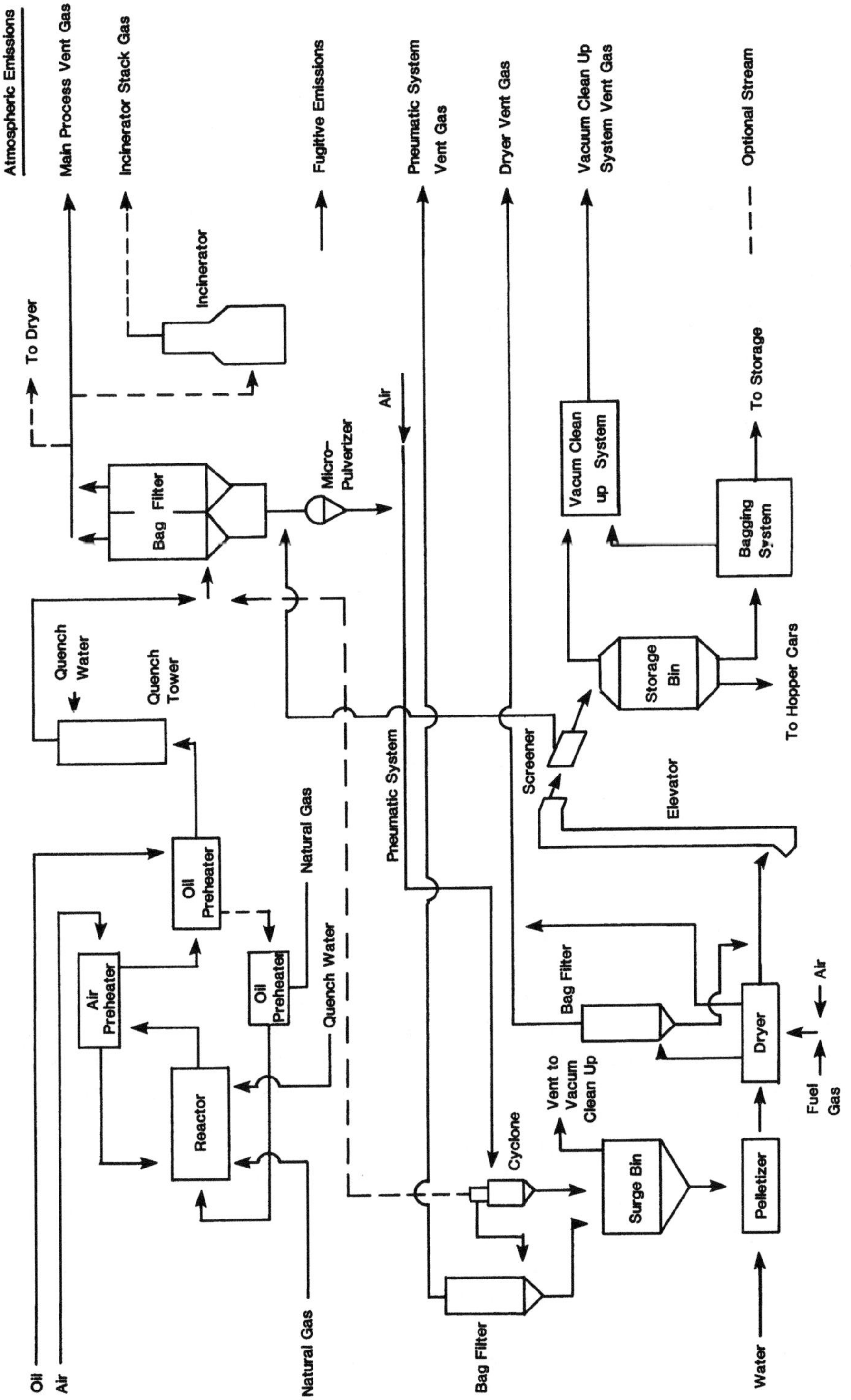

Fig. 3. Schematic diagram of the oil furnace carbon black process

Table 5. Emissions for US oil furnace process plants in 1974[a]

Material	Source[b]	Mean emission factor, g/kg	Emission from all plants[c] 10^3 mt/yr
Nitrogen	MPV	8,200	
Water	MPV, DV	7,300	
Carbon dioxide	DV	900	
Carbon monoxide	MPV	1,400	2,100
Hydrogen	MPV	100	150
Sulfur compounds		70.2	150
Hydrogen sulfide	MPV	30	
Carbon disulfide	MPV	30	
Carbonyl sulfide	MPV	10	
Sulfur dioxide	DV	0.2	
Hydrocarbons		50.7	76
Acetylene, methane etc.	MPV	50	
Feedstock vapor	OST	0.7	
Polycyclic aromatic	MPV	0.002	0.003
Nitrogen oxides		0.9	1.4
	DV	0.3	
	MPV	0.6	
Carbon black		0.6	0.9
	PSV	0.3	
	MPV	0.1	
	DV	0.1	
	FE, VSV	0.1	
Trace elements	MPV	< 0.25	< 0.4

[a] [8, 9]
[b] MPV = main process vent, DV = dryer vent, OST = oil storage tanks, PSV = pneumatic system vent, FE = fugitive emissions, VSV = Vacuum system vent
[c] Based on production of 1.5×10^6 mt

Physical and Chemical Properties

Carbon exists in two crystalline forms; diamond – a tetrahedrally hybridized face-centered cubic latice, and graphite which usually consists of parallel layers of trigonally hybridized carbon in a hexagonal network. Most industrial carbons are based on an imperfect graphite structure but are devoid of long range order and are thus considered to be amorphous. Electron microscope and X-ray diffraction analyses of carbon black aggregates suggest a roughly parallel alignment of distorted, small graphite layers oriented concentrically around growth centers. Figure 4 is a high resolution electron micrograph of a typical particle or nodule of an aggregate of a rubber grade carbon black. It shows the arrangement of layers relative to a principal growth center and several subsidiary growth centers. Surface layers appear to be more defective than those in the bulk due partly to the effect of oxidation. Relatively high surface defect concentrations are corroborated by X-ray and chemical analyses [10] and may indicate the presence of non-extractable layer fragments equivalent to large polynuclear aromatic molecules.

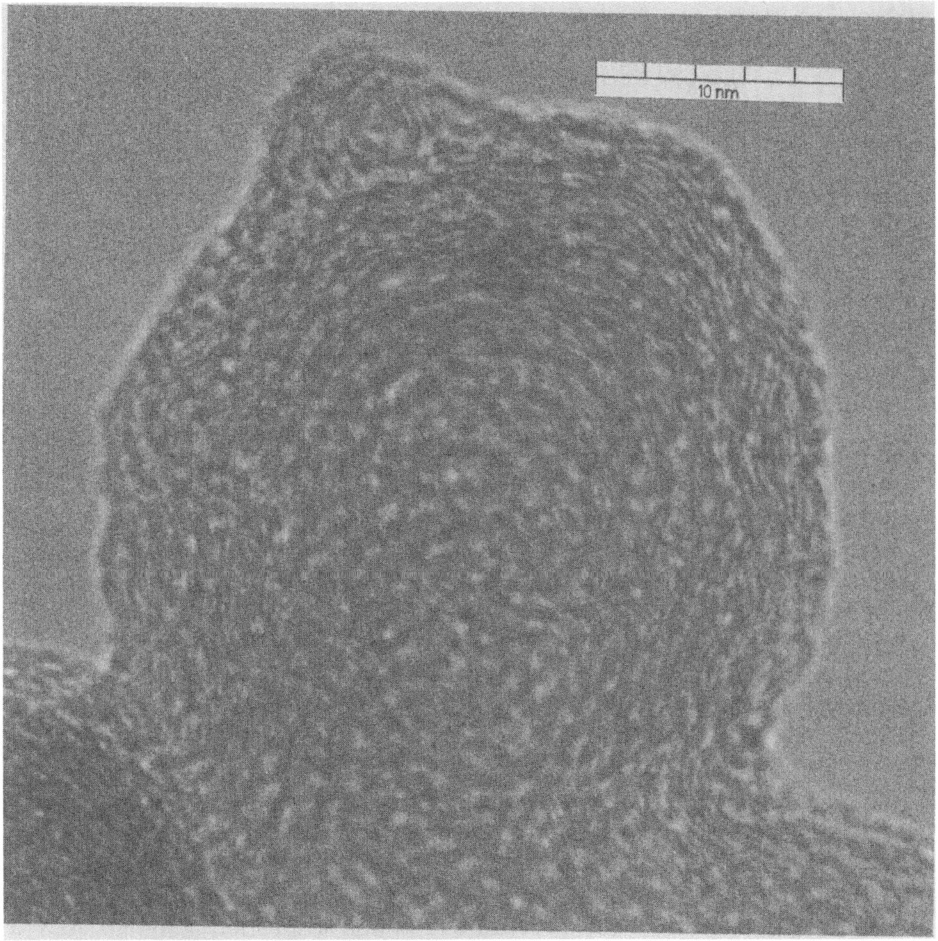

Fig. 4. Microstructure of a carbon black "particle"

Non-colloidal industrial carbons (e.g. coke, char, charcoal) prepared from condensed phase precursors tend to have less graphitic order and higher impurity levels than carbon black.

Unique products are obtained from each carbon black process although there is some overlap in analytical properties as shown in Table 6. Data from post-treated grades modified by oxidation are included in the table, which greatly broadens the observed range in properties. Oxidation increases surface area, oil absorption and volatile content, and decreases pH.

Carbon black grades were originally classified according to performance or property characteristics, but as the number of grades multiplied, the inadequacies of this procedure became obvious. In 1965, ASTM Committee D-24 adopted a revised nomenclature (ASTM D2516) and classification system (ASTM D1765) for grades used in rubber applications. The ASTM system uses a letter (N for normal cure rate, S for slow cure rate), and three digits to identify each grade. The

Table 6. Typical analytical properties of carbon blacks

Property	Furnace	Channel or impingement	Thermal	Acetylene	Lampblack
Particle diameter, nm	13 – 80	9 – 29	150 –500	35 – 50	30 –200
Surface area, m^2/g	20 –250	100 –1,000	5 – 15	60 – 70	15 – 95
DBP absorption, ml/100 g	60 –195	60 – 475	30 – 45	250 –350	110 –250
Volatile material, %	0.3 – 5.0	3.5– 16	0.1 – 0.5	0.3– 0.7	0.5 – 15
pH	3.5 – 9	3 – 6	7 – 9	5 – 7	3 – 7
Inorganic impurities, %	0.3 – 1.0	≦ 0.3	0.05– 0.4	≦ 0.01	0.01– 0.15
Sulfur, %	0.1 – 1.5	≦ 0.2	≦ 0.3	≦ 0.2	0.01– 10
Organic impurities, %	0.01– 0.3	≦ 0.1	0.02– 1.7	≦ 0.01	0.01– 1.5

first number indicates the approximate particle size decade (e.g. 1 = 11–19 nm) and the remaining numbers are assigned arbitrarily to denote such property characteristics as morphology and surface area. No comparable system exists for carbons used in inks, paints, plastics, and other non-rubber applications. These grades are classified according to performance; particularly in terms of imparted color intensity, rheology (flow), and conductivity.

All applications of carbon black depend to varying extents on surface area, aggregate morphology, and surface chemistry. A number of sophisticated procedures have been developed to evaluate these properties accurately [10] but a satisfactory overview is afforded by the simpler analytical methods listed in Table 6.

Number average particle size determined from electron micrographs is a characteristic parameter of more historic than practical importance. It is, however, inversely related to external area and thus gives an estimate of porosity when used in conjunction with a direct measurement of surface area. The latter is usually measured by the nitrogen BET procedure although adsorption of iodine (ASTM Method D1510) or the surfactant, cetyltrimethyl ammonium bromide, is used for quality control purposes.

Aggregate bulkiness or structure is an important determinant of rheological and reinforcement behaviour. It is related to void volume which is determined routinely by vehicle demand. The original manual procedure using linseed oil has been replaced by automated measurement of dibutylphthalate absorption (ASTM Method D2414).

Differences in surface chemistry can be inferred from the values for gravimetric pyrolysis at 910 °C (volatile content) and the pH of aqueous slurries of carbon black (ASTM Method D1512). Both analyses provide indirect information on the concentration of oxygen-containing functional groups. Evolution of carbon monoxide and carbon dioxide from thermal decomposition of surface groups accounts for most of the weight loss at temperatures ≦1,000 °C, for pre-dried carbon blacks. Aqueous pH is dependent primarily on the concentration of carboxylic acids, with a secondary contribution from phenols which are normally more prevalent. An ash-free carbon black devoid of acidic groups exhibits pH >9 which decreases to pH 7–9 when phenols, but not carboxylic acids, are present.

Table 7. Trace metal content of furnace black[a]

Element[b]	ppm	Element	ppm
Aluminum	30 –50	Molybdenum	0.5– 1
Arsenic	0.02– 0.3	Nickel	0.5– 3
Barium	0.4 – 1.6	Phosphorus	2 – 8
Bismuth	0.03– 0.6	Potassium	10 – 40
Boron	$\leqq$ 0.1	Rubidium	$\leqq$ 0.1
Calcium	20 –60	Selenium	$\leqq$ 0.3
Chromium	0.4 – 2	Silicon	80 –150
Cobalt	0.1 – 0.5	Silver	$\leqq$ 0.5
Copper	0.2 – 2	Sodium	450 –650
Iron	15 –40	Strontium	0.5– 5
Lead	0.6 – 1	Tin	$\leqq$ 0.2
Lithium	0.3 – 0.7	Titanium	5 – 8
Magnesium	12 –18	Vanadium	$\leqq$ 2
Manganese	0.2 – 3	Zinc	0.2– 2
Mercury	$\leqq$ 0.1		

[a] Analysis of fifteen oil furnace blacks from three manufacturers. The observed range of impurity contents is shown

[b] Not included are twenty elements which are below the detection limit of 0.05 ppm. These are: Au, Be, Cd, Cs, Ga, Hf, Ir, Nb, Os, Pa, Pt, Rh, Ru, Sb, Sc, Ta, Te, Tl, W, Zr

At greater acidities, there is a linear correlation between pH and carboxyl content [11].

Salts of alkali and alkaline earth metals derived mainly from quench and process water are the major impurities in commercial carbon black. Total inorganics are reported as ash after combustion at 550 °C in air (ASTM Method D1506). A comprehensive trace element analysis for a representative furnace black is listed in Table 7.

Increases in the average sulfur content of crude petroleum in recent years is reflected as a higher sulfur content in furnace blacks and lampblacks produced from petroleum based feedstocks. Up to one half of the feedstock sulfur is retained in a furnace black, mostly as combined sulfur, but with appreciable amounts of free elemental sulfur under certain process conditions.

Trace organic impurities in commercial carbon black are predominantly PAH with smaller amounts of other polynuclear aromatic compounds (PNA), such as nitrogen-or sulfur-heterocyclics, and oxidized PAH derivatives, though they are prevalent on some types of carbonaceous particles. PNA are strongly adsorbed on carbon black and require prolonged extraction with an aromatic solvent for quantitative removal. A correction must be made for elemental sulfur which is often present in the extract.

Carbon black aerosols are not ignited by a spark source [12] but with difficulty, can be caused to explode at dust concentrations > 50 g/m^3 by high intensity chemical igniters [13]. The maximum explosion pressure is 10 bar with a maximum rate of pressure rise between 30 and 100 bar/s. In laboratory furnace tests, the minimum fire ignition temperature for carbon black dust clouds in air exceeds

700 °C [12] although a dust layer ignites in air at temperatures above 315 °C [13] and will burn slowly with a dull red glow in the absence of an external flame. Carbon monoxide is the main product of combustion. Fires in storage areas may not be readily detected because of the thermal insulating properties of carbon black and the absence of flame and smoke. Fires may be extinguished by purging with carbon dioxide or other inert gas or by copious water spray.

Sorption of PAH on Carbon Black

PAH are not only the major organic impurity on carbon blacks but an important adsorbate on most types of particulate carbon. Maximum adsorption of PAH from the vapor phase is determined by the packing density of the adsorbed molecule and the available surface area of the carbon. PAH interact strongly with carbon surfaces mainly through van der Waals and dipole attractive forces. The surface of a non-crystalline carbon is energetically heterogenous so that adsorption energy usually diminishes slowly with increasing surface coverage (θ) but a marked attenuation due to adsorbate shielding occurs when the monolayer capacity ($\theta = 1$) is attained.

For adsorption from solution, incremental energy decreases with increasing concentration to a minimum which corresponds to a stable surface configuration of solute and solvent molecules. The composition of this layer depends on entropic factors and the relative strength of the solvent and solute surface interactions. It contains solvent molecules predominantly except for adsorption at very high solute concentration from poorly interacting solvents. Desorption of PAH from carbon depends also on the competitive interaction of solvent and adsorbate molecules with the surface as well as on the solubility of the adsorbate in the extraction fluid. For example, negligible PAH desorption occurs in simple aqueous systems due to the low affinity of water for both the adsorbate and the carbon surface.

Isosteric heats of adsorption obtained by gas chromatography are a convenient means of comparing the relative strength of molecular interactions with carbon. Adsorption enthalpy data for solvents used in carbon sorption studies in Table 8 clearly show the increase in surface affinity in going from water to aromatic hydrocarbons and also from graphitic to more heterogenous, oxidized surfaces. Solvent interaction with commercial carbon black is probably close to the level for the oxidized heat treated sample and is thus similar to that observed for the diesel particulate sample. The reported enthalpies are maximum values which have been shown to decrease drastically with increasing surface coverage as fewer unoccupied high-energy adsorption sites become available to approaching adsorbate molecules [16].

Studies of the isothermal adsorption of pyrene or BaP from dilute solutions in cyclohexane provide considerable information about the interaction of these PAH's with carbon [17–19]. The net heat of adsorption of pyrene calculated from the Langmuir adsorption isotherm is 23 kJ/mol on the heterogenous surface of N339 furnace black. In all cases, the true enthalpy of adsorption is greater by an amount equivalent to the solvent-solute interaction energy and the endothermic

Table 8. Heat of adsorption on particulate carbon[a]

Adsorbate	ΔH, kJ/mol			
	CB-1	CB-2	CB-3	Diesel
Naphthalene	72	–	–	51
Toluene	–	47	58	–
Benzene	41	39	52	54
Cyclohexane	36	34	39	39
Dichloromethane	–	26	–	21
Water	23	–	–	29

[a] Identification of carbons: CB-1; Graphitized i.e. (heat-treated at 2,500–3,000 C) Furnace Black – [14]. CB-2; Graphitized thermal black (GT) – [15]. CB-3; Oxidized GT having 0.9 wt% surface oxygen – [15]. Diesel; Diesel particulate after removal of endogenous organic adsorbates – [16]

desorption of initially adsorbed solvent. Maximum adsorption of pyrene at the isotherm plateau is 14 mg/100 m^2 on graphitized black and 12 mg/100 m^2 on furnace black. This corresponds to a mixed adsorption layer in which two-thirds of the surface is covered by solvent, setting an upper limit for the area occupied by high-energy sites capable of preferentially adsorbing pyrene.

PAH adsorbed on carbon are oriented parallel to the basal layer and interact mainly with the pi-electron cloud at the surface but there may be an additional Lewis acid-base or dipole-dipole contribution for polar surfaces [20]. Thus the heat of adsorption of pyrene on graphite increases with surface oxidation [17] and the limiting adsorption of BaP from cyclohexane increases from 8 mg/100 m^2 to 24 mg/100 m^2 in the approximate order of surface oxygen content for seven furnace and channel carbon blacks [19]. The highest BaP adsorption capacity is about 50% of the calculated close-packed monolayer which indicates that preferential adsorption of PAH's is limited to a portion of the surface even on highly oxidized carbons.

Competitive adsorption measurements under non-equilibrium conditions have also been used to estimate the affinity of PAH's for carbon in the presence of solvents. Partition ratios for PAH's on graphitized channel black in various solvents are given in Table 9a [21] and in Table 9b for different PAH's on a fur-

Table 9a. Adsorption ratios for PAH on graphitized carbon black[a]

PAH	Toluene	Benzene	Dichloromethane	Hexane	Water
Fluoranthene	0.0091	0.0095	0.020	0.20	∞
Benzo(k)fluoranthene	0.058	0.23	6.3	9.0	∞
Benzo(ghi)perylene	0.21	3.4	–	–	∞
Indeno(1,2,3-cd)pyrene	0.21	3.6	–	–	∞

[a] Partitioning between 200 mg carbon and 100 ml solvent at 25 °C; (PAH/carbon black)/(PAH/solvent)

Table 9b. Adsorption ratios for PAH on N326 furnace black[a]

PAH	Toluene	Benzene	Dichloromethane	Cyclohexane
Phenanthrene	0.0057	0.011	0.20	0.59
Benzo(a)pyrene	0.56	1.4	13	1.1×10^2

[a] PAH (500 μg)/Carbon (1 g)/Solvent (5 ml) at 25 °C; [(PAH (μg)/Carbon (g))/(PAH (μg)/Solvent (ml))] × 100

nace black [22]. Relative adsorption of PAH increases with molecular weight and varies inversely with solvent adsorption enthalpy. Solubility parameters also influence these ratios, but to an unknown extent.

Determination of PNA Content

Extraction

Individual components of complex mixtures of PNA adsorbed on carbon must be desorbed prior to quantitation by current methods. Extraction is carried out conveniently in a Soxhlet apparatus using high boiling liquids which interact strongly with carbon and are good solvents for PNA. Toluene is the preferred extractant for PAH and sulfur-PNA whereas chlorobenzene and more polar solvents such as aromatic hydrocarbon-alcohol mixtures are preferred for removing oxygen- and nitrogen-containing PNA from oxidized carbon black [23–28].

Relatively high temperatures and/or long extraction periods are required for quantitative removal of ≧ five-ring PAH adsorbed on carbon black and other high surface area carbons. Benzene extraction of three- and four-ring PAH from furnace black is complete in 50 h but at least 150 h is required for extraction of six- and seven-ring PAH [29]. Similarly, in 40 h benzene extracts nearly all of the five-ring PAH but only fifty percent of the six- and seven-ring PAH from a lampblack, whereas more complete extraction of the > six-ring compounds is obtained in 6 h at a higher temperature using naphthalene at reduced pressure [30a].

Proper selection of solvent and extraction conditions is important even for removal of high loadings of weakly or partially adsorbed PAH from environmental carbonaceous particulates. For example, only 6% of added BaP could be recovered from a coal-derived fly ash by ultrasonic extraction in benzene at 25 °C although 90%–95% of added two- and three-ring PAH were removed [31]. Sonification is more effective for extraction of large quanties of BaP added to diesel particulate in that, ultrasonic extraction by toluene, dichloromethane, chloroform, and acetone give 82%, 69%, 62%, and 28%, respectively, of the BaP recovered by Soxhlet extraction with toluene [32]. Lower molecular weight PAH are preferentially eluted from carbonaceous solids at short contact times, especially with poor extraction solvents. This phenomenon is demonstrated in Table 10 for extraction of four- and five-ring PAH from diesel soot equilibrated for 1 h with boiling solvent [33].

Table 10. PAH extraction from diesel soot in one hour

PAH (M.W. in daltons)	Extraction efficiency[a], %				
	Methanol	Acetone	Cyclohexane	Xylene	Toluene
Benz(a)anthracene/chrysene (228)	58	70	70	83	90
Benzofluoranthenes (252)	20	44	46	81	92
Benzopyrenes (252)	20	43	46	81	89
Indeno(1,2,3-cd)pyrene (276)	0.5	8	21	72	91
Benzo(ghi)perylene (276)	0.5	9	18	62	89

[a] Proportion of PAH in first extraction when followed by subsequent extraction with toluene for one hour

In summary, toluene and chlorobenzene are the solvents of choice for Soxhlet extraction of PAH and most PNA on carbon blacks and other carbonaceous solids. Essentially all PAH $\leq$300 daltons are extracted from furnace blacks in 48 h with toluene [27] or in 24 h with chlorobenzene [28].

Organic extract on commercial carbon black is estimated routinely by ASTM Standard Test Method D1618 which measures the percent transmittance at 425 nm of a toluene supernatant equilibrated with carbon black for 1 min at ambient temperature. In an alternative ASTM procedure, Standard Recommended Practice D3392, orthodichlorobenzene is substituted for toluene and absorbance is determined at several wavelengths between 300 nm and 600 nm. These methods give, at best, an approximation of the amount of weakly bound aromatic impurities, but neither is suitable for determining total PNA content, especially for low extract carbon blacks. A gravimetric analysis based on ultrasonic extraction in cyclohexane proposed by NIOSH [30] is also unsuitable in that it involves preferential desorption of non-aromatic impurities and low molecular weight PNA. A reliable gravimetric determination of extractable organic impurities on carbon black has been developed by the Environmental Health Association of the Carbon Black Industry. This procedure employs a 48 h Soxhlet extraction with toluene. It has been validated by testing within the Industry and is undergoing a round-robin evaluation by the ASTM.

Analysis

Unlike solvent extracts from most carbonaceous particulates which are rich in PNA including alkyl-PAH, carbon black extracts contain mainly nonalkylated PAH which can be analyzed by high performance liquid chromatography (HPLC) or gas chromatography (GC) without prior sample fractionation. Other chromatographic techniques, such as thin layer chromatography, are sometimes used to isolate individual components for identification by ultraviolet absorption or fluorescence spectroscopy.

Fractionation is an essential step, however, in the analysis of complex extract mixtures from soots and atmospheric particulates. In current practice, liquid-liquid partitioning or column chromatography is most often used to separate non-

PNA components and to prepare chemically homogeneous PAH and other PNA sub-fractions for subsequent analysis [33].

HPLC and high resolution capillary GC in conjunction with sensitive, high specificity detectors have become the standard methods for analysis of PNA in carbon black extracts. Some recent applications of these techniques are discussed below, more comprehensive surveys of carbon extract analyses are available elsewhere [18, 34].

Capillary GC affords excellent resolution but is difficult to use for PNA with > six rings because of their low vapor pressure at reasonable column temperatures. Larger PAH and sulfur-PNA have been identified in a furnace black extract by temperature programmed mass spectrometry, but this method gives lower resolution than GC [35]. Both high resolution and high sensitivity to large PNA are attainable, however, with reverse-phase HPLC which employs a chemically bonded hydrophobic stationary phase and a programmed gradient polar or aqueous mobile phase. With this technique, a total of fifty-three PAH and sulfur-PNA having molecular weights from 202–448 daltons (four-eleven rings) are resolved in a furnace black extract [36]. An even higher resolution analysis of the same extract has been demonstrated recently using a capillary column adaptation of reverse phase HPLC [37]. A total of 114 PNA with an overall molecular range of 228–400 daltons were detected by mass spectrometry but few compounds were positively identified.

There is considerable interest in the development of improved high specificity detectors for PNA. For example, negative ion chemical ionization mass spectrometry coupled with capillary GC was used to identify nitro-PNA in an extract from a pre-1980 oxidized furnace black [38] and PNA-ketones in the moderately polar fraction of a soot extract [39]. Shpol'skii spectroscopy which is based on low temperature quasilinear fluorescence, appears to be a powerful technique for the analysis of PAH, and heterocyclic nitrogen- and sulfur-PNA. In a recent application, an alkylated PAH (methyl coronene) and eight pericondensed thiophenes were identified in HPLC fractions of a carbon black extract [40]. The thiophenes appear to be sulfur adducts of four- and five-ring PAH also present in the extract. Shpol'skii spectroscopy can be used to identify the major components in unfractionated extracts but does not seem to offer any advantage over the usual HPLC or GC procedures [41].

PNA Composition in Carbon Black Extract

PNA impurities adsorbed on carbon black may range in size up to the dimensions of carbon basal layers but molecules having > eleven-rings are not detected due to limitations in extraction and analysis. Solvent extracts of carbon black contain mainly unsubstituted four- to seven-ring PAH of which several are animal carcinogens in their non-adsorbed form. Concentration ranges for ten compounds which account for most of the PAH fraction in extracts from non-aftertreated carbon blacks are given in Table 11. The content of individual PAH and their sum on modern furnace and thermal blacks, are near the lower end of the range shown.

Table 11. PAH in carbon black extract, ppm range

Carbon black type	Channel (3)[a]	Oil furnace (20)	Thermal (5)
Fluoranthene	0.2–0.5	0.1 – 82	1 –197
Pyrene	0.1–0.3	0.1 –400	1 –603
Benzofluoranthenes	–	<0.01–102	0.02– 20
Cyclopenta(cd)pyrene	<0.2	0.1 –145	–
Benzo(a)pyrene	0.1–0.2	0.1 – 25	0.01–186
Indeno(1,2,3-cd)pyrene	0.1	<0.3 – 35	–
Benzo(e)pyrene	0.1–0.2	0.1 – 40	0.1 –145
Benzo(ghi)perylene	0.5–0.7	0.6 –166	0.2 –217
Anthanthrene	–	<0.01– 26	0.2 –299
Coronene	0.4–0.6	0.1 –219	0.3 –800
References	[18]	[18, 27, 29, 42, 43]	[33, 42, 44–46]

[a] Number of different grades analyzed

Nitro aromatics are present on certain aftertreated carbon blacks but nitrogen-PNA have not been detected on non-oxidized carbon blacks. Sulfur heterocyclics present to a small extent in most furnace black and some channel black extracts, probably form by reaction of PAH radicals with sulfur compounds during the carbon formation process. Oxygenated PNA produced mainly by oxidative degradation of adsorbed PAH are a major component of extracts from channel and oxidized furnace blacks. The effect of oxidation on PNA composition was demonstrated for a series of aftertreated impingement blacks [24]. PAH and a smaller amount of quinones were in the benzene/methanol extract of the original black (4% O) whereas carboxylic acids and anhydrides predominated in the extract from a moderately oxidized carbon (12% O). Phthalic acid was the major extract component for the most heavily oxidized carbon in the series (19% O). This sequence in a less exaggerated form is generally relevant to soot and environmental carbon.

Soots

Characterization

For historical reasons, carbon black is sometimes regarded as a form of soot, and in fact, many languages use the same word to designate both materials. This comparison is tenuous at best, and is not valid for most environmental soots, which are mixtures of various forms of particulate carbon with organic tars, resins, and refractory inorganics. They exhibit variable composition and properties reflecting their origin as uncontrolled by-products of the partial combustion or pyrolysis of carbonaceous matter. Some environmental soots contain more extractable organic tar (ca. 50 wt%) than particulate carbon, with most deposited soots having very little carbon in colloidal form [27].

In contrast, carbon blacks are high purity, colloidal carbons having a unique morphology defined as aciniform carbon (AC), characterized by a turbostratic arrangement of graphitic layers, colloidal size and cluster morphology [47].

Aciniform carbon is prevalent in chimney smoke, and engine exhaust but not in soot deposits. The particulate portion of diesel soot is almost exclusively aciniform carbon. Unlike carbon black, diesel aggregates have a very wide particle size distribution suggesting that particles or small aggregates formed in different flame zones have been cemented together by subsequent deposition of a layer of carbon.

Soot deposits recovered from chimneys, fireplaces, and flues, etc. often contain less than 1% aciniform carbon in the presence of several other forms of carbonaceous material [47]. These include fragments of pyrolyzed solid fuel (coke and char), carbon cenospheres from liquid fuel sprays, and crosslinked resins. The latter may act as a binder for aciniform carbon particles to produce carbonaceous microgel (CM). This material, shown in Fig. 5 is composed of microscopic entities in which pseudo-spheroidal particles of colloidal dimensions are embedded in a carbon or carbonaceous matrix.

In addition to insoluble carbonaceous material, soots also contain extractable tars and other organic compounds which make up the soluble organic fraction (SOF), and various inorganic components. Soots generally have much greater organic extract and/or ash content than does carbon black. Extensive chemical, physical, and microscopic characterization is required for complete description of

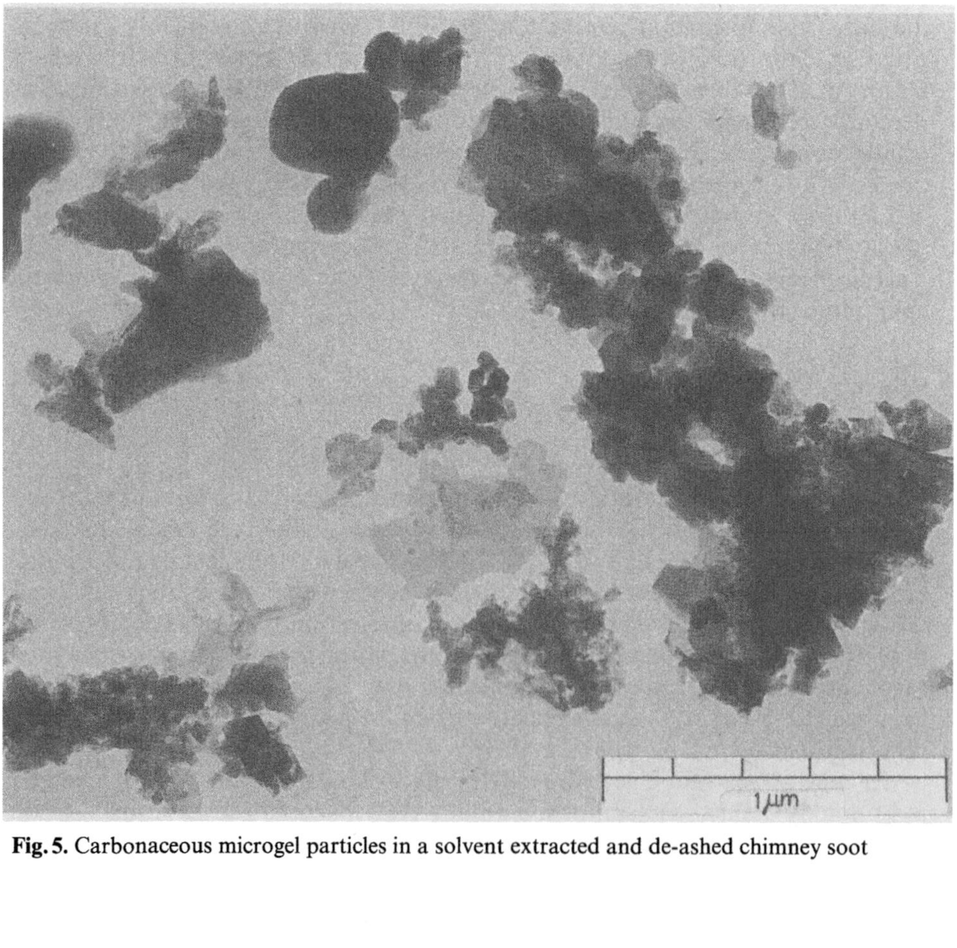

Fig. 5. Carbonaceous microgel particles in a solvent extracted and de-ashed chimney soot

Table 12. Description and composition of carbon blacks and soots[a]

Material	N_2 S.A., m^2/g	Extract, %		ASH, %	Atomic ratio H/C	Aciniform carbon, %	Physical description
		SOF	Water				
Furnace process CB							
N472	198	0.08	0.80	0.92	0.015	99	Black pellets
N351	73	0.15	0.90	0.27	0.040	99	Black pellets
N762	29	0.04	0.70	0.28	0.031	99	Black pellets
Domestic chimney soots							
Furnace "Soot Box" (oil)	32	0.6	61	54	ND	0.8	Brownish-black powder
Fireplace (coal)	21	0.9	9.3	74	1.18	1.4	Black powder; light brown and white particles
Combined flues (coal)	17	36	19	25	1.21	0.04	Oil black powder
Fireplace (hardwood)	3	16	14	22	1.08	0.02	Black flakes and chunks
Industrial boiler							
"Cold return" (oil)	93	0.2	87	28	ND	0.2	Sticky black powder with white flakes
Power plant flyash	20	0.07	29	56	0.44	0.4	Dark gray powder
Flue (coal)	1.3	0.008	8.4	92	ND	ND	Gray-black powder
Diesel soots							
Single cyl-steady state	72	51	3.6	2.2	ND	45	Fine black powder
EPA mode 11	ND	43	> 8[b]	ND	ND	ND	Fine black powder
Urban dust SRM 1648	29	2.9	ND	65	1.67	0.5	Fine dark-gray powder

[a] Taken from [27, 47]
[b] Mainly sulfuric acid

Table 13. Composition of non-alkylated PAH in environmental soots

Source PAH, ppm/particulates	Urban air[a]						Diesel engine exhaust[b]			Gasoline engine exhaust[c]			
	1	2	3	3		4							
Fluoranthene	9	3	8	9	5	50	NR	175	346	1,750	91	253	972
Pyrene	9	5	7	5	6	41	35	217	310	1,325	115	383	1,770
Benzofluoranthenes	14	3	3	NR	8	2	NR	17	100	NR	NR	NR	54
Cyclopenta(cd)pyrene	NR	NR	NR	1	NR	NR	NR	2	74	< 20	NR	NR	NR
Benz(a)anthracene	5	1	7	3	3	1	NR	96	NR	230	≤280	NR	131
Chrysene	12	1	3	NR	4	15	NR	98	36	NR	70	NR	175
Benzo(a)pyrene	5	0.8	3	3	2	1	3	5	NR	34	42	38	73
Benzo(e)pyrene	14	2	NR	NR	4	NR	3	4	NR	45	119	47	422
Indeno(1,2,3-cd)pyrene	6	0.8	5	3	3	5	10	NR	NR	167	NR	NR	NR
Benzo(ghi)perylene	43	5	6	5	4	7	16	NR	NR	191	63	165	NR
Anthanthrene	2	0.3	NR	NR	NR	NR	NR	NR	NR	< 10	NR	5	NR
Coronene	30	5	NR	5	NR	NR	9	32	NR	62	18	68	338
References	[50]	[50]	[51]	[52]	[51]	[51]	[52]	[49]	[53]	[52]	[55]	[54]	[56]

[a] 1) Los Angeles – heavy traffic
2) Los Angeles area – light traffic
3) St. Louis air particulate – NBS SRM 1648
4) Washington, DC air particulate – NBS SRM 1649

[b] Average of eight automobiles

[c] Condensed and adsorbed PAH

a soot mixture. However, a useful comparison of environmental samples can be obtained from a microscopic evaluation in conjunction with limited elemental and compositional analyses as shown in Table 12.

Diesel soots consist mainly of AC and SOF, whereas domestic and industrial soots and fly ash contain very little AC and frequently more inorganic material than particulate carbon. The presence of even minor amounts of CM in airborne soot implicates a domestic source, while the presence of char or coke fragments and substantial ash is indicative of a chimney soot. Soots emitted by high temperature combustion systems (e.g. power plants, industrial boiler flues) have low SOF and high ash due to extensive oxidation.

Further characteristic differences between soots are apparent from an examination of their trace metal content and certain components of the SOF, especially PNA [48]. Some indication of the variability in PNA content of soots is given in Table 13, which shows the concentration of non-alkylated PAH in urban air and engine exhaust particulates. As a class, PAH are the most intensively studied PNA, but they constitute only a small minority of the PNA fraction of a soot SOF. For example, fractionation of automotive diesel SOF on a normal-phase HPLC silica column [49] gave the following composition; aliphatics and < two-ring aromatics: 46%, acids: 31%, heterocyclic PNA: 13%, oxy-PAH (carbonyl, carboxyl, hydroxyl): 7%, and PAH: 4%. Approximately half of the extract is raw or partially converted diesel fuel and most of the remainder is made up of pyrolysis products which have reacted with combustion gases in the engine or in the collection system. Oxidation also has an important effect on the SOF of aged, airborne soot (e.g. urban particulate) in that adsorbed PAH is converted to polar PNA by thermal and photochemical processes under ambient conditions (see "Atmospheric Chemistry" section).

Comparison of the major SOF sub-fractions in carbon black to those in several soots (Table 14) indicates that the trace organic impurity in carbon black is mainly PAH whereas the soot SOF contain predominantly polar PNA, although their absolute PAH content is still much greater than that of carbon black. It should also be noted that there are almost no alkylated PAH in carbon black, but a relatively high proportion in soots which reflects their lower temperature of formation.

Table 14. Extract composition: carbon black and soots [a]

Sample	N351 carbon black	Wood soot	Coal soot	Diesel
Carbon [b], particulate/aciniform, %	99/99	50/0.02	23/0.4	45/50
Methylene chloride extract, wt. %	0.05	15	35	51
Main HPLC fraction, %				
Non-aromatic	1.3	0.5	3.4	35
PAH (>3 ring)	70	3.9	11	14
Polar PNA	18	92	73	23

[a] [57]

[b] Particulate carbon is material remaining after aqueous and organic solvent extraction, corrected for inorganic ash. Aciniform carbon determined by light absorption – dilute dispersion in xylene

Health Effects of Soot

It was observed over 200 years ago that the high incidence of scrotal cancer in London chimney sweeps could be attributed to exposure to soot [58]. During the next 75 years, several reports confirmed that chimney soot is a human skin carcinogen [59]. About 50 years ago, epidemiologic studies of Japanese men employed in coal gas production indicated a possible link between lung cancer and occupational exposure to soot [60]. Subsequent studies of gas workers [61] and coke-over workers [62] suggested that an increased risk of lung and other respiratory cancer is more likely associated with PAH fumes than with soot.

Experimental carcinogenesis provided insight into the nature of the carcinogenic fraction of soot, in that topical application of an organic extract of coal soot induced skin tumors in mice [63]. Similar experiments with benzene extracts of air particulate implicated a number of PAH as the putative carcinogens [64]. More extensive subcutaneous injection studies with chromatographically fractionated extracts from air particulate samples showed that most of the carcinogenicity resided in the PAH fraction. Fewer tumors were produced by the oxygenated PNA fractions, whereas the aliphatic fractions were inactive [65]. Although BaP is often used as a standard indicator of PAH content, the tumor inducing ability of the total extract or its PAH fraction showed no relationship with BaP content. For a series of air particulate extracts fractionated by a different scheme, it was estimated that no more than 50% of the tumorigenicity was due to BaP [66].

Non-carcinogenic, toxic effects have also been attributed to soot. Ambient soot concentration was highly correlated with the incidence of bronchitis and pneumonia in England and Wales [67], and with chronic bronchitis in Japan [68]. In an occupational setting, British gas plant workers who were heavily exposed to soot and fumes experienced a significant increase in mortality due to bronchitis [61]. A recent review of the literature on health effects of ambient air particulate suggests that morbidity rates due to chronic bronchitis generally increase with the level of sulfur dioxide and particulate air pollutants [69]. However, the contribution of soot as opposed to other respirable particulates is not determined.

There is also one reported instance of severe neurologic effects immediately following topical application of a diesel soot extract in C57 black mice [70]. Pathological conditions of both the liver and urinary system were observed.

Atmospheric Chemistry

Carbonaceous aerosol particles from incomplete combusion of fossil fuel are a major constituent of airborne particulate matter. Ambient soot not only effects visibility but adsorbs toxic air pollutants and is an efficient heterogeneous catalyst for some important atmospheric reactions. These include reactions of sulfur oxides, nitrogen oxides, ammonia, and adsorbed PAH, observed under actual or simulated atmospheric conditions.

In 1974, Novakov and co-workers reported that carbon catalysed the oxidation of SO_2 to sulfate in air at room temperature [71]. Subsequent work by this group demonstrated that key steps of the reaction take place in a film of adsorbed water surrounding the carbon particle and that the oxidation rate is independent of pH at pH <7.6 indicating that all S(IV) species, $SO_2 \cdot H_2O$, HSO_3^- and $SO_3^=$, have comparable reactivity [72–74]. A saturation limit observed as the oxidation proceeds is attributed to poisoning of active sites by adsorbed sulfate.

Kinetics of the oxidation of SO_2 in the presence of hydrated carbon can be represented by a Langmuir-Hinshelwood model [75, 76] in which oxygen dissociatively adsorbed on the carbon surface reacts in the rate controlling step with adsorbed SO_2 to yield SO_3. A mechanism involving surface oxygen groups has also been proposed [77], whereby SO_2 dissolved in the adsorbed water film is oxidized to sulfate at a diffusion controlled rate by reaction with a functional group. In a slower reaction, the surface group is regenerated by oxygen. In both mechanisms, the overall oxidation rate is dependent on the active site concentration rather than on total mass or surface area of carbon. A correlation between electron paramagnetic resonance spin concentration and oxygen chemisorption or SO_2 oxidation has been noted for carbons of different origin and thermal history [78]. The unpaired spin concentration of aged air particulate samples is lower than that of fresh carbons which may explain the lower catalytic activity of an ambient carbonaceous aerosol relative to activated charcoal [76].

Carbon also catalyses the reaction of SO_2 with oxidants other than oxygen, particularly NO_x and ozone [79, 80]. Oxidation of SO_2 adsorbed on carbon black decreased in the order: $NO_2 \sim O_3 > N_2O \sim O_2$ [81]. After 2.5 h, the weight of adsorbed sulfuric acid produced by 5 ppm NO_2/air or O_3/air exceeded the weight of the original carbon sample. Overall stoichiometry of the surface reaction of SO_2 with NO_2 [75] and with O_3 [81] is similar:

$$SO_2 + NO_3 = SO_3 + NO$$

$$SO_2 + O_3 = SO_3 + O_2 .$$

In a separate study [76], the reaction of SO_2 with ammonia was found to occur at a rate comparable to that with NO_2 or O_3. This process is consistent with a Langmuir-Hinshelwood model in which SO_2 and NH_3 adsorbed on carbon in the presence of water form ammonium sulfite which remains adsorbed and slowly oxidizes to ammonium sulfate.

When carbon-free mixtures of SO_2, NO_2, and propylene in air are exposed to sunlight, a primary photochemical aerosol is produced [82]. Analysis of the aerosol particles by X-ray photoelectron spectroscopy (ESCA) indicated the presence of ammonium sulfate and nitrate. On aging, the nitrate concentration decreases, probably by volatilization of nitric acid produced via proton exchange from adsorbed sulfuric acid: $NO_3^- + H_2SO_4 \rightarrow HNO_3 + HSO_4^-$. The relative importance of photochemical vs. catalytic oxidation processes is not established, but the occurrence of an inverse relationship between sulfate and nitrate content of Los Angeles air particulate samples has been reported [82].

At 25 °C, propane soot catalyses the reaction of NO with NH_3 in air to yield mainly ammonium nitrate as determined by ESCA [83]. Above 150 °C, these gases interact with the carbon surface to produce stable nitrogenous surface species identified as amines, amides, and nitriles [83, 84]. The same groups were observed in ambient aerosols [83].

Carbon readily reacts with NO_2 at room temperature to form both oxygen- and nitrogen-containing surface groups [85]. The presence of C-NO_2, C-ONO, and C-NNO was shown by FT-IR analysis. NO_2 addition follows Langmuir-Hinshelwood kinetics with an apparent activation energy of 10.2 kJ/mol.

Carbon catalyses both the oxidation and reduction of NO, yielding NO_2 in air at ambient temperatures [86] and nitrogen by reaction with NH_3 at elevated temperatures [87]. A sulfuric acid activated carbon at 150 °C, generated almost 80% of the theoretical N_2 in ten minutes, whereas a ten-fold lower rate was observed with an untreated carbon. A detailed mechanism for this reaction has not been determined.

Oxidized and substituted PAH are major components of the organic extracts of most soots, airborne carbonaceous matter, and commercial oxidized carbon blacks. Nitro-derivatives have been studied extensively because of their wide occurrence and high mutagenicity but sulfur- and oxygen-substituted PAH are formed also under environmental conditions [88].

Differences in reactivity between free and adsorbed PAH range from large in the case of NO_x to quite small for SO_2 and photolysis in air. Nitration rates appear to be substrate-dependent as shown for adsorbed pyrene exposed to 1.33 ppm NO_2 in air at 20 °C. Conversion to nitropyrene was quantitative on silica gel, decreasing to 57% on fly ash, 2.7% on alumina and 0.4% on carbon [89]. Under similar conditions, PAH adsorbed on silica gel, fly ash, or alumina did not react with NO [90]. Addition of 0.12 ppm gaseous HNO_3 during high volume sampling of ambient air led to a higher conversion of PAH than did addition of 1 ppm NO_2 (e.g. BaP-95% vs 35%, BaA-55% vs 20%, perylene −55% vs 35%) [91]. There was no reaction of PAH upon addition of 0.2 ppm O_3 or 0.1 ppm HNO_2 to the air sample. Similarly, no PAH loss was observed when BaP or perylene deposited on glass or teflon filters loaded with fly ash, diesel, or airborne particles was exposed for 3 h at high-volume sampling velocities to; humid air, 0.1 ppm NO_2, 0.1 ppm O_3, or 0.1 ppm SO_2 [92]. At slightly higher NO_2 concentrations, conversion of BaP to 1-, 3-, and 6-nitro BaP in 8 h on glass fiber filters is 18% at 0.25 ppm NO_2 + 3 ppb HNO_3 and 40% at 1 ppm NO_2 + 11 ppb HNO_3 [93].

Nitration of PAH is an aromatic electrophilic substitution process, and as such, its rate depends strongly on reaction medium and PAH molecular structure. Half-lives for nitration of PAH adsorbed on carbon and for the same PAH in aqueous solution are shown in Table 15. Both media give a similar order of PAH reactivity although the aqueous system produces a 10^5 greater range of reactivities. As mentioned previously, ring-substituted nitro-PAH are the major nitration products, but oxy-PAH are also formed in these oxidizing mixtures [94]. In strong sunlight, additional PAH nitration may occur by a non-ionic radical mechanism, whilst some nitro-PAH's are photolysed to hydroxy derivatives and quinones, via intermediate rearrangement to nitrite with subsequent elimination of NO [95].

The high light absorptivity of carbon should limit photoreactions of adsorbed PAH, but surprisingly, photolysis has been reported for PAH adsorbed on soot [96] whereas PAH adsorbed on coal fly ash appear to be resistant to photodegradation [97]. Other investigations, however, produced the expected result. Upon irradiation in air for two days, four- and five-ring PAH exhibit $\leq$10% deg-

Table 15. Half-lives for nitration of PAH at 25 °C

PAH	$T_{1/2}$, days	
	Ads. carbon [a]	Solution [b]
Anthanthrene	3.7	0.0015
Benzo(a)pyrene	7	0.14
Benzo(ghi)perylene	8	5.5
Benz(a)anthracene	11	34
Pyrene	14	5.7
Benzo(e)pyrene	24	>100
Chrysene	26	24
Fluoranthene	27	>400
Coronene	29	23
Phenanthrene	30	> 70
Indeno(1,2,3-cd)pyrene	–	>800

[a] [100] – PAH (72–4,717 ppm) adsorbed on ethylene soot and exposed to 10 ppm N_2O_3

[b] [94] – PAH (0.1–3 ppm) dissolved in water-methanol-dioxane containing 0.16 M fuming nitric acid and 0.016 M sodium nitrite

radation when adsorbed on soot as compared to $\leqq 44\%$ degradation for the same compounds deposited on filter paper [98]. Furthermore, a recent kinetic study of the oxidation of pyrene under simulated solar radiation demonstrated a strong protective effect of carbon [99]. In the absence of light, pyrene adsorbed on carbon or coal fly ash had a half-life ~25 h in air at 24 °C. Upon irradiation, the half-life of pyrene adsorbed on fly ash decreased to ~7 h but the stability of pyrene on carbon was unchanged. PAH are also susceptible to photochemically induced hydroxylation via reaction of an intermediate radical cation with oxygen [88].

Experiments with pre-oxidized thermal carbon black show that alterations in surface chemistry have no significant effect on the rate of photo-oxidation of adsorbed PAH [101]. This insensitivity to surface polarity and the observation of endoperoxide formation from anthracene derivatives is consistant with a photo-oxidation mechanism involving reaction with singlet O_2 ($^1\Delta_g$) [88]. It is suggested [102] that irradiation of PAH adsorbed on carbonaceous particles sensitizes production of singlet oxygen which then oxidizes proximate PAH:

$$PAH + O_2 \xrightarrow{h\nu} PAH* + {}^3O_2 \rightarrow PAH + {}^1O_2 \rightarrow oxy\text{-}PAH$$

PAH adsorbed on some non-carbonaceous surfaces are readily oxidized by ozone [103]. For example, in air containing 0.2 ppm O_3, the half-lives of BaP and perylene adsorbed on silica are 2.6 min and 132 min, respectively [104]. Although ozonolysis is not light induced, PAH degradation in the presence of ozone is accelerated by simulated sunlight presumably by parallel reaction with singlet oxygen [105].

Of the various potential environmental oxidants, SO_2 seems to be least reactive towards PAH. No reaction is observed at SO_2 concentrations relevant to combustion emissions or atmospheric pollution [90, 100]. At a very high concentration (10%) of SO_2 in air, however, pyrene and BaP adsorbed on fly ash and alumina yielded the expected mono- and di-sulfonated derivatives [106].

Exposure to Carbonaceous Dusts

Airborne Particulate Matter

Air samples from thirteen sites in Los Angeles County were analyzed for total dust and extractable PAH [107]. Average annual dust loadings were 66–128 μg/m^3 with little variation in benzene extract (9% ± 1%) despite a three-fold range in automobile traffic density. Individual PAH concentrations, especially that of coronene, did correlate well with automobile usage.

Dust samples collected in the Tsuburano tunnel in Tokyo using an Anderson low-volume cascade impactor showed approximately 70% of the PAH to be on particles with diameters <0.5 μm [108]. The average dust level was 0.7 mg/m^3 consisting mainly of soot generated by diesel trucks. Measured PAH concentrations (μg/g) were:

Pyrene	– 124
Benz[a]anthracene	– 15
Chrysene	– 20
Benzo[a]pyrene	– 13
Perylene	– 3
Benzo[ghi]perylene	– 37

Wood burning is an often overlooked but important source of airborne particulates. A study of residential fireplace emissions found an average of 10 g of particles/kg of wood burned [109]. At these levels, it is estimated that wood burning accounts for 20%–30% of the Denver winter time air particulate. Analysis of the emitted particles indicate a mass median diameter of 0.2 μm and a BaP content of 3–120 ppm depending on the wood fuel. Particles emitted from softwood fires were found to contain 33% elemental carbon and 38% organic carbon whereas those from hardwoods had 8% elemental carbon and 46% organic carbon.

Particle emission rates from wood-burning stoves are similar to those from fireplaces [110]. Extremely high benzene extracts (42–67 wt%) are observed, however, with a total PAH content of up to 25%.

Burning of agricultural residua, including slash burning after logging operations is a significant source of atmospheric carbonaceous aerosols and PAH. A recent study indicates that PAH emission factors are comparable for slash burning and for wood-burning stoves [111]. Based on data from four slash burn sites, the estimated BaP deposition in western Oregon alone is 2.9 t/year, all of which is associated with carbonaceous particles.

Occupational Exposure

Concentrations of PAH and dust in the workplace can be 2–3 orders of magnitude greater than concentrations in urban air. A review of ambient BaP levels for industries using processes or products associated with hydrocarbon pyrolysis indicated potential exposures approaching 1 mg/m^3 [112]. Results of this survey are summarized in Table 16.

Table 16. Ambient BaP levels in selected industries

Industry description	BaP concentration range ($\mu g/m^3$)
Aluminum reduction	0.01 –975
Coke production	0.01 –161
Carbon impregnation	0.80 – 84
Coal tar pitch roofing	<0.06 – 27
Coal liquefaction	0.01 – 19
Petroleum refining	<0.01 – 9.3
Coal gas works	1.4 – 4.8
Steel mill	0.04 – 4.2
Hot Forging	1.6 – 2.9
Ferrous foundry	<1.1 –<1.4
Airport fueling	<0.08 – 0.36
Tire manufacturing	0.002– 0.057

Worker exposure to total PAH is undoubtedly much greater as demonstrated by measurements in the aluminum [113] and the coke [113, 114] industries where particulate PAH (benzene soluble) was 10–30 fold ambient BaP. In coke oven effluent, 98%–99% of the PAH is adsorbed on respirable particles [113, 114].

There is no federal standard for occupational exposure to PAH in the United States, but there are OSHA limits for 8-h time-weighted average exposure to the benzene soluble fractions of coal tar pitch volatiles (0.2 mg/m^3) and coke oven emissions (0.15 mg/m^3). Although not strictly applicable in many cases, these indirect standards frequently are used to regulate occupational exposure to selected or total PAH.

Determination of Carbon in Ambient Dust

Of various methods used to determine the carbon content of air particulate material, those based on optical properties or combustion are preferred. Elemental carbon is predominantly responsible for the absorption of visible light by atmospheric aerosols and can be quantitated by comparing measured optical properties (e.g. reflectance, refractive index or absorption coefficient) of filter samples to that of a carbon powder standard [115, 116]. Combustion methods are most frequently used to determine total carbon but can be selectively applied to the analysis of elemental carbon [117].

Combined optical and combustion analyses of collected dust samples indicate that urban aerosols are 10%–30% carbon, of which 25%–50% is elemental carbon [117–119]. Similar, but more variable, results are obtained by combustion analysis only, if the sample is pre-treated to remove inorganic carbonates and organic carbon. Pre-treatments are rarely quantititive, however, as in the case of nitric acid oxidation [120] or pyrolysis [121] which partially carbonize susceptible organics.

Estimation of carbon black dust exposure in an occupational setting depends on the nature of the workplace. In a carbon black plant or storage area, a simple gravimetric procedures such as NIOSH method S262 [122] provides a reasonable approximation of exposure to carbon black. At the other extreme, in a rubber factory for example, ambient dust could contain; organic oils, resins, polymers, accelerators, stabilizers, inorganic oxides and carbonates, sulfur and other materials in addition to carbon black. The collected dust would appear black if several percent carbon black were present. A thermogravimetric pyrolysis procedure for the analysis of the carbon black content of vulcanized rubber [123] appears to be applicable to such dust samples. A modified combustion method developed specifically for the determination of carbon black in rubber factory dust was published recently [124]. In this method, dust samples collected on silver membrane filters are solvent extracted to remove organic components, then heated to 550 °C in air to combust elemental carbon. Inorganic content is estimated by difference and insoluble sulfur is removed in a two-step process using acidified, alcoholic thiourea. With minor modifications, this general procedure should be useful for analysis of carbon black dusts in facilities producing inks, paints, plastics, etc.

Carbon Black Production – USA

A comprehensive survey of ambient dust levels was carried out in 24 plants using personal filter samples [125]. Total dust exposure by area of employment as summarized in Table 17 is indicative of good industrial hygiene practice. Archival photographs and employee interviews support the view that much higher exposures were probable in channel and early furnace plants. In addition, average dust levels of 23 mg/m^3 and 33 mg/m^3 for two plants in Texas were reported in 1948 [126].

Ambient levels of PAH and CO in USA carbon black plants are believed to be within atmospheric limits but no published reports are available.

Table 17. Total dust exposure in the USA carbon black industry, mg/m^3

Area of employment	Geometric mean TWA	95% Confidence limits
Administration	0.01	0.00–0.01
Laboratory	0.04	0.02–0.07
Production	0.44	0.36–0.53
Maintenance	0.59	0.50–0.70
Material handling	1.45	1.28–1.63

Carbon Black Production – USSR and Europe

PAH in high-volume air samples were adsorbed on charcoal then analysed by GC in a Rumanian plant which produced carbon black by channel, furnace, and thermal processes [127]. BaP levels ranged from 52–510 $\mu g/m^3$ and the concentration of four- and five-ring PAH was 500–3,500 $\mu g/m^3$. Anthracene oil used as feedstock generated anthracene air levels of 1.25 mg/m^3 at the unloading dock.

A 1973 survey of a USSR furnace black plant found carbon black dust levels of 90–196 mg/m^3 and comparably high levels of flue gas [128]. Stack emissions of 50,000 m^3/h from this plant were composed of: N_2, 60%–70%; H_2, 14%–20%; CO, 12%–15%; H_2O, 6%–12%; CO_2, 3%–6%; hydrocarbons, 0.7%; and negligible amounts of sulfur gases. Also in 1973, a survey of three USSR furnace black plants [129] found 25% of the ambient dust samples > 10 mg/m^3 and most of the CO measurements above the USSR permissible exposure level. Blood samples from exposed workers had an average carboxyhemoglobin content of 10.2% vs. 5.7% for non-exposed controls. A subsequent study of four USSR carbon black plants [130] confirmed the occurrence of extremely high CO and dust levels, with CO ranging from 150%–500% of the permissible level and ambient dust generally above the MPC of 10 mg/m^3.

In contrast, measurement of BaP associated with ambient dust in and around a German furnace black plant indicated that average air concentrations (1.6 ng/m^3) were less than those due to auto exhaust (4.6 ng/m^3) in the center of a nearby city [131].

Industrial Use of Carbon Black

Fewer workers are potentially exposed to carbon black during its manufacture, than are occupationally exposed in some of the major industries using this product. Two industries which have been the subject of extensive industrial hygiene surveys are tire manufacturing and printing.

Tire compounds are very complex mixtures of polymers, plasticizers, extenders, curatives, antioxidants, antiozonants, and reinforcing agents. Thus, it is not surprising that in a tire retreading factory, 58 organic compounds were identified in air samples from the vulcanization area and 45 compounds were identified in the extrusion area [132]. Aromatic hydrocarbons, mainly benzene derivatives, accounted for 30% of a vulcanization air filtrate (0.81 mg/m^3) and 62% of an extrusion area sample (0.19 mg/m^3). In this study, no PAH were observed and the sample collection procedure did not allow separate analysis of particulates. Total dust, PAH and BaP levels were reported, however, for thirteen British tire factories [133] and three USA tire factories [134]. Arithmetic mean exposures to dust, benzene extractible compounds and BaP are summarized by area of employment in Table 18. Dust levels are highest in the compounding and mixing areas where carbon black and other powder ingredients are handled, whereas ambient organics and BaP are prevalent at the later processing stages. It is estimated that < 1% of the airborne PAH and BaP is contributed by carbon black [126].

Carbon black appears also to be a minor contributor to ambient BaP in newspaper press rooms. Dust samples collected in pressrooms are mainly ink mist

Table 18. Occupational exposure in tire factories

Area of employment	U.K.		USA	
	Dust (mg/m³)[a]	Benzene extract (μg/m³)[a]	Dust (mg/m³)[a]	Benzo(a)pyrene (μg/m³)[b]
Compounding	4.5 (0.4–36)[c]	30	3.1 (1.3–5.0)[d]	–
Mixing	3.6 (0.6–26)	30	1.9 (1.4–5.8)	≦32
Milling	1.7 (0.4– 8.5)	<20– 800	0.8 (0.2–1.3)	≦15
Extrusion and calendering	0.6 (0.2– 3.9)	–	0.6 (0.3–0.9)	–
Tire Building	0.5 (0.1– 1.7)	20– 680	–	–
Curing	0.5 (0.1– 1.1)	10–1,980	1.3 (0.6–2.1)	≦ 8.8

[a] Personal filter samples
[b] High-volume filter samples
[c] Range of experimental value
[d] Range of plant mean concentrations

droplets and cellulose paper fibers. An industrial hygiene survey of ten British pressrooms yielded total dust levels of 0.65–2.16 mg/m³ and respirable dust levels of 0.05–0.9 mg/m³ [135]. Airborne BaP concentrations of 5.2–18 mg/m³ (8–8.3 ppm on total dust) correlate well with the BaP content of the inks used in the plants. British letterpress newsinks containing 10–12 wt% carbon black in mineral oil are reported to have a mean BaP concentration of 8.3 ppm, most of which is derived from the oil [135].

Ink mist concentrations vary greatly with printing technology and plant design. For instance, in the 1960's, pressmen in a New York City newspaper plant were exposed to a time weighted average ink mist concentration of 7.2 mg/m³ (mass median diameter ca. 15 μm) of which 16% was in the respirable size range [136], whereas a more recent study in an Italian pressroom found a time weighted average ink mist concentration of 1.4 mg/m³ with a respirable fraction of only 5% [137].

Acute Toxicity

Animal Studies

Particulate carbon, especially charcoal, is well known in veterinary and medical practice as a safe and effective adsorbent to counteract the effect of ingested toxins. For this reason it is not surprising that there are few studies of the acute toxicity of carbon itself. A lower limit for the acute oral toxicity of carbon black was derived from a study where aqueous dispersions of furnace or channel black were administered by gavage to Sprague Dawly mice [138]. There was no evidence of incompatibility up to the maximum dose, which corresponds to $LD_{50} > 10$ g/kg body weight.

Low extract, colloidal carbon was not significantly toxic to Chbb rats when tested by several methods of exposure [139]. Intravenous administration of a suspension of carbon dust (≦8 μm diameter) in saline resulted in carbon particle deposition in large arteries and alveolar capillaries within 12 h after injection. Some deposition was noted also in the liver and kidneys, with the latter clearing in 48 h. Intraperitoneal injection caused a mild, temporary irritation and rapid phagocytosis of the carbon particles. After 24 weeks, particles were agglomerated in granulomas close to the serosa. Carbon transported by lymphatics deposited in all parenchymal organs with no evidence of foreign-body reaction. Articular implantation of carbon dust did not cause acute inflammation. Intra-articular deposits cleared within 1 week with no effect on the cartilage. The authors concluded that tissue tolerance to carbon is excellent.

Similar deposition was observed in rabbits following intravenous injection of an aqueous dispersion of carbon black [140]. Nephrectomy and renal biopsy showed aggregated carbon particles in the glomerular mesangium and other areas of the kidney within 1 h and persisting for 42 days after injection. Carbon particles did not penetrate the glomerular or peritubular capillary basement membranes and were not observed in epithelial cells and urinary spaces. No inflammation was noted.

Large doses of carbon or other readily phagocytosed particles partially block the reticuloendothelial systems' ability to produce antibodies and eliminate antigens. Studies in mice show that the normal immune response to the introduction of sheep red blood cells is reduced temporarily by prior injection of colloidal carbon [141–143]. When administered intravenously, the carbon preferentially impairs antigen phagocytosis in the liver by competing with the erythrocytes for a limited number of hepatic phagocytic cells. Antigen is removed at a slower rate, mainly in the spleen and bone marrow. A similar reduction in immune response is observed upon intraperitoneal injection, but apparently without major diversion of antigen from the liver to the spleen.

Skin painting studies with oil or water suspensions of furnace blacks [144] or thermal blacks [145] did not cause dermal irritation in mice, rabbits or monkeys.

Human Studies

Workers in a Rumanian factory producing channel and furnace carbon black, and high-extract lamp black were found to have no skin cancer and relatively little serious skin disease [146]. A majority of the lamp black workers exhibited mild dermatoses such as hyperkeratosis. No skin disease was associated with exposure to channel or furnace black. Dermatitis arising from soap irritation and other factors not related to carbon black was noted.

There are reports of carbon deposition beneath the eyelids and mild folliculosis after long-term use of eye makeup cosmetics containing carbon black (i.e. liquid eyeliner, mascara) [147, 148]. The carbon particles are believed to be phagocytosed then transported via the lymphatic system before deposition in nodules in the subconjunctival tissue. No eye irritation occurs as supported by in vitro testing with human corneal endothelial cells whereby carbon black was non-toxic at all doses (0.1–10 mg/ml) [149]. It should be noted that this ocular tissue test is more discriminatory than the standard Draize procedure.

Inhalation Toxicology

Deposition and Clearance of Insoluble Particles

Inhaled particles deposit in the respiratory tract mainly by inertial impaction, gravitational sedimentation and diffusion [150, 151]. To a first approximation; particles having aerodynamic diameters > 5 μm deposit by impaction in the nose and head; particles between 1 and 5 μm deposit primarily in the tracheobronchial region by impaction and sedimentation, whereas particles between 1 and 2 μm deposit in the bonchioles and alveoli by sedimentation and diffusion. A deposition minimum is observed at a particle diameter of about 0.5 μm, at which size sedimentation ceases whereas diffusion into the alveoli increases for smaller particles.

Insoluble particles deposited in the extrathoracic region (nose, pharynx, and larynx) are ingested with mucus within minutes of entrainment. Particles in the ciliated portion of the respiratory tract are transported via the mucociliary escalator to the larynx then ingested. This process has a maximum clearance half-time of days. Particles which penetrate into the alveolar region are rapidly engulfed by macrophages. Most are eventually moved to the gastrointestinal tract by the mucociliary apparatus. A few particles are carried through the alveolar wall into the interstitial substance and thence via the lymphatics to the regional lymph nodes where they can enter the systemic blood and be circulated to other organs or excreted. Alveolar clearance half-times usually are measured in months although retention times are much shorter for soluble dusts. Dogs exposed to slightly soluble inorganic dusts exhibit an increased alveolar clearance rate and ratio of urinary to fecal excretion with increasing solubility [152]. Some insoluble particles are retained for years in the lung and lymphoid tissue, with tracheobronchial lymph nodes becoming the major repository after several years [153].

Impurities and adsorbates on insoluble particles can be dissolved and enzymatically converted to metabolites in macrophages. Some of these products are toxic or genotoxic, for example, the lung cytochrome P450 system can convert PAH's to carcinogenic intermediates. There is an apparent threshold effect in that malignant transformations are not observed for materials containing very small amounts or strongly adsorbed PAH's.

Deposition and Clearance of Carbonaceous Particles

Human subjects exposed to high levels (25–50 mg/m^3) of respirable carbon dust [154] or xerographic toner [155] undergo marginal changes in mucociliary clearance, airway resistance or respiratory function. Successive inhalation of teflon dust followed by carbon dust caused ca. 20% acceleration in clearance of the teflon ($T_{1/2} \sim 1$ h) due to increased tracheobronchial mucus secretion. This should be compared to a bronchial clearance half-time of 1.9 h for a similar sized ferric oxide dust [156]. Inhalation of the carbon-containing toner caused only slight discomfort with the subjects complaining of dryness in the nose and pharynx. Measurement of toner deposition in the nasal airways showed that about half of $\geqq 2$ μm particles and more than 80% of $\geqq 9$ μm particles were retained in the nose.

The darkly pigmented lungs of city dwellers is often assumed to be evidence of soot accumulation although there are no quantitative studies of the retention of carbonaceous aerosols in human lungs. In a retrospective study of 146 randomly selected urban males [157], lung tissue samples analysed soon after autopsy were found to contain about 0.1 wt% of particulate carbon. Accumulation increased with age at death but showed no correlation with occupation, cigarette smoking, or lung disease.

A strong decrease in alveolar clearance rate was demonstrated for smokers, however, in a controlled experiment where male subjects were exposed to a single dose of 2.8 μm ferric oxide dust [158]. After 11 months, smokers retained half of the original dose, whereas only about 10% of the initially deposited ferric oxide remained in the lungs of the nonsmokers (alveolar clearance half-time ~70 days). Smoking also decreased short-term clearance rates but not as dramatically.

Animal Inhalation of Diesel Exhaust

Most information regarding the effect of carbonaceous aerosols on the mammalian respiratory system has come from animal inhalation studies with diluted diesel exhaust. In interpreting the results of these studies, it should be noted that diesel exhaust contains high concentrations of combustion products including nitrogen oxides in addition to respirable soot particles. Nevertheless, valid conclusions concerning the fate of inhaled carbon particles appear to be attainable.

Immediately following acute exposure to 6 mg/m^3 of 0.2 μm radiolabeled diesel particulate, about 16% of the inhaled dose was retained in the lungs of male Fischer 344 rats [159]. Clearance rates showed three components [159, 160] with approximate half-times of; 1 day, 8 days, and 70 days. These were interpreted, respectively, as due to; mucociliary clearance of particles deposited in the tracheobronchial tree, transport of material deposited in the proximal bronchioles, and removal of particles from the alveolar region by endocytosis, absorption and metabolism. Almost 40% of the initial deposit was eliminated in the feces via the gastrointestinal tract within 4 days. There was longer term retention in the hilar and mediastinal lymph nodes (6% in 28 days) and in the lung (27% after 105 days), but no activity was detected in the urine, liver, spleen or kidneys [159].

Rats and male Hartley guinea pigs exposed for 45 min to diluted diesel exhaust at a particle concentration of 7 mg/m^3 had comparable deposition efficiencies and tracheobronchial clearance rates [161]. After 10 days, in contrast to the normal lung clearance and lymphatic retention in rats, the guinea pigs exhibited very little alveolar clearance and almost no carbon in the lymph nodes. This indicates the need for caution in extrapolating data from a single animal species to humans.

Alveolar clearance rates decrease in rats at diesel particle lung burdens above 1 mg, and almost cease at lung burdens >12 mg which occur after 7 weeks continuous exposure at 6 mg/m^3 [162]. This phenomenon may be explained by particle sequestering in immobile macrophage aggregates which accumulate in peribronchial and subpleural regions in the lungs [163]. A similar process appears to occur to a small extent in guinea pigs at lower doses, with no evidence of cytotoxicity or pathological changes such as fibrosis or emphysema [164].

In order to determine whether chronic exposure to diesel exhaust gas is a contributing factor to the impairment of alveolar clearance, rats were subjected to labeled diesel particles after long term (1–11 weeks) exposure to carbon black dust at a concentration of 6 mg/m^3 [165]. A carbon black lung burden of 4 mg (1 week exposure) had little effect on clearance rate, but there was a marked slowing at 10–16 mg (3–5 weeks) and extremely slow clearance at a lung burden of 30 mg (11 weeks). It follows, that the impairment in pulmonary clearance is due mainly to the retention of carbon particles rather than to other components of diesel exhaust.

Chronic inhalation of diesel exhaust causes minor changes in pulmonary function which may vary with species and exposure level. After 4 weeks inhalation of diluted diesel exhaust at a particle concentration of 6.8 mg/m^3, guinea pigs experienced an increase in pulmonary airflow resistance [166]. Noninvasive pulmonary function tests on rats exposed at a concentration of 1.5 mg/m^3 for up to 612 days, however, showed no change from controls during the first year [167]. A small increase in lung volume (FRC) and a decrease in airflow resistance in the small airways (MEF_{20}) developed near the end of the exposure period. These changes are not consistant with any known human lung disease.

Animal Inhalation of Carbon Black

In a very extensive study, Nau and coworkers [168] exposed monkeys, guinea pigs, hamsters, and mice to channel and furnace black aerosols at dust levels of 85 mg/m^3 and 57 mg/m^3, respectively. Exposures were for 7 h per day, 5 days per week up to the life span for some animals. Radiographic changes attributed to pulmonary carbon deposition were detected in mice and monkeys. There was no significant fibrosis. Electrocardiography of monkeys exposed for 1,000–10,000 h indicated right atrial and right ventricular strain. In a subsequent study [145], the same species were exposed for almost 6,000 h to a carbon black aerosol at a concentration of 53 mg/m^3. No impairment in respiratory function was observed but some Rhesus monkeys developed right ventricular hypertrophy. The authors concluded that the heart effects, noted in these studies at extremely high exposure levels, were not carbon black specific but were a secondary effect of congestion caused by inert dust accumulation in the lungs.

There were no pathologic changes in the larynx and trachea of golden hamsters after inhalation of thermal black for 172 days at a concentration of 110 mg/m^3 [169]. Tracheal edema in some members of a group of hamsters exposed to 58 mg/m^3 for 236 days was taken to represent a mechanical, rather than toxic, effect of excessive dust inhalation. Several cases of eosinophilia, which is suggestive of an allergic response, also were observed in the lower exposure group.

The absence of major respiratory pathology including proliferative or fibrotic changes was reported also in other studies with rabbits and rats at medium to high exposure levels [170, 171].

Carbon black suspensions were intratracheally instilled in rats as controls in a study of the effects of silica [172]. After six months, the lymph nodes of rats treated with carbon black showed no fibrosis, but lymph nodes of rats exposed to free silica or mixtures of carbon black and 10% or 20% silica exhibited extensive fibrosis.

Cellular and Immunological Response to Carbonaceous Aerosols

Increased production of scavenger cells is the primary cellular defense to the introduction of exogenous material into the alveolar region of the lung. Thus, increased macrophage output in Swiss albino mice occurred in less than 4 h following intratracheal instillation of a carbon suspension (4 mg of India ink) [173]. Within one day, macrophage production had increased tenfold and no free carbon particles were present in the alveoli [174].

Vostal et al. [175] summarized findings at General Motors on the effect of diesel exhaust inhalation on the murine pulmonary defense system. A threshold effect was observed in that increased macrophage production was not detected below an exposure of 0.75 mg/m^3, but was dose related at higher concentrations. Focal proliferation of Type II cells and extracellular deposits of cholesterol arising from macrophage agglomerates were noted after prolonged exposure at high doses. Inhaled diesel particles did not appear to be cytotoxic and had no marked effect on the immune system.

Decreased immune response, especially after long-term exposure to carbon has been reported in other studies. Guinea pigs exposed to E. coli aerosol after 4 weeks inhalation of carbon black at 15 mg/m^3 exhibited the same pulmonary bacteria count as controls but with a greater fraction of viable bacteria [176]. In a more extensive, chronic study of male CDI mice, intranasal inoculation with K. pneumoniae after 20 weeks inhalation of thermal carbon black (1.5 mg/m^3) alone or combined with sulfuric acid mist (1.4 mg/m^3) led to a similar increase in viable bacteria which was attributed to a decrease in bactericidal activity due to an accumulation of phagocytized carbon particles [177]. A biphasic change in the response of the lymphatic system was also observed, with the spleen producing higher antibody levels after 4 weeks inhalation but depressed levels after 12 weeks. Carbon dust did not decrease resistance to influenza virus, but with the acid mixture an increase in pulmonary consolidation and mortality was observed.

BALB/C mice exposed to an E. coli aerosol after inhalation of carbon dust (0.6 mg/m^3) for 102–192 days showed a progressive decrease in antibody formation in the spleen but biphasic production in the mediastinal lymph nodes [178]. Prior exposure to carbon had no consistant effect on immune response if the antigen was administered by intraperitoneal or intravenous injection rather than by inhalation [179].

Under some conditions, carbon may localize the effect of toxic air pollutants. Inhalation of charcoal dust caused no anatomic abnormalities of the lung in Swiss albino mice, but when the carbon was saturated with NO_2, some focal destructive pulmonary lesions occurred [180].

Occupational Lung Diseases

Although human exposure to carbonaceous dusts may occur via ingestion or dermal contact, inhalation is by far the most important route of contact. A number of lung diseases are associated with occupational exposure to certain airborne fine particles. Among these, are; chronic bronchitis – characterized by oversecretion of mucus due to irritation of the epithelial surface of the respiratory tract, emphysema – characterized by a loss in elasticity of the alveolar membrane leading to rupture and subsequent air space enlargement, pulmonary fibrosis – in which alveolar membranes lose elasticity and thicken due to accumulation of collagen which results in lower gas permeability and decreased response to gas pressure changes, occupational asthma – often a hypersensitivity reaction occurring as muscle spasms and increased mucus production in the small bronchi and bronchioles, lung cancer – most frequently in the tracheobronchial region of the respiratory tract, and pneumoconiosis – dust accumulation in the lungs (usually lower lobes) sometimes accompanied by an adverse tissue reaction. Emphysema and chronic bronchitis known collectively as chronic obstructive pulmonary disease show changes in spirometry manifested by a reduction in expiratory air flow combined with increased airway resistance. A loss of lung recoil, also may occur with emphysema. Asthma causes obstructive impairment also, mainly by airway narrowing due to muscle spasms. Fibrosis often results in a restrictive impairment of lung function, characterized by a reduction in lung volume, ventilatory capacity, and gas exchange, with an increase in lung recoil. Pneumonconioses may be detected radiographically in the absence of symptoms or adverse physiological responses. In more severe cases, pneumoconioses can be associated with obstructive or restrictive lung impairment, depending upon the nature of the dust.

Decreased blood oxygenation and increased flow resistance due to lung disease can cause the heart to increase pumping rate and pressure. As a result, the heart may become enlarged and diseased (cor pulmonale), leading eventually to heart failure.

Carbon Black Workers

Human exposure studies are by necessity statistical evaluations of cause-effect relationships. Ideally, exposure levels, and other important variables should be known and controlled prior to determination of health effects, but in most studies of workers exposed to carbon black dust these parameters are crudely estimated, if not ignored.

Large-scale epidemiologic mortality studies are discussed in the section "Epidemiology". Here we will review respiratory morbidity investigations which unfortunately are fraught with statistical and methodological problems. Almost all of these reports emanate from the USSR or Eastern Europe and lack critical information necessary for proper interpretation. Often several of the following are omitted: 1) size and composition (sex and age distributions) of exposed group and controls if any; 2) protocol, methodology, and diagnostic criteria; 3) carbon black exposure level and duration; 4) other major occupational exposures; 5) confounding factors, such as smoking, health history, and non-occupational determinants.

From the meager details available, it appears that working conditions in USSR and Eastern European carbon black plants were extremely harsh. Many of the reports note that workplace ventilation was poor or nonexistent and that ambient concentrations of dust, CO, and hydrocarbon vapors were far in excess of their occupational limits. Workers in these plants exhibit a high incidence of pneumoconiosis and minor respiratory disease (Table 19). It is not known

Country	Year	Carbon black	Cohort-size	Exposure conditions	Respiratory effects			Ref.
					Pneumo-coniosis	Decreased pulmonary function (FVC, FEV)	Other	
USSR	1961	Gas	89	N.S.[a]	4/89 (4%)	–	Emphysema-1	[181]
USSR	1965	Furnace, lampblack	> 80	Dust (pkg. dept); 10–1,000 mg/m^3, CO; 0.005–0.13 mg/l, hydrocarbons; 0.16–0.67 mg/l	–	–	2-fold increase in respiratory and general morbidity	[182]
USSR	1968	Furnace	43 women, 23 men } 66	N.S.	7/52 (13%)	General	Pneumoconiosis from pulmonary function-18/52 (35%)	[183]
USSR	1973	Furnace	643	Dust; 75% > 10 mg/m^3 CO; 74% over limit hydrocarbons; 14% over limit	7/51 (14%)	34/51 (67%)	Bronchitis-193/643 (30%)	[129]
USSR	1975	Channel, furnace, thermal	357	Extremely high dust levels	17/357 (5%)	–	Bronchitis-12/357 (3.5%)	[130]
Rumania	1969	Channel, furnace, thermal	82	Very high PAH levels (e.g. BaP $\leqq 0.5\ mg/m^3$, BghiF $\leqq 8\ mg/m^3$	15/72 (21%)	14/82 (17%)	–	[127]
Rumania	1976	Furnace	143	N.S.	29/143 (20%)	–	Emphysema (20%)	[184]
Czechoslovakia	1970	Channel, furnace	52	Dust; 8–30 mg/m^3	9/52 (17%)	–	–	[185]
Yugoslavia	1975	N.S.	35	N.S.	3/35 (9%)	General	Interstitial fibrosis-6/35 (17%)	[186]
Yugoslavia	1980[b]	N.S.	35	Dust; 13 mg/m^3 (arith. mean)	11/35 (31%)	General	Chronic bronchitis (50%)	[187]
U.K. and USA	1979	Furnace	512	6 dust exposure categories-TWA $\leqq 4.4\ mg/m^3$, V. low CO and hydrocarbon exposure	1/466 (0.2%)	f (smoking)	Emphysema-2 Bronchitis f (smoking)	[188]

[a] Not specified

whether these health effects are due to exposure to carbon black or other agents, smoking, or personal factors. Some insight into this question comes from a well-designed study of U.K. and USA workers also in Table 19. Here respiratory morbidity as a function of dust exposure was determined using spirometry, chest X-rays, physical examinations, and detailed questionairres. Bronchitic symptoms and decreased pulmonary function were prevalent in the 53% of the workers who smoked cigarettes but did not correlate with exposure level.

The high incidence of respiratory abnormalities in USSR and Eastern European carbon black workers is probably due, in part, to the effect of smoking compounded by excessive occupational exposure to CO, dust and organic compounds. This hypothesis is likely, but will remain untested until a statistically valid study of these workers is completed.

Genetic Toxicology

Bioavailability of Adsorbed PAH

Activation and detoxification of PAH takes place in mammals via enzymatic oxidation by microsomal mixed function oxygenase. An important member of this group of enzymes, aryl hydrocarbon hydroxylase (AHH), is found in the liver but is present also in macrophages and other mammalian tissues. Induction of AHH activity is a measure of PAH bioavailability although it is not necessarily a predictor of neoplasticity.

The absence of detectable AHH induction or increase in lung or liver microsomal cytochromes was demonstrated in rats after long-term inhalation of diesel exhaust at a particle concentration of 1.5 mg/m^3 [189, 190]. Direct intraperitoneal administration of solvent extract from diesel particles produced a slow, selective increase in AHH activity in the lung, indicating poor absorption of the extract into the lung circulation with minimal distribution to the liver. It is estimated that a minimum pulmonary dose of 6 mg of diesel extract per kilogram of body weight is required to obtain measurable AHH induction [191].

Ingestion of diesel particulate induces AHH activity in the liver and lung of rodents, attaining maximum activity in 6 weeks [192]. In contrast, two generations of mice fed large quantities of N375 carbon blacks (1 g/g body weight/year) for 26 weeks showed no increase in AHH levels in liver or lung tissue [193].

Induction of AHH in the gastrointestinal mucosa was determined in rats 24–72 h after an intragastric dose of BaP (50 mg/kg bodyweight), alone or mixed with fly ash or powdered charcoal (1 g/kg bodyweight) [194]. Enzyme levels in the small intestine increased ten-fold 24 h after BaP ingestion and decreased to background levels within the intestinal transit time of 48 h. Similar behavior was observed for the mixture of BaP and fly ash. No increase in mucosal AHH level occurred in the presence of charcoal presumably due to reduced bioavailability of BaP adsorbed on carbon.

PAH Elution by Biological Systems

PAH and other adsorbates on carbon presumably require prior elution in order to become available for enzymatic transformation. Water is a poor solvent for PAH. Aqueous solubility at 25 °C is ca. 1–10 mg/l for three-ring PAH's, ca. 0.001 mg/l for five-ring compounds and generally decreases with molecular weight [195]. Elution of PAH in body fluids is largely dependent, therefore, on the presence of natural surfactants such as the phospholipid, dipalmitoyl-L-α-phosphatidylcholine (DPPC), in lung fluid [196] and albumin in blood serum [197]. DPPC is a major component (~50%) of the pulmonary surfactant in alveolar cells [198].

Most attempts to desorb PAH from carbon blacks or carbon black-filled rubber vulcanizates using edible oils and surfactant-free aqueous eluants have been unsuccessful [145, 199, 200]. Incomplete, slow desorption was observed, however, for several low surface area, high extract carbon blacks [45, 201]. Blood and lung fluids which contain surfactants are only slightly better eluants in vitro for PAH on commercial furnace blacks. Limiting desorption of BaP is attained in 24 h at 40 °C with human plasma, swine serum, swine lung homogenate, and swine lung washings, but maximum elution (lung homogenate) is below 0.005% of the BaP extractable with toluene [193].

As might be expected, greater elution of PAH by tissue fluids occurs at or above the monolayer adsorption capacity of carbon black. In the case of thermal black containing exogenous PAH, human blood plasma elutes about 80% of added four-ring compounds and 47% of added BaP in 4 h at 25 °C in vitro [202]. Bovine serum appears to be a less efficient extractant, in that it was found to elute ≦20% of supramonolayer quantities of BaP on several carbon blacks with the rate of desorption inversely related to the adsorption area of the carbon black [203].

Electrophoresis and extraction experiments with human serum proteins reveal that BaP is desorbed from carbon black by albumin but not globulin fractions [203]. Similarly, the water soluble albumin-like fraction of cytoplasm was shown to be the active eluant in rat lung homogenate [204]. Evidence for PNA solubilization by protein complexes was obtained in studies of the effect of prior equilibration with serum or lung cytosol on the mutagen content of diesel particles [205]. Proteins cause a large reduction in the extractable mutagenic activity, which can be reversed by incubation with protease. The existence of a strong interaction with protein implies that PNA which desorb in the presence of proteinaceous surfactants may be less bioavailable than the same compound in solution.

Transfer of an adsorbed compound from a particle surface to a cell membrane is a crucial determinant of bioavailability. For a series of PAH the rate of the reverse process, that is, transfer from a phospholipid membrane (phosphatidylcholine vesicle) to water decreases with increasing molecular volume and partition coefficient between hydrocarbon and polar phases [206]. Solvation of the transfer species at the interface between water and the phospholipid surface appears to be rate-limiting. In comparing transfer rates for PAH adsorbed on different particles, it is likely that surface coverage and especially, adsorbate-surface interaction energy are dominant factors.

Table 20. BaP transfer from particles to DPPC vesicles or liver microsomes

Particle	BaP loading μg/m^2	Transfer to DPPC vesicles, % BaP uptake in 30 min	Transfer to liver microsomes, rate constant, min^{-1}
Anthophyllite asbestos	25	90	0.60
Amosite asbestos	53	85	–
Porous glass	0.98	83	–
Chrysotile asbestos	11	81	0.18
Crocidolite asbestos	36	74	–
Talc	25	70	–
Partisil silica	0.75	65	–
Amorphous silica	0.79	54	0.071
Hematite	38	53	0.014
Glass beads (< 10 mm)	–	50	–
Titanium dioxide	48	27	–
Free BaP microcrystals	–	21	0.006
Furnace carbon black	9.6	0	0.0

Kinetics of BaP transfer from high surface area inorganic particles to either rat liver microsomes [207–210] or DPPC vesicles [208, 211] was determined at 25–55 °C in aqueous suspension. Rates of transfer at a nominal BaP loading of 300 ppm are summarized in Table 20 for both the lipid bilayer vesicle and liver microsome systems. BaP adsorbed on inorganic oxide surfaces is taken up more readily by the membranes than BaP from a microcrystalline dispersion. Although adsorption of BaP on these surfaces is relatively weak, enhanced rates are not observed for simple mixtures of BaP microcrystals and particles. Transfer does not occur, however, if BaP is strongly bound to the particle, as with carbon black at sub-monolayer surface coverage. Desorption of BaP into the aqueous phase appears to be the rate-limiting transfer step. This process is facilitated by the presence of weakly bound, highly dispersed BaP. The affinity of N375 carbon black for BaP was estimated from equilibrium partition coefficients at 37 °C to be over 200 – fold greater than that of DPPC vesicles and 70-fold greater than that of liquid crystalline vesicles of dimyristoyl-L-α-phosphatidylcholine (DMPC) [212].

In order to promote transfer, BaP enriched carbon blacks (2.7 ± 0.5 μg/m^2) were equilibrated at 37 °C with a large weight excess of vesicles (wt. vesicles/wt. carbon black = 100) as compared to earlier studies which employed a weight ratio of 0.6. BaP transfer to vesicles under this forcing condition allows differentiation between carbons, as shown for four furnace blacks in Table 21. The extent of transfer is greater with the liquid crystalline membrane and depends on the nature of the carbon but not its surface area. The authors conclude that the PAH impurity on carbon black does not transfer readily to membranes in vivo because of preferential affinity for the carbon surface, the low endogenous concentration on commercial blacks and the low effective ratio of biological eluant to particle. As a corrollary to this, it would seem that the bioavailability of PAH added to carbon particles in amounts near or above the surface adsorption capacity cannot be extrapolated to the endogenous levels on carbon blacks and other low extract content carbons.

Table 21. BaP transfer from carbon black to phospholipid vesicles

Carbon black type	BET surface area, m^2/g	% Transfer in 1 h	
		DPPC	DMPC
N234	128	nd[a]	nd[a]
N375	101	35	45
N339	90	26	31
N351	70	23	50

[a] No detectable transfer

Pulmonary clearance rates for BaP intratracheally instilled in animals are much slower when the compound is administered as an intimate mixture with inert carrier particles, rather than as a free suspension [213]. The effect has been confirmed for a wide variety of particles and generally is treated as an adsorption-related phenomenon [214]. This interpretation is questionable, however, because the particle surface can accommodate only a few percent of the admixed BaP at customary loadings.

Experiments with nutshell charcoal indicate that carbon increases BaP retention more than does similarly sized particles of iron oxide or aluminum oxide [215]. Pulmonary clearance of BaP as a function of carbon particle size is shown for two species in Table 22. For relatively large particles, BaP and carbon are cleared at comparable rates but for the more highly adsorbing sub-micron carbon particles, BaP is cleared much faster than carbon.

High concentrations of proteinaceous exudate in the bronchi and alveoli of mice infected with PR8 influenza virus have no significant effect on carbon clearance whereas BaP desorption is accelerated by the surfactant or by cellular changes in the infected lungs.

Pulmonary clearance of BaP is dependent on surface elution when the compound is at sub-monolayer coverage on carbon or other particles, but desorption is probably not rate-controlling at the high BaP concentrations used in these intratracheal studies. In this case, it is likely that the decrease in BaP clearance rate

Table 22. Pulmonary retention of BaP and nutshell charcoal

Species	BaP/ carbon, wt/wt	Clearance half-time, h				Ref.
		BaP[a]			Carbon	
		Alone	15–30 μm	0.5–1 μm		
Syrian golden hamster	1:1	–	60	10	–	[215]
DBA/2BD mouse[a]	1:1	1.5	≳100	36	~140	[216]
BPA/2BD mouse[b]	2:1	1.5	–	~ 9	~125	[217]

[a] Alone or with particles of specified size range
[b] Intratracheal instillation 7 days after infection by PR8 influenza virus

is an indirect result of the increase in the number of pulmonary scavenger cells and other physiologic changes which occur in response to the introduction of foreign particles. Thus for example, BaP and carbon particles may be co-incorporated in macrophages with only limited release of BaP or its metabolites prior to cell lysis. This mechanism could explain the increased retention of BaP in the presence of large, porous charcoal particles as arising from the low accessibility of material trapped in pores and inclusions to cellular enzymes.

Mutagenicity

Carbon Black. Short-term mutagenicity assays are used for the preliminary evaluation of genotoxicity as a relatively inexpensive alternative to animal carcinogenicity studies. For example, many workers have applied the Ames test, which is based on modified strains of Salmonella typhimurium bacteria, to extracts from carbon black and other carbonaceous particles [18, 48]. In contrast, there is only one published report of a short-term assay of whole carbon black [218]. In an extensive study, N339 furnace black having 294 ppm of toluene extractable PAH was subjected to a battery of five assays for genetic activity. These are the Ames S. typhimurium test (metabolically activated and non-activated), sister chromatid exchange in hamsters, mouse lymphoma and cell transformation tests, and detection of chromosome damage in fruit flies. Carbon black did not exhibit significant genetic activity in any of the tests.

Solvent extracts of non-aftertreated carbon black are mutagenic after metabolic activation, in both the Ames reverse mutation test [27] and in an S. typhimurium forward assay [219]. Direct acting mutagens were detected in extracts of an aftertreated carbon black used in xerographic toners [220, 221].

HPLC fractionation of the extract led to the identification of five nitropyrene isomers, one of which, 1,8-dinitropyrene, accounted for most of the mutagenicity. The manufacturing process was subsequently altered to achieve a more than 20 fold reduction in nitropyrene content so that the mutagenic activity in extracts of xerographic toners containing the modified black is below normal detection limits [26, 221].

The average mutagenic activity of extracts from carbon black and environmental soot are comparable, although in accord with its lower mutagen content, the activity per unit weight of particulate sample is about 100 fold lower for carbon black [27]. Mutagenic properties of some of the major types of soot are described below.

Air Particulate. The mutagenicity of extracts and extract fractions from a wide range of atmospheric carbonaceous particulates have been evaluated using Ames S. typhimurium bacterial strains [48]. Most extracts contain both direct acting mutagens (e.g. polar PNA) and also promutagens which require activation by mammalian liver enzymes (e.g. PAH). In almost all cases, direct acting mutagenicity predominates. When the effect of particle size was examined, it was found that mutagens were associated mainly with the smallest, most highly adsorbing particles ($\leqq 2$ μm diameter).

Fossil Fired Soot and Fly Ash. Ames strain TA 538 was used to determine the mutagenicity of extracts from soot/fly ash collected from the effluent of a coal burning electric power plant [222, 223]. Of three extractants, horse serum eluted more mutagen than cyclohexane, whereas saline was almost completely ineffective. Enzymatic activation increased mutagenicity about 30% for the serum extract and 150% for the cyclohexane extract. Extracts from particles $\leqq 3$ μm diameter were mutagenic, with relatively little mutagenicity associated with larger diameter particles.

Atmospheric particulates and fly ash contain variable quantities of carbonaceous material making it difficult to ascertain the extent of mutagen adsorption on soot. Furthermore, the conversion of combustion-formed PAH to polar PNA derivatives is strongly dependent on time-temperature history and the concentration of environmental reactants. Thus, ambient suspended particles and fly ash show considerable variation in properties and mutagen content depending on the conditions and location of collection. These materials provide very limited information about the potential health effects of carbon black but, conversely, carbon black can serve as an excellent model for predicting the binding and reactivity of PAH and other adsorbates on the surfaces of carbonaceous particles.

Soots formed in laboratory hydrocarbon flames and in automotive engines are often similar to carbon black in morphology and surface properties. These particles are usually associated with much greater quantities of extractible organic matter than is found on commercial carbon blacks. PAH are generally the most common adsorbate on fresh laboratory soots but are mixed with unburned fuel and oxidized PNA on automotive soots. For example, when a dichloromethane extract of a kerosene soot was examined in an S. typhimurium forward mutation assay all mutagenicity was attributed to two PAH, fluoranthene and cyclopenteno[c,d]pyrene [219]. The same extract also mutates human lymphoblasts when activated with rat liver enzymes [224].

Automotive Exhaust Particulate. Diesel particulate extracts are substantially more mutagenic than extracts from particles formed in gasoline engines. Diesel engines have a higher rate of soot emission and produce mainly direct acting mutagens while PAH are the dominant mutagens from internal combustion engines [48]. Nitroaromatics, especially nitropyrenes, formed by post-combustion reaction of NO_x with PAH are reponsible for most of the mutagenicity of diesel exhaust extracts assayed in S. typhimurium systems [225]. Nitropyrenes are extremely active in the Ames assay with 1,8-dinitropyrene being one of the most potent bacterial mutagens known [221]. However, this compound is only moderately active in the L5178y mouse lymphoma mutation assay [226] which suggests that the bacterial mutagenicity of diesel extracts may exaggerate their potential genotoxic effects on mammalian cells.

Desorption of Mutagens. Mutagen assays have been used to monitor physiologic desorption of adsorbed PNA in vivo or in vitro. The fate of the desorbed species is determined by metabolic processes and complexing phenomena in tissues and body fluids. In the special case of alveolar carbon deposits, the most likely routes for elimination of adsorbed mutagen's are elution and metabolism within macrophages, and solubilization by phospholipids in lung surfactants [191].

Alveolar macrophages obtained by lavage from rats which had inhaled diesel particles, were solvent extracted and tested for mutagenicity as a function of post-exposure time [227]. Extracts of macrophages collected immediately and one day after exposure were mutagenic to S. typhimurium tester strain TA98. No mutagenicity was detected in macrophages obtained two or more days after exposure. A similar result was observed in vitro after incubation of rabbit aveolar macrophages with diesel particles [228]. Only 3% of the original mutagenicity was detected in the extract after 40 h incubation. These findings indicate that the known ability of alveolar macrophages to metabolize PAH [229] is applicable at high surface loadings to molecules adsorbed on carbon, including nitro-PNA's which are important mutagens in diesel extract.

The presence of carbon does not appear to alter the metabolic pathway for mutagen detoxification. Both free BaP and BaP adsorbed on furnace blacks yield 3-hydroxy BaP in the presence of guinea pig macrophages or liver microsomes in vitro [230]. This metabolite is the major enzymatic detoxification product from BaP and is not a carcinogen or procarcinogen [231].

Carcinogenicity

There have been numerous attempts to produce tumors in animals by administration of carbon black in various carriers via ingestion, intratracheal instillation, intraperitoneal injection, subcutaneous and/or intramuscular injection, or skin painting. No evidence of tumorigenesis was observed in monkeys, rabbits, hamsters, guinea pigs or rats [2, 18]. The same is true for studies in mice with one possible exception where subcutaneous administration of furnace black suspended in tricaprylin produced local sarcomas although no significant tumor formation occurred upon injection of a suspension of furnace and channel black or injection of furnace black without tricaprylin [232]. These findings were not corroborated by a subsequent subcutaneous study in mice in which furnace and thermal blacks were not tumorigenic [200].

In contrast, solvent extracts of carbon black are carcinogenic when administered by skin painting and also produce tumors, but at lower yield, when administered by ingestion or subcutaneous injection [2, 18]. Tumor incidence is reduced greatly when the extract is co-administered with carbon black [2], presumably due to decreased bioavailability of the re-adsorbed extract components. Similarly, it has been shown in oral and subcutaneous studies that the carcinogenicity of individual PAH is reduced or eliminated by adsorption on carbon [19, 200].

Comprehensive health effect studies have been carried out also on one type of environmental soot, diesel particulate. The conclusion that can be drawn from extensive biomedical and animal exposure data is that, in spite of a high extract and PNA loading, diesel particles are not carcinogenic by any means of administration. As with carbon black, organic solvent extracts of diesel particulate test positive in bacterial mutagenicity assays but are not readily bioavailable in vivo [191].

It is generally assumed that respiratory tumors are a potential result of the inhalation of PAH or PAH contaminated materials in man but experiments in animals show that inhalation of PAH does not produce lung tumors even at extremely high exposures [191].

Table 23. Respiratory tumors from intratracheal instillation of BaP and carbon[a]

Species	Dose, mg BaP	Frequency	BaP/C	BaP	C	BaP+C	References
Duck	20–50 (saline)[b]	Daily, 1–4 times	<1:1	1/34	0/38	0/ 32	[239]
Rat	0.5 (infusine)	Biweekly, 36 weeks	1:1	24/48	1/16	15/ 51	[236]
Hamster	4 (gel-saline)	Weekly, 25 weeks	1:1	–	0/50	93/190	[240]
Hamster	3 (saline)	Weekly, 15 weeks	1:1	2/48	0/48	13/ 48	[234]

[a] Total respiratory tumors/total animals, except for results with rats which are reported as animals with tumors/total animals
[b] Liquid medium

An alternate approach to the evaluation of potential respiratory tract carcinogens is the injection of liquid suspension or solutions of PAH's directly into lung tissue, usually by intratracheal instillation. This method results in an abnormally high non-uniform concentration of PAH localized in the lower bronchi and alveoli, which is quite different from the normal conditions of human exposure. Nevertheless, PAH demonstrate low carcinogenic potency when administered intratracheally except when mixed with certain high surface area powders.

Enhanced formation of lung tumors has been observed for BaP in combination with particles of aluminum oxide, arsenic trioxide, asbestos, hematite, lead oxide, magnesium oxide, nickel sulfide, silica, talc, and titanium dioxide [214]. Results for mixtures of BaP and carbon are inconclusive with some workers finding synergism [233], others finding little increase in BaP tumorigenicity [234], and still others finding deactivation of BaP [235, 236]. Data from publications which provide a description of exposure conditions and quantitative tumor yields are summarized in Table 23. The only consistant result from these studies is that carbon black is not a tumorigen. However, the role of adsorption cannot be ascertained due to the wide variation in relative tumor yields for BaP carbon mixtures, the use of different species and treatment conditions, and most importantly, the use of BaP loadings ($BaP/C \geqq 1$) well in excess of the adsorption capacity of the carbon.

Epidemiology

Over a thirty year period Ingalls and co-workers have published three retrospective cohort studies of North American carbon black workers. The first described the morbidity and mortality experience of employees of a single large producer of carbon black, based on insurance claims [237]. No significant excess of cancer, respiratory or circulatory disease was seen. Much the same result was obtained in a follow-up mortality study of the same population for the period 1939–1956 [238].

In 1980, an expanded study including a majority of North American carbon black producers compared the observed mortality of employees with that expected on the basis of population death rates in the states or provinces in which the plants were located [241].

Occupational exposures ranged from 1 year to 25 years with a total of almost 35,000 person-years at risk. Again, the observed incidence of mortality in major disease categories did not differ significantly from the expected incidence.

Carbon black dust exposure is the major variable in a recently completed case-control study of US carbon black workers [242]. A group of white male workers with major circulatory, malignant or respiratory morbidity was carefully matched by age and years of service with two separate groups of healthy controls and then the total occupational exposure to carbon black was determined for individuals in each group. No correlation was found between carbon black exposure and morbidity.

Many more workers are exposed to carbon black dust in user industries than in the carbon black industry itself, however, few of these workers are exposed solely to carbon black. The principal user of carbon black, the tire industry, has been the subject of many large-scale occupational epidemiology studies [126].

Inspection of cause-specific mortality data shows an elevated risk for leukemia and bladder cancer attributable to exposure to benzene and 2-naphthylamine, and weaker evidence for an excess of stomach and lung cancer in British but not US workers. Both types of cancer are most prevalent in workers involved in the late stages of tire manufacturing, rather than in raw material handling, compounding and milling, where carbon black exposure is most likely.

For all work categories, the incidence of cardiovascular and respiratory mortality is significantly lower than expected. As regards respiratory morbidity; bronchitis, decreased pulmonary function and other symptoms of chronic obstructive lung disease appear to be related to cigarette smoking and exposure to fumes and solvent vapors, but not to dust exposure. Pneumoconiosis was absent in occupationally exposed workers and although emphysema was the major cause of pulmonary disability retirement it was not particularly common for workers exposed primarily to dust (carbon black, silica, talc, etc.) These findings indicate that exposure to carbon black is not a major determinant of mortality or morbidity in rubber workers but a definitive conclusion is not possible.

Epidemiologic data for other carbon black user industries is much less complete and its interpretation also suffers from the problem of mixed exposures and confounding factors. For example, a recent study of workers in a Nigerian dry cell battery factory found that a majority of the employees exhibited symptoms of bronchitis and that there was a significant decrease in pulmonary function for workers exposed to high levels of dust (31 ± 2 mg/m^3 in the molding department and 22 ± 2 mg/m^3 in the mixing department) [243]. It was assumed that these effects were due to inhalation of carbon black because the ambient dust was black although the actual dust exposure was to a powder mixture containing aluminum chloride, carbon black, graphite, manganese dioxide, mercuric chloride, starch, zinc chloride and possibly other components.

On balance, the most reliable information regarding the overall occupational health effects of exposure to carbon black comes from the series of studies of North American carbon black workers. These studies employ sound epidemiologic protocols and avoid the complication of overriding exposure to other agents. They are not as strong statistically as some rubber industry studies, for instance, due to the relatively small size of the exposed cohort. Taken collectively, the carbon black industry studies argue strongly for the noncarcinogenicity of carbon black and the absence of major occupationally related diseases.

Standards and Regulations

Occupational exposure to carbon black is regulated in the USA by the Occupational Safety and Health Administration (OSHA). The current OSHA permissible exposure limit or PEL of 3.5 mg/m^3 (8 h TWA) is a direct descendent of a 1965 recommendation by the Threshold Limit Value (TLV) Committee of the ACGIH [244]. When the TLV Committee set the exposure limit for carbon black

Table 24. Legislation on carbon black in food contact applications[a]

Country	Publication	Date	Details
Austria[b]	61 Farbenverordnung	1975	German BGA regulations apply
Belgium[b]	Arrête Royal (Publ. 9/24/76)	8/25/76	Transmission of benzene or toluene extract >70% at 390 nm. Aromatic amines <500 ppm
Canada	B23.001	1/8/81	No specific regulations, general health safety requirement
Czechoslovakia[b]	Instruction 49/1978	1978	BaP content 0.1 ppm max for use in plastics
Denmark		1980	No government regulations but USA (FDA) and German regs are used
Finland		1980	No government regulations
France[b]	Circular 176	12/1/80	Benzene extract 0.1% max. No detectable BaP (<7 ng). Aromatic amines <500 ppm
West Germany[b]	B. Gesundh. B1.15, 268	7/1/72	Toluene extract (DIN 53553) 0.15% max. Absorbance of cyclohexane extract 0.1 max at 386 nm
West Germany[b]	B. Gesundh. B1.21, 330	7/1/78	Carbon black content 2.5% max except in polyolefin fittings (3% max) and rubber articles in contact with milk (30% max). No limit in flexible containers for solid powders, etc.
	B. Gesundh., 163	12/1/83	Primary aromatic amines <500 ppm (not including carboxy and sulfo derivatives). Benzidine, B-naphthylamine, and 4-amino phenol forbidden. Applies only to plastics
Italy[b]	Ministerial Decree (Gazette Ufficiale 104)	4/20/73	Benzene extract 0.1% max. Extract absorption (1 cm cell); 0.15 max at 280–289 nm, 0.12 max at 290–299 nm, 0.08 max at 300–359 nm, 0.02 max at 360–400 nm. Primary aromatic amines <500 ppm
Netherlands[b]	Exec. Reg. CIII-55		Toluene extract 0.15% max after 8 h. Aromatic amines <500 ppm

Table 24 (continued)

Country	Publication	Date	Details
Norway		1980	No government regulations but USA (FDA) and German BGA used
Spain [b]	30787 Resolucion (corrected by Off. Gazette)	12/28/82	Transmission of benzene or toluene extract >70% at 390 nm. Primary aromatic amines <500 ppm. Applies to plastics, cellulose ethers and esters
Sweden [b]		1980	No government regulations, but USA (FDA) and German BGA used
Switzerland			German BGA regulations apply
U.K. [b]			No government regulations. USA (FDA) and German BGA used. EEC directives enforced by Statutory Instrument
	Stat. Inst. 1927 (EEC Directive 893)	1978	Materials and articles in contact with food must be made by good manufacturing practice so that they do not transfer their constituents to food in such a way as to endanger health or deteriorate the food
	BS4991 BS1972	1974 1967	For use in pipes for drinking water; toluene extract 0.1% max, volatiles (950 °C, 7 min) 9% max, particle size 10–25 nm
	BFP Code of Practice	1978	Carbon black 5 wt% max in plastics
USA [b]	Code of Federal Regulations Title 21-Food and Drugs		
	175.105		Only channel black to be used in adhesives
	175.300 175.320		Only channel black to be used in resinous and polymeric coatings
	176.170 176.180		Only channel black to be used in paper and paper board
	177.1210		Only channel black to be used in closures with sealing gaskets for food containers
	177.1460		Only channel black to be used in melamine formaldehyde resins for molded articles
	177.2410		Only channel black to be used in phenolic resins for molded articles
	177.2600		Channel black or furnace black 50 wt% max in rubber articles for repeated use (10 wt% max for use in contact with milk or edible oils)

[a] Compiled by the European Committee for Biological Effects of Carbon Black (October 1984)
[b] Trace metal limits in Table 25

Table 25. Maximum allowed metal content of carbon black[a, b]

Country	Antimony	Arsenic	Barium	Cadmium	Chromium	Copper	Lead	Mercury	Nickel	Selenium	Silver	Tin	Uranium	Zinc
Austria	2,000	200	4,000	4,000		4,000	2,000	400				4,000	4,000	4,000
Belgium	2,000	50	100	2,000	1,000		100	50		100				
Czechoslovakia		50	100	100	100		1,000	50		100				2,000
France		50	100	1,000			100	50		100				2,000
Germany[c]	50	100	100	1,000			100	50		100				
Italy		50	100	2,000			100	50		100				2,000
Netherlands[d]	2,000	100	100	1,000	1,000		100	50	100	100				
Spain	2,000	100	1,000	2,000	1,000		100	50		100				2,000
Sweden	← Nil →													
U.K.[e]	250	250	500	100	250		2,500	100						
USA[f]		100	2,000	20	100		100	4		20	100			
Europe (EN71)														
Plastic/Paper	250	100	500	100	100		250	100						
Paints	250	100	500	100	250			100						

[a] Compiled by the European Committee for Biological Effects of Carbon Black (October 1984)
[b] ppm metal extracted by 0.1 N HCl
[c] DIN Method 53770
[d] Extracted by 0.1 N H_2SO_4
[e] Colour Manufacturers Association
[f] RCRA EP toxicity limits at pH 5 (acetic acid if necessary) in Sect. 261.24 and Appendix II; Federal Register 45 (98), May 19, 1980
[g] Ash analysis

lower than that for nuisance dusts they indicated that this was based on housekeeping considerations (i.e. highly visible, soiling, black dust) and did not result from evidence of adverse health effects.

After converting the TLV to a PEL, OSHA justified the exclusion of carbon black from the nuisance dust category by assigning it health codes 2 and 3 in the substance toxicity table; which are "suspect carcinogen" and "cumulative heart damage" [245]. The suspect carcinogen designation was removed in 1984 following an in-depth study of the pertinent literature which convinced OSHA that PAH trace impurities adsorbed on carbon black are not bioactive and do not cause cancer in animals or humans. In the most recent edition of OSHA's Industrial Hygiene Technical Manual, carbon black is assigned health codes 3 and 10; "cumulative heart damage" and "cumulative lung damage" [246]. OSHA has indicated that they are reviewing these health effect designations which were determined originally on the basis of two animal inhalation studies at extremely high dust levels (53–85 mg/m^3) [145, 168].

A number of countries have set 3.5 mg/m^3 as the occupational exposure limit for carbon black, these include Australia, Belgium, Denmark, Finland, Italy, The Netherlands. Sweden, and the U.K. [247]. Higher dust exposure limits exist in Germany (6 mg/m^3) and Switzerland (8 mg/m^3).

Aside from occupational exposure, carbon black is also regulated for use in food, drugs, and cosmetics. Although not used as a direct food additive in many countries it generally is approved for food contact applications. Relevant government specifications are summarized in Table 24 for North America and Western Europe and corresponding limits for extractable trace metals on carbon black are listed in Table 25. Countries which have no general regulations concerning the use of carbon black usually approve specific applications on an individual basis.

In Japan, the government supports industry specifications (Jishu Kisei) which require, for example, carbon blacks having $\leqq 0.1\%$ benzene extract for use in olefin and styrene plastics, only channel black in PVC, and any carbon black at a loading $\leqq 50$ wt% in rubber.

References

1. Kastein, B.: Rubber World *188,* 44 (1982)
2. IARC Monographs on The Evaluation of the Carcinogenic Risk of Chemicals to Humans, Vol. 33, Part 2, Lyon 1984
3. Infochimie *239,* 77 (1983)
4. Chem. Rept., 2, March 19, 1984
5. Stokes, C.A., Guercio, V.J.: Oil & Gas J. 62–68, March 12, 1984
6. Beck, M.R., Hebermehl, H.W.: Rubber World *189,* 84 (1983)
7. Neoh, K.G., Howard, J.B., Sarofim, A.F.: In: Particulate Carbon: Formating During Combustion, Siegla, D.C., Smith, G.W., eds. 261–282, Plenum Press, New York 1981
8. Rawlings, G.D., Hughes, T.W.: Spec. Conf. Emiss. Inventories, 173 (1978)
9. Gerstle, R.W., Richards, J.R.: U.S. E.P.A. Rept. 600/2-77-023 (1977)
10. Medalia, A.I., Rivin, D.: In: Characterization of Powder Surfaces, Parfitt, G.D., Sing, K.S.W., eds. Chap. 7 Academic Press, London 1976
11. Rivin, D.: Rubber Chem. Technol. *44,* 307 (1971)
12. Nagy, J., Dorsett, H.G., Cooper, A.R.: BuMines Rept. Inv. 6597 (1965)
13. Cox, R.A.F.: Proc. P.R.I. *13*/1 (1984)

14. Kiselev, A.V., Yashin, Y.I.: Gas Adsorption Chromatography, Plenum Press, New York 1969
15. Elkington, P.A., Curthoys, G.: J. Phys. Chem. *73*, 2321 (1969)
16. Ross, M.M., Risby, T.H.: Environ. Sci. Technol. *16*, 75 (1982)
17. Groszek, A.J.: J. Chem. Soc., Faraday Disc. *59*, 109 (1975)
18. Rivin, D., Smith, R.G.: Rubber Chem. Technol. *55*, 707 (1982)
19. Von Haam, E., Titus, H.L., Caplan, I., Shinowara, G.Y.: Proc. Soc. Exp. Biol. Med. *98*, 95 (1958)
20. Daisey, J.M., Low, M.J.D., Tascon, J.M.D.: In: Polynuclear Aromatic Hydrocarbons, Cooke, M., Dennis, A.J., eds. 307–315, Battelle Press, Columbus 1984
21. Lagana, A., Petronio, B.M., Rotatori, M.: J. Chromatogr. *198*, 143 (1980)
22. Zoccolillo, L., Liberti, A., Coccioli, F., Rochetti, M.: Chromatogr. *288*, 347 (1984)
23. Kamata, K., Yamazoe, R., Harada, H.: Ann. Rep. Tokyo Metr. Res. Lab. P.H. *29*, 124 (1978)
24. Fitch, W.L., Smith, D.H.: Environ. Sci. Technol. *13*, 341 (1979)
25. Taylor, G.T., Redington, T.E., Bailey, M.J., Buddingh, F., Nau, C.A.: Am. Ind. Hyg. Assoc. J. *41*, 819 (1980)
26. Sanders, D.R.: In: Polynuclear Aromatic Hydrocarbons, Cooke, M., Dennis, A.J., eds., 145–158, Battelle Press, Columbus 1981
27. Medalia, A.I., Rivin, D., Sanders, D.R.: Sci. Total Environ. *31*, 1 (1983)
28. Giammerise, A.T., Evans, D.C., Butler, M.A., Murphy, C.B., Kiriazides, D.K., Marsh, D., Mermelstein, R.: In: Polynuclear Aromatic Hydrocarbons, Cooke, M., Dennis, A.J., eds. 325–334, Battelle Press, Columbus 1982
29. Locati, G., Fantuzzi, A., Consonni, G., Ligotti, I., Bonomi, G.: Am. Ind. Hyg. Assoc. J. *40*, 644 (1979)
30. Criteria for a Recommended Standard: Occupational Exposure to Carbon Black, DHEW (NIOSH) Publ. 78–204, 1978
30a Fitch, W.L., Everhart, E.T., Smith, D.H.: Anal. Chem. *50*, 2122 (1978)
31. Griest, W.H., Yeatts, L.B., Jr., Caton, J.E.: Anal. Chem. *52*, 199 (1980)
32. Lee, F.S., Harvey, T.M., Prater, T.J., Paputa, M.C., Schuetzle, D.: ASTM STP 721, 92–110, American Society for Testing and Materials, 1980
33. Grimmer, G.: In: Environmental Carcinogens: Polycyclic Aromatic Hydrocarbons, Grimmer, G. ed., Chap. 2, CRC Press, Boca Raton 1983
34. Lee, M.L., Novotny, M.V., Bartle, K.D.: Analytical Chemistry of Polycyclic Aromatic Compounds, Academic Press, New York 1981
35. Lee, M.L., Hites, R.A.: Anal. Chem. *48*, 77 (1976)
36. Peadon, P.A., Lee, M.L., Hirata, Y., Novotny, M.: Anal. Chem. *52*, 2268 (1980)
37. Hirose, A., Wiesler, D., Novotny, M.: Chromatographia *18*, 239 (1984)
38. Ramdahl, T., Urdal, K.: Anal. Chem. *54*, 2256 (1982)
39. Ramdahl, T.: Environ. Sci. Technol. *17*, 666 (1983)
40. Colmsjo, A.L., Ostman, C.E.: In: Polynuclear Aromatic Hydrocarbons, Cooke, M., Dennis, A.J., Fisher, G.L., eds. 201–210, Battelle Press, Columbus 1982
41. Drake, J.A.G., Jones, D.W., Causey, B.S., Kirkbright, G.F.: Fuel *57*, 663 (1978)
42. Novricik, J., Novrocikova, M., Frycka, J.: Intl. Polymer Sci. Technol. *10*, T/106 (1983)
43. Matsushita, H.: Paper 01G01, PAC CHEM84, Honolulu Dec. 16–21, 1984
44. Dewiest, F.: J. Pharm. Belg., *35*, 253 (1980)
45. Falk, H.L., Miller, A., Kotin, P.: Science *127*, 474 (1958)
46. Stenberg, U.R., Alsberg, T.E.: Anal. Chem. *53*, 2067 (1981)
47. Medalia, A.I., Rivin, D.: Carbon *20*, 481 (1982)
48. Barfknecht, T.R.: Prog. Energy Combust. Sci. *9*, 199 (1983)
49. Lee, F.S.C., Schuetzle, D.: In: Handbook of Polycyclic Aromatic Hydrocarbons, Bjorseth, A., ed. Chap. 2, Marcel Dekker, New York 1983
50. Gordon, R.J., Bryan, R.J.: Environ. Sci Technol. *7*, 1050 (1973)
51. May, W.E., Wise, S.A.: Anal. Chem. *56*, 225 (1984)
52. Medalia, A.I., Rivin, D., Sanders, D.R.: Rubber Div. A.C.S., Oct. 1981
53. Yergey, J.A., Risby, T.H., Lestz, S.S.: Anal. Chem. *54*, 354 (1982)
54. Hangebrauck, R.P., Lauch, R.P., Meeker, J.E.: Am. Ind. Hyg. Assoc. J. *27*, 47 (1966)

55. Lee, F.S.C., Prater, T.J., Ferris, F.: In: Polynuclear Aromatic Hydrocarbons, Jones, P.W., Leber, P., eds. Ann. Arbor Science, Ann Arbor 1979
56. Kraft, J., Lies, K.H.: Paper No. 810082, SAE, Detroit, March 1981
57. Rivin, D., Medalia, A.I.: In: Soot in Combustion Systems, Lahaye, J., Prado, G., eds., 25–35, Plenum Press New York 1983
58. Pott, P.: Chirurgical Observations, Hawes, Clarke and Cullings, London 1775
59. Paget, J.: Lancet *2,* 265 (1850)
60. Kuroda S., Kawahata, K.: Z. Krebsforsch. *45,* 35 (1936)
61. Doll, R., Fisher, R.E.W., Gammon, E.J., Gunn, W., Hughes, G.O., Tyrer, F.H., Wilson, W.: Br. J. Ind. Med. *22,* 1 (1965)
62. Redmond, C.K., Strofino, C.R., Cypess, R.H.: Ann. N.Y. Acad. Sci. *271,* 102 (1976)
63. Passey, R.D., Caster-Braine, J.: J. Path. Bact. *28,* 133 (1925)
64. Kotin, P., Falk, H.L., Mader, P., Thomas, M.: Arch. Ind. Hyg. Occup. Med. *9,* 153 (1954)
65. Hueper, W.C., Kotin, P., Tabor, E.C., Payne, W.W., Falk, H., Sawicki, E.: Arch. Path. *74,* 89 (1962)
66. Wynder, E.L., Hoffmann, D.: J. Air Pollut. Contr. Assoc. *15,* 155 (1965)
67. Stocks, P.: Br. J. Cancer *14,* 29 (1960)
68. Nose, Y.: Proc. First Intl. Clean Air Conf., London 1960
69. Goldsmith, J.R., Friberg, L.T.: In: Air Pollution, Vol. II, Stern, A.C., ed., Academic Press, New York 1977
70. Kotin, P., Falk, H.L., Thomas, M.: Arch. Ind. Hyg. Occup. Med. *9,* 164 (1954)
71. Novakov, T., Chang, S.G., Harker, A.B.: Science *186,* 259 (1974)
72. Brodzinsky, R., Chang, S.G., Markowitz, S.S., Novakov, T.: J. Phys. Chem. *84,* 3354 (1980)
73. Benner, W.H., Brodzinsky, R., Novakov, T.: Atmos. Environ. *16,* 1333 (1982)
74. Chang, S.G., Toossi, R., Novakov, T.: Atmos. Environ. *15,* 1287 (1981)
75. Britton, L.G., Clarke, A.G.: Atmos. Environ. *14,* 829 (1980)
76. Harrison, R.M., Pio, C.A.: Atmos. Environ. *17,* 1261 (1983)
77. Novakov, T.: Sci. Total Environ. *36,* 1 (1984)
78. Siedlewski, J.: Int. Chem. Eng. *5,* 297 (1965)
79. Cofer, W.R., Schryer, D.R., Rogowski, R.S.: Atmos. Environ. *14,* 571 (1980)
80. Boulos, B.M.: In: Polynuclear Aromatic Hydrocarbons, Cooke, M., Dennis, A.J., eds. 215–225, Battelle Press, Columbus 1984
81. Cofer, W.R., Schryer, D.R., Rogowski, R.S.: Atmos. Environ. *15,* 1281 (1981)
82. Harker, A.B., Richards, L.W., Clark, W.E.: Atmos. Environ. *11,* 87 (1977)
83. Chang, S.G., Novakov, T.: Atmos. Environ. *9,* 495 (1975)
84. Chang, S.G., Brodzinsky, R., Gundel, L.A., Novakov, T.: In: Particulate Carbon: Atmospheric Life Cycle, Wolff, G.T., Klimisch, R.L., eds., 159–179, Plenum Press, New York 1982
85. Akhter, M.S., Chughtai, A.R., Smith, D.M.: J. Phys. Chem. *88,* 5334 (1984)
86. Rao, M.N., Hougen, O.H.: Chem. Eng. Progr. Symp. Series *48,* 110 (1952)
87. Mochida, I., Komatsubara, Y., Fujitsu, H., Ida, S.: 16th Biennial Conf. on Carbon, Ext. Abs. 345–346, 1983
88. Nielsen, T., Ramdahl, T., Bjorseth, A.: Environ. Health Perspect. *47,* 103 (1983)
89. Jager, J., Hanus, V.: J. Hyg. Epidemiol. Microbiol. Immunol. *24,* 1 (1980)
90. Hughes, M.M., Natusch, D.F.S., Taylor, D.R., Zeller, M.V.: In: Polynuclear Aromatic Hydrocarbons, Bjorseth, A., Dennis, A.J., eds., 1–8, Battele Press, Columbus 1980
91. Lindskog, A.: Environ. Health Perspect. *47,* 81 (1983)
92. Grosjean, D., Fung, K., Harrison, J.: Environ. Sci. Technol. *17,* 673 (1983)
93. Pitts, J.N., Van Cauwenberghe, K.A., Grosjean, D., Schmid, J.P., Fitz, D., Belser, W.L., Knudson, G.B., Hynds, P.M.: Science *202,* 515 (1978)
94. Nielsen, T.: Environ. Sci. Technol. *18,* 157 (1984)
95. Ioki, Y.: J. Chem. Soc., Perkin II, 1240 (1977)
96. Tebbens, B.D., Thomas, J.F., Mukai, M.: Am. Ind. Hyg. Assoc. J. *27,* 415 (1966)
97. Korfmacher, W.A., Natusch, D.F.S., Taylor, D.R., Mamantov, G., Wehry, E.L.: Science *207,* 765 (1980)
98. Falk, H.L., Markul, I., Kotin, P.: Arch. Ind. Health *13,* 13 (1956)

99. Daisey, J.M., Lewandowski, C.G., Zorz, M.: Environ. Sci. Technol. *16*, 857 (1982)
100. Butler, J.D., Crossley, P.: Atmos. Environ. *15*, 91 (1981)
101. De Wiest, F.: Proc. Eur. Symp. Phys. Chem. Behav. Atmos. Pollut. 185–193, 1980
102. Eisenberg, W.C., Taylor, K., Cunningham, L.B., Murray, R.W.: In: Polynuclear Aromatic Hydrocarbons, Cooke, M., Dennis, A.J., eds. 395–410, Battelle Press, Columbus 1984
103. Pitts, J.N., Jr., Lokensgard, D.M., Ripley, P.S., Van Cauwenberghe, K.A., Van Vaeck, L., Shaffer, S.D., Thill, A.J., Belser, W.L., Jr.: Science *210*, 1347 (1980)
104. Wu, C.-H., Salmeen, I., Niki, H.: Environ. Sci. Technol. *18*, 603 (1984)
105. Katz, M., Chen, C., Tosine, B., Sakuma, T.: In: Polynuclear Aromatic Hydrocarbons, Jones, P.W., Leber, P., eds. 171–189, Ann Arbor Science Publ., Ann Arbor 1979
106. Jager, J., Rakovic, M.: J. Hyg. Epidemiol. Microbiol. Immunol. *18*, 137 (1974)
107. Gordon, R.J.: Environ. Sci. Technol. *4*, 370 (1976)
108. Handa, T., Kato, Y., Yamanaura, T., Ishii, T., Suda, K.: Environ. Sci. Technol. *14*, 416 (1980)
109. Dasch, J.M.: Environ. Sci. Technol. *16*, 639 (1982)
110. Butcher, S.S., Sorenson, E.M.: J. Air Pollut. Contr. Assoc. *29*, 724 (1979)
111. Sullivan, T.J., Mix, M.C.: Bull. Environ. Contam. Toxicol. *31*, 208 (1983)
112. Wallingford, K.M., Que Hee, S.S.: In: Polynuclear Aromatic Hydrocarbons, Cooke, M., Dennis, A.J., eds. 1369–1384, Battelle Press, Columbus 1984
113. Bjorseth, A.: In: Polynuclear Aromatic Hydrocarbons, Jones, P.W., Leber, P., eds. 371–382, Ann Arbor Science Publishers, Ann Arbor 1979
114. Broddin, G., Van Vaeck, L., Van Cauwenberghe, K.: Atmos. Environ. *11*, 1061 (1977)
115. Gerber, H.E.: Particulate Carbon: Atmospheric Life Cycle, Wolff, G.T., Klimisch, R.L., eds., 145–157, Plenum Press, New York 1982
116. Rosen, H., Hansen, A., Dod, R.L., Novakov, T.: Science *208*, 741 (1980)
117. Stevens, R.K., McClenny, W.A., Dzubay, T.G., Mason, M.A., Courtney, W.J.: In: Particulate Carbon: Atmospheric Life Cycle, Wolff, G.T., Klimisch, R.L., eds. 111–126, Plenum Press, New York 1982
118. Delumyea, R.G., Chu, L.C., Macias, E.S.: Atmos. Environ. *14*, 647 (1980)
119. Novakov, T.: NSF/NASA Conf. Heterogeneous Catalysis, Albany 1981
120. Pimenta, J.A., Wood, G.R.: Environ. Sci. Technol. *14*, 556 (1980)
121. Cadle, S.H., Groblicki, P.J., Mulawa, P.A.: Atmos. Environ. *17*, 593 (1983)
122. NIOSH, Manual of Analytical Methods, Vol. 3, Part II, 2nd Ed., DHEW, 1977
123. Charsley, E.L., Dunn, J.G.: Plast. Rubber Proc. Appl. *1*, 3 (1981)
124. Mehler, J.E., Lankenau, R.O., Stevens Mees, F.: Am. Ind. Hyg. Assoc. J. *43*, 908 (1982)
125. Smith, R.G., Musch, D.C.: Am. Ind. Hyg. Assoc. J. *43*, 925 (1982)
126. Rivin D.: Dangerous Props. Ind. Mat. Rep. *5*, 2 (1985)
127. Gabor, S., Raucher, C., Stefanescu, A., Ossian, A., Cornea, G.: Igiena *28*, 57 (1969)
128. Spodin, Y.N.: Gig. Tr. *9*, 22 (1973)
129. Komarova, L.T.: Gig. Tr. Prof. Zbl. *17*, 38 (1973)
130. Troitskaya, N.A., Velichkovskii, B.T., Bikmullina, S.K., Sazhina, T.G., Corodnova, N.V., Andreyeva, T.D.: Gig. Tr. Prof. Zbl. *17*, 32 (1975)
131. Deimel, M., Dulson, W.: VDI-Ber. *358*, 139 (1980)
132. Cocheo, V., Bellomo, M.L., Bombi, G.G.: Am. Ind. Hyg. Assoc. J. *44*, 521 (1983)
133. Health and Safety Executive, Rubber NIG, HMSO, London 1981
134. Williams, T.M., Harris, R.L., Arp, E.W., Symons, M.J., Van Ert, M.D.: Am. Ind. Hyg. Assoc. J. *41*, 204 (1980)
135. Casey, P., Hagger, R., Harper, P.: Ann. Occup. Hyg. *27*, 127 (1983)
136. Lippmann, M., Goldstein, D.H.: Arch. Environ. Health *21*, 591 (1970)
137. Soma, R., Confortini, C., Peruzzo, G.F., Maddalon, G., Alessio, L.: Med. Lav. *69*, 507 (1978)
138. Leuschner, F.: Rept. Hamburg 1984
139. Helbing, G., Burri, C., Mohr, W., Neugebauer, R., Wolter, D.: In: Evaluation of Biomaterials, Winter, G.D., Leroy, J.L., deGroot, K., eds. Chap. 38, Wiley, New York 1980
140. Ishiyama, T.: Jpn. J. Nephrol. *25*, 397 (1983)
141. Sabet, T.Y., Friedman, H.: Immunology *17*, 535 (1969)
142. Sabet, T.Y., Newlin, C., Friedman, H.: Immunology *16*, 433 (1969)

143. Souhami, R.L.: Immunology *22,* 685 (1972)
144. Nau, C.A., Neal, J., Stembridge, V.A.: Arch. Ind. Health *18,* 511 (1958)
145. Nau, C.A., Taylor, G.T., Lawrence, C.H.: J. Occup. Med. *18,* 732 (1976)
146. Capusan, I., Mauksch, J.: Berufsdermatosen *17,* 28 (1969)
147. Sugar, H.S., Kobernick, S.: Am. J. Opthal. *62,* 146 (1966)
148. Haddad, R., Zehetbauer, G.: Klin. Mbl. Augen. *177,* 829 (1980)
149. Douglas, W.H.J., Spilman, S.D.: In: Alternative Methods in Toxicology, Vol. I, Goldberg, A.M., ed., 205–229, M.A. Liebert Inc., New York 1983
150. Heyder, J., Gebhart, J., Rudolf, G., Stahlohfen, W.: J. Aerosol Sci. *11,* 505 (1980)
151. Wright, G.W.: In: Patty's Industrial Hygiene and Toxicology, Clayton, G.D., Clayton, F.E., eds., Chap. 7 Interscience, New York 1977
152. Kilburn, K.H.: Int. Rev. Cytology *37,* 153 (1974)
153. Menzel, D.B., McClellan, R.O.: In: Toxicology, Casarett, L.J., Doull, J., eds., Second Edition, Chap. 12, MacMillan, New York 1975
154. Camner, P., Helstrom, P.A., Philipson, K.: Arch. Environ. Health *26,* 294 (1973)
155. Andersen, I.B., Lundqvist, G.R., Proctor, D.F., Swift, D.L.: Am. Rev. Resp. Dis. *119,* 619 (1979)
156. Wilkey, D.D., Lee, P.S., Hass, F.J., Gerrity, T.R., Yeates, D.B., Lourenco, R.V.: Arch. Environ. Health *35,* 294 (1980)
157. Lewis, G.P., Coughlin, L.: Atmos. Environ. *7,* 1249 (1973)
158. Cohen, D., Arai, S.F., Brain, J.D.: Science *204,* 514 (1979)
159. Chan, T.L., Lee, P.S., Hering, W.E.: J. Appl. Toxicol. *1,* 77 (1981)
160. Vostal, J.J., Schreck, R.M., Lee, P.S., Chan, T.L., Soderholm, S.C.: In: Toxicological Effects of Emissions from Diesel Engines, Lewtas, J., ed. 143–159, Elsevier, New York 1982
161. Lee, P.S., Chan, T.L., Hering, W.E.: GM Res. Publ. GMR-4313 (1983)
162. Chan, T.L., Lee, P.S., Hering, W.E.: Fund. Appl. Toxicol. *4,* 634 (1984)
163. White, H.J., Garg, B.D.: J. Appl. Toxicol. *1,* 104 (1981)
164. Barnhart, M.I., Chen, S.T., Salley, S.O., Puro, H.: J. Appl. Toxicol. *1,* 88 (1981)
165. Lee, P.S., Gorski, R.A., Hering, W.E., Chan, T.L.: GM Res. Publ. GMR-4683, FASEB, April 1984
166. Wiester, M.J., Iltis, R., Moore, W.: Environ. Res. *22,* 285 (1980)
167. Gross, K.B.: J. Appl. Toxicol. *1,* 116 (1981)
169. Snow, J.B.: Laryngoscope *8,* 267 (1970)
170. Blenkinsopp, W.K.: J. Path. Bact. *96,* 297 (1968)
171. Borchardt, H.: Virchows Arch. Path. Anat. Phys. *271,* 366 (1929)
172. Sano, T., Osanai, H., Sato, K., Tanaka, T., Ito, M., Kobayashi, T., Suzuki, K.: Rodo Kagaku (Tokyo) *10,* 700 (1959)
173. Bowden, D.H., Adamson, I.Y.R.: Lab. Invest. *38,* 422 (1978)
174. Adamson, I.Y.R., Bowden, K.H.: Lab. Invest. *38,* 430 (1978)
175. Vostal, J.J., White, H.J., Strom, K.A., Siak, J.S., Chen, K.C., Dziedzic, D.: In: Toxicological Effects of Emissions from Diesel Engines, Lewtas J., ed. 201–221, Elsevier, New York 1982
176. Rylander, R.: Arch. Environ. Health *18,* 551 (1969)
177. Fenters, J.D., Bradof, J.N., Aranyi, C., Ketels, K., Ehrlich, R., Gardner, D.E.: Environ. Res. *19,* 244 (1979)
178. Zarkower, A.J.: Arch. Environ. Health *25,* 45 (1972)
179. Zarkower, A., Morges, W.: Infect. Immun. *5,* 915 (1972)
180. Boren, H.G.: Arch. Environ. Health *8,* 119 (1964)
181. Kareva, A.I., Kollo, R.M.: Vestn. Rent. Radiol. *36,* 40 (1961)
182. Komarova, L.T.: Nauchn. Tr. Omsk. Med. Inst. *61,* 115 (1965)
183. Komarova, L.T., Rapis, B.L.: Nauchn. Tr. Omsk. Med. Inst. *86,* 144 (1968)
184. Cocarla, A., Cornea, G., Dengel, H., Gabor, S., Milea, M., Papilian, V.V.: Int. Arch. Occup. Environ. Health *36,* 217 (1976)
185. Slepicka, J., Eisler, L., Mirejovsky, P., Simecek, R.: Prac. Lek. *22,* 276 (1970)
186. Nau, C.A., Neal, J., Stembridge, V.A., Cooley, R.N.: Arch. Environ. Health *4,* 415 (1962)
186. Valic, F., Beritic-Stahuljak, D., Mark, B.: Intl. Arch. Arbeitsmed. *34,* 51 (1975)

187. Beritic-Stahuljak, D., Cigula, M., Rubala, D., Valic, F., Mark, B.: Acta Med. Iug. *34*, 363 (1980)
188. Crosbie, W.A., Cox, R.A.F., Leblanc, J.V., Cooper, D., Thomas, W.C.: Akron Rubber Grp. Proc. (1979)
189. Navarro, C., Charboneau, J., McCauley, R.: J. Appl. Toxicol. *1*, 124 (1981)
190. Chen, K.C., Vostal, J.J.: J. Appl. Toxicol. *1*, 127 (1981)
191. Vostal, J.J.: Environ. Health Perspect. *47*, 269 (1983)
192. Lee, I.P., Suzuki, K., Lee, S.D., Dixon, R.L.: Toxicol. Appl. Pharm. *52*, 181 (1980)
193. Buddingh, F., Bailey, M.J., Wells, B., Haesemeyer, J.: Am. Ind. Hyg. Assoc. J. *42*, 503 (1981)
194. Kivela-Ikonen, P., Hanninen, O., Kalliokoski, P., Koivusaari, U.: Environ. Res. *32*, 1 (1983)
195. May, W.E., Wasik, S.P., Freeman, D.H.: Anal. Chem. *50*, 997 (1978)
196. King, R.J.: Fed. Proc. *33*, 2238 (1974)
197. Anghilari, L.J.: Eur. J. Pharmacol. *3*, 153 (1968)
198. King, R.J.: J. Appl. Physiol. *53*, 1 (1982)
199. Neal, J., Thornton, M., Nau, C.A.: Arch. Environ. Health *4*, 598 (1962)
200. Nau, C.A., Neal, J., Stembridge, V.A.: Arch. Environ. Health *1*, 512 (1960)
201. Medvedev, V.I.: Vopr. Onkol. *20*, 49 (1974)
202. DeWiest, F.: J. Pharm. Belg. *35*, 333 (1980)
203. Kutscher, W., Tomingas, R., Weisfeld, H.P.: Arch. Hyg. *151*, 646 (1967)
204. Kutscher, W., Tomingas, R., Weisfeld, H.P.: Arch. Hyg. *151*, 656 (1967)
205. King, L.C., Kohan, M.J., Austin, A.C., Claxton, L.D., Lewtas Huisingh, J.: Environ. Mut. *3*, 109 (1981)
206. Plant, A.L., Pownall, J.H., Smith, L.C.: Chem. Biol. Interact. *44*, 237 (1983)
207. Lakowicz, J.R., Bevan, D.R.: Biochem. *18*, 5170 (1979)
208. Lakowicz, J.R., Bevan, D.R.: In: Effects of Mineral Dusts, Brown, R.C., Chamberlain, M., Davies, R., eds. 69–175, Academic Press, London 1980
209. Lakowicz, J.R., Bevan, D.R.: In: Polynuclear Aromatic Hydrocarbons, Bjorseth, A., Dennis, A.J., eds. 879–898, Battelle Press, Columbus 1980
210. Lakowicz, J.R., Bevan, D.R.: Chem. Biol. Interact. *29*, 129 (1980)
211. Lakowicz, J.R., Bevan, D.R., Riemer, S.C.: Biochim. Biophys. Acta *629*, 243 (1980)
212. Bevan, D.R., Worrell, W.J.: In: Polynuclear Aromatic Hydrocarbons, Cooke, M., Dennis, A.J., eds. Battelle Press, Columbus 1984
213. Saffiotti, U., Borg, S.A., Grote, M.I., Korp, D.B.: Chicago Med. Sch. Bull. *24*, 10 (1964)
214. Bevan, D.R.: Haz. Assess. Chem. *3*, 141 (1980)
215. Henry, M.C., Kaufman, D.G.: J. Nat. Cancer Inst. *51*, 1961 (1973)
216. Creasia, D.A., Poggenburg, J.K., Nettesheim, P.: J. Toxicol. Environ. Health *1*, 967 (1976)
217. Creasia, D.A.: J. Toxicol. Environ. Health *3*, 1003 (1977)
218. Kirwin, C.J., Leblanc, J.V., Thomas, W.C., Haworth, S.R., Kirby, P.E., Thilager, A., Bowman, J.T., Brusick, D.J.: J. Tox. Environ. Health *7*, 973 (1981)
219. Kaden, D.A., Hites, R.A., Thilly, W.G.: Cancer Res. *39*, 4152 (1979)
220. Lofroth, G., Hefner, E., Alfheim, I., Moller, M.: Science *209*, 1037 (1980)
221. Rosenkranz, H.S., McCoy, E.C., Sanders, D.R., Butler, M., Kiriazides, D.K., Mermelstein, R.: Science *209*, 1039 (1980)
222. Chrisp, C.E., Fisher, G.L., Lammert, J.E.: Science *199*, 73 (1978)
223. Fisher, G.L., Chrisp, C.E., Raabe, O.G.: Science *204*, 879 (1979)
224. Skopek, T.R., Liber, H.L., Kaden, D.A., Hites, R.A., Thilly, W.G.: Nat. Cancer Inst. *63*, 309 (1979)
225. Rosenkranz, H.S.: Mutat. Res. *101*, 1 (1982)
226. Cole, J., Arlett, C.F., Lowe, J., Bridges, B.A.: Mutat. Res. *93*, 213 (1982)
227. Vostal, J.J., White, H.J., Strom, K.A., Siak, J.S., Chen, K.C., Dziedzic, D.: In: Toxicological Effects of Emissions from Diesel Engines, Lewtas, J., ed., 201–221, Elsevier, New York 1982
228. King, L.C., Loud, K., Tejada, S.B., Kohan, M.J., Lewtas, J.: Environ. Mut. *5*, 577 (1983)
229. McLemore, T.L., Warr, G.A., Martin, R.R.: Cancer Lett. *2*, 327 (1977)

230. Tomingas, R., Lange, H.U., Beck, E.G., Manojlovic, N., Dehnen, W.: Zbl. Bakt. *155*, 148 (1971)
231. Thakker, D.R., Yagi, H., Whelan, D.L., Levin, W., Wood, A.W., Conney, A.H., Jerina, D.M.: In: Environmental Health Chemistry, McKinney, J.D., ed. Chap. 19, Ann Arbor Science Publ., Ann Arbor 1981
232. Steiner, P.E.: Cancer Res. *14*, 103 (1954)
233. Pylev, L.N.: Proc. 10th Intl. Cancer Congr. *2*, 441 (1971)
234. Stenback, F., Rowland, J., Sellakumar, A.: Oncology *33*, 29 (1976)
235. Sellakumar, A., Stenback, F., Rowland, J.: Eur. J. Cancer *12*, 313 (1976)
236. Davis, B.R., Whitehead, J.K., Gill, M.E., Lee, P.N., Butterworth, A.D., Roe, F.J.R.: Br. J. Cancer *31*, 443 (1975)
237. Ingalls, T.H.: Arch. Ind. Hyg. Occup. Med. *1*, 662 (1950)
238. Ingalls, T.H., Risquez-Iribarren, R.: Arch. Environ. Health *2*, 429 (1961)
239. Rigdon, R.H., Neal, J.: Tex. Rept. Biol. Med. *23*, 494 (1965)
240. Farrell, R.L., Davis, G.W.: In: Experimental Lung Cancer: Carcinogenesis & Bioassays, Karbe, E., Park, J.F., eds., 186–198, Springer, New York 1977
241. Robertson, J.M., Ingalls, T.H.: Arch. Environ. Health *35*, 181 (1980)
242. Robertson, J.M., Ingalls, T.H.: Submitted to Arch. Environ. Health
243. Oleru, U.G., Elegbeleye, O.O., Enu, C.C., Olumide, Y.M.: Environ. Res. *30*, 161 (1983)
244. Documentation of the Threshold Limit Values, American Conference of Governmental Industrial Hygienists, 4th ed. 1980
245. OSHA Industrial Hygiene Manual, Occup. Safety Health Rep. *77*, 8039 (1981)
246. OSHA Industrial Hygiene Technical Manual, Occup. Safety Health Rep. *77*, 8701 (1984)
247. Occupational Safety and Health Series, No. 37, International Labour Office, Geneva 1977

Creosote

G. Sundström

National Environmental Protection Board, Emission and Product Control Laboratory, Special Analytical Section
Box 1302, S-171 25 Solna, Sweden

Å. Larsson and M. Tarkpea

National Environmental Protection Board, Emission and Product Control Laboratory, Brackish Water Toxicology Section
Studsvik, S-611 82 Nyköping, Sweden

Summary

Creosote is a complex chemical mixture with major use as wood preservative. Recent chemical and biological investigations have shown that creosote contains a number of highly biologically active organic components and exerts toxic effects, especially against aquatic organisms. Many components are also chemically and biologically stable and environmental contamination from creosote persists for many years. The general use of creosote as a pesticide has accordingly been questioned and its use been restricted in some countries.

Introduction

Distillation products from wood and coal have been used as wood preservatives for many hundred, even thousand, years. Already in the Bible such material is mentioned when Noah was told:

*"Make thee an ark of gopher wood; room shalt thou make in the ark, and shalt pitch it within and without with pitch**" (Genesis 6:14)

* pitch = black substance made from coal tar

During the nineteenth century, the scientific background behind the use of coal tar for wood preservation was established and the method of pressure or vacuum impregnation was introduced (British Patent 7.731).

Coal tar creosote is an extremely complex chemical mixture obtained by distillation of coal tar produced by coking, i.e. the high temperature carbonization of coal. The distillation range of creosote for impregnation of wood is usually about 200–350 °C. The liquid has a very characteristic sharp odour (naphthalene, phenol) and a yellow to black colour. A number of coal derived oils and tar products with other distillation ranges, as well as with different types of physical/chemical treatments, are also produced. Many investigations of such products, and those of petroleum oils, are relevant also for creosote and its components, and will be included in this overview when appropriate.

A product which should not be confused with coal tar creosote is *wood* creosote. This product is used as an antiseptic and expectorant, nowadays mostly for veterinary use, and is described as a mixture of mainly phenols obtained from wood tar. Wood creosote will not be dealt with in this overview.

Coal tar creosote has recently attracted considerable attention both from environmental and health effects reasons. Chemical investigations have shown the presence of a number of highly biologically active components and the release of creosote into the environment have given both acute and long-term effects. The present report is mainly devoted to the potential *environmental* effects of coal tar creosote. The *health* effects have been reviewed recently by the US Environmental Protection Agency [99, 100], in order to possibly restrict the use of creosote, as well as some other wood treatment agents, and by the International Agency for Research on Cancer (IARC) [45].

IARC has classified creosote as a human carcinogen belonging to group 2A, i.e. there is sufficient evidence that creosote is carcinogenic in animals and limited evidence from epidemiologic studies that creosote is carcinogenic in humans.

Production and Use

The production and use of coal tar products in the Western World were revieved by IARC [45] and is only summarized here. From the data given by IARC it can be assumed that the total World production of creosote probably amounts to more than one million tonnes per year, 90% of which is used for wood treatment. During the recent years more than 10,000 tonnes coal tar creosote has been used yearly in Sweden, and this constitutes 60% of the total use of pesticides in this country.

Although the major use of creosote is as an impregnating agent to protect wood from rot and worms, other minor uses are known [99]:

1. Herbicide
   Ornamental flowering plants and gardens
   Product storage yards
   Agricultural premises and highway rights-of-way
2. Fungicidal use
   Canvas and rope

3. Disinfectant
   Seed potato storage premises and equipment
   Livestock premises
   Poultry premises
   Home and institutional uses
   Transportation vehicles
   Tree wound dressings
   Metal working fluids
4. Uses as larvicide, insecticide, repellent
   Mosquito larvicide
   Insect repellent
   Animal repellent
   Bird repellent
   Screw worm control
   Insect control
   Gypsy moth control
   Animal dip

It is thus evident that creosote might have been used in several areas where food, feed, animals etc. can become contaminated. However, it is estimated that in the US the abovementioned uses only constitute about 2% of the total amount creosote produced. No doubt the largest risk for environmental contamination arise from the large scale use of creosote as a wood impregnating agent.

Environmental Contamination

Extensive contamination of the environment by creosote has been reported from several countries, old wood treatment facilities often being the source. One case of contamination has been described in detail by Black et al. [7, 8]. In 1978 fish in a small river (Michigan, USA) started to taste like "medicine". Investigations of the bottom sediments revealed creosote residues emanating from an impregnating plant that had been in use between 1902 and 1949. Thus, residues were still present more than twenty years after the closing of the plant and the creosote could be traced 4–5 km downstream. As chemical-analytical tracers were used some polycyclic aromatic hydrocarbons (PAH) which are typical for creosote (phenanthrene, 1,2-benzanthracene and benzo[*a*]-pyrene). The analyses were performed using high pressure liquid chromatography with fluorescence detection.

Similar investigations in Sweden [95] and Norway [47] of the soil of old impregnating plants and sediments in neighbouring areas such as lakes and brooks have likewise shown that creosote residues are very persistent in the environment. Biological tests have also shown that creosote residues still after many years exert a considerable toxicity (see below).

In St. Louis Park, Minnesota, USA, investigations have been made of an aquifer contaminated from coal tar wastes [44, 76, 82]. Creosote and coal tar from a coal tar distillation and wood preserving facility, in use between 1918 and 1982, have entered the ground water by infiltration from waste water ponds and runoff from the plant. Already in 1932 a well drilled 1,000 m from the plant yielded water with a coal tar taste. A large number of aromatic compounds were identified in

Table 1. Aromatic compounds identified in coal-tar and in ground-water contaminated from the tar [82]

Compound	H_2O	Tar	Compound	H_2O	Tar
1-Ethyl-3-methylbenzene	+	−	3-Methylphenanthrene	+	+
Benzofuran	+	+	2-Methylphenanthrene	+	+
1,2,3-Trimethylbenzene	+	+	2-Methylanthracene	+	+
2,3-Dihydro-1H-indene	+	+	C_1-9H-Carbazole	+	−
Indene	+	+	4H-Cyclopental[*def*]phenanthrene	+	+
C_1-2,3-Dihydro-1H-indene	+	−	C_1-9H-Carbazole	+	−
C_1-Benzofuran	+	−	2-Methyl-9H-carbazole	+	+
2-Methylbenzofuran	+	+	9-Methylanthracene	+	+
C_1-2,3-Dihydro-1H-indene	+	−	2-Phenylnaphthalene	+	+
C_1-2,3-Dihydro-1H-indene	+	−	3,6-Dimethylphenanthrene	+	+
C_1-1H-Indene	+	−	Fluoranthene	+	+
C_1-1H-Indene	+	−	Acephenanthrylene	−	+
Naphthalene	+	+	Phenanthro[4,5-*bcd*]thiophene	+	+
Benzo[*b*]thiophene	+	+	Pyrene	+	+
C_2-Benzofuran	+	−	Benzonaphthofuran isomer	+	+
C_2-Benzofuran	+	−	Benzonaphthofuran isomer	+	+
C_2-Benzofuran	+	−	Benzo[*b*]naphtho[2,3-*d*]furan	+	+
2-Methylbenzol[*b*]thiophene	+	+	Benzonaphthofuran isomer	+	+
2-Methylnaphthalene	+	+	C_1-Fluoranthene/Pyrene isomer	+	+
3-Methylbenzol[*b*]thiophene	+	+	9H-Fluorene-2-carbonitrile	+	−
1-Methylnaphthalene	+	+	11H-Benzo[*a*]fluorene	+	+
Biphenyl	+	+	11H-Benzo[*b*]fluorene	+	+
3-Methyl-1H-indole	+	−	Benzo[*c*]phenanthrene	+	+
2-Ethylnaphthalene	+	+	C_2-Pyrene isomer	−	+
2,6-(2,7-)Dimethylnaphthalene	+	+	C_2-Pyrene isomer	+	+
1,3-(1,6-)Dimethylnaphthalene	+	+	Benzo[*b*]naphtho[2,1-*d*]thiophene	+	+
Diphenylmethane	+	−	Benzo[*ghi*]fluoranthene	+	+
2,3-Dimethylnaphthalene	+	+	Benzo[*b*]naphtho[1,2-*d*]thiophene	+	+
Acenaphthylene	+	+	Benzo[*b*]naphtho[2,3-*d*]thiophene	+	+
1,2-Dimethylnaphthalene	+	−	Benz[*a*]anthracene	+	+
Acenaphthene	+	+	Benzo[*a*]carbazole	+	−
1-Cyanonaphthalene	+	+	Triphenylene	+	+
Dibenzofuran	+	+	Benzocarbazole isomer	+	−
C_3-Naphthalene	+	−	Chrysene	+	+
2,3,6-Trimethylnaphthalene	+	−	Benzophenanthrene isomer	+	+
9H-Fluorene	+	+	C_1-Benzo[*b*]naphtho[2,1-*d*]thiophene	−	+
C_1-Acenaphthene	+	−	C_1-Benz[*a*]anthracene	+	+
C_1-Acenaphthene	+	+	Benzo[*b*]carbazole	+	+
9-Methyl-9H-fluorene	+	+	11-Methylbenz[*a*]anthracene	+	+
C_1-Acenaphthene	+	+	1-Methylbenz[*a*]anthracene	+	+
C_1-Dibenzofuran	+	+	9-Methylbenz[*a*]anthracene	+	+
C_1-Dibenzofuran	+	+	2,2′-Binaphthalene	+	+
9H-Xanthene	+	+	Binaphthalene isomer	+	+
9,10-Dihydroanthracene	+	+	Benzo[*b*]fluoranthene, or [*j*]	+	+
9,10-Dihydrophenanthrene	+	−	Benzo[*k*]fluoranthene	+	+
2-Methyl-9H-fluorene	+	+	Benzo[*e*]pyrene	+	+
1-Methyl-9H-fluorene	+	+	Benzo[*a*]pyrene	+	+
C_1-9H-fluorene	+	+	Perylene	+	+
Dibenzothiophene	+	+	Dinaphtho[2,1-*b*:1′,2′-*d*]thiophene	−	+
Phenanthrene	+	+	Pentacene	−	+

Table 1 (continued)

Compound	H_2O	Tar	Compound	H_2O	Tar
Anthracene	+	+	Anthanthrene	−	+
9H-Carbazole	+	+	Indeno[1,2,3-*cd*]pyrene	+	+
4-Methyldibenzothiophene	+	+	Benzo[*ghi*]perylene	+	+
1-Phenylnaphthalene	+	+	Dibenzofluoranthene/pyrene	−	+
2-Methyldibenzothiophene or 9H-Thioxanthene	+	+			

the ground water and in the corresponding coal tar residues, Table 1. A specific study of organic bases contaminating the ground water was also performed and is referred to below.

The contribution of PAHs from coal tar products to the general pollution of hydrocarbons in a highly industrialized area – Elizabeth River, Norfolk, Virginia, USA – was investigated by Merrill and Wade [63]. Thereby the relative amounts of PAH, alkanes and other organic constituents in carbonized coal products (creosote, coal tar, roofing tar and creosoted wood) were compared to that of soot and petroleum products. By the use of a gas chromatographic fingerprinting technique, after fractionation of extracts in aromatics and alkane components by thin layer chromatography, these data were further compared to those of sediment extracts. It was concluded that the sediments of the river were contaminated with hydrocarbons both from coal and petroleum products. The PAHs often had point sources associated with old wood preserving facilities.

Environmental contamination by creosote have so far mostly been traced by the analysis of selected PAHs. The presence of elevated levels of PAH in aquatic environments have often been attributed to the presence of creosote treated wood in e.g. harbour constructions [22, 23, 89, 109]. Mussels thereby can be used as indicators as suggested in the "Mussel watch" Program in the USA [64] as these organisms does not seem to metabolize PAHs. An overview of PAHs in the aquatic environment – formation, sources, effects etc. – have been published by the National Research Council Canada [65].

The extensive contamination of sediments by creosote has resulted in the need of methods for decontamination and disposal of such materials. A field method has been developed by the US EPA [2, 101] which includes sedimentation of the excavated material and subsequent coagulation, filtering and active coal treatment of the water. The sediment and coagulated material is disposed of on landfills and the water is returned to the river. The effectiveness of the treatment is monitored by biotests using *Daphnia,* crayfish and fish. A highest level of 5,000 ppm hexane soluble creosote residues in the sediment was considered an acceptable level for the protection of the aquatic fauna. The total cost of this treatment was estimated to about 100 US dollar per meter of the river investigated.

Chemical Composition

Creosote and other oil products produced from coal are very complex mixtures of organic compounds. Originally therefore such oils have been characterized by

Table 2. Example of a specification of coal tar creosote for wood impregnation (Sweden)

Property	Requirement
Density at +20/4 °C	1.04–1.12 g/cm^3
Water content	< 1.0%
Distillation range	
Distillate below 210 °C	< 4%
Distillate below 235 °C	<20%
Distillate below 355 °C	Must be reported
Acids (phenols)	< 3%
Toluene insolubles	< 0.2%
Transparency at 23 °C	Completely clear

Table 3. Compounds expected to be present in creosote as suggested by Nestler [67, 68] from the compilation of Rhodes and Woolridge in 1945 on compounds identified from the carbonization of coal [80]

Aliphatic hydrocarbons

Undecane
Nonadecane
Heneicosane
Docosane
Tricosane
Tetracosane
Pentacosane
Hexacosane
Heptacosane
Octacosane

Aromatic hydrocarbons

1,2,3-Trimethylbenzene
Cymene
Indane
Indene
3,4-Dimethylbenzene

1,2,4,5-Tetramethylbenzene
1,2,3,5-Tetramethylbenzene
4-Methylindene
Tetrahydronaphthalene
Naphthalene

Dimethylindene
2-Methylnaphthalene
1-Methylnaphthalene
2-Ethylnaphthalene
Biphenyl

2,6-Dimethylnaphthalene
2,7-Dimethylnaphthalene
1,7-Dimethylnaphthalene
1,6-Dimethylnaphthalene
1,5-Dimethylnaphthalene

2,3-Dimethylnaphthalene
1,2-Dimethylnaphthalene
3-Methylbiphenyl
4-Methylbiphenyl
1,3,7-Trimethylnaphthalene

Acenaphthene
2,3,5-Trimethylnaphthalene
2,3,6-Trimethylnaphthalene
3,4′-Dimethylbiphenyl
4,4′-Dimethylbiphenyl

Fluorene
4,5-Benzindane
2-Methylfluorene
3-Methylfluorene
Phenanthrene

Tetramethylbiphenyl
Anthracene
3-Methylphenanthrene
9-Methylphenanthrene
1-Methylphenanthrene

2-Phenylnaphthalene
Naphthacene
2-Methylanthracene
2,7-Dimethylanthracene
1,2,3,4-Tetrahydrofluoranthene

Truxene (di-Indane)
Fluoranthene
Pyrene
Retene (8-Methyl-2-isopropylphenanthrene)
1,2-Benzofluorene

2,3-Benzofluorene
Naphtho-2′,3′-1,2-anthracene
1,2-Benzonaphthacene

Table 3 (continued)

Chrysene
Triphenylene
4,5-Phenanthrylenemethane

Sulfur compounds

2,3-Benzothiophene
Methylthionaphthalene
Dibenzothiophene
Dibenzothionaphthalene
Sulfur

Miscellaneous Compounds

Benzonitrile
1-Naphthonitrile
2-Naphthonitrile
Acetophenone
Phenanthridone

Acids

Benzoic acid

Inorganic compounds

Ammonium thiocyanate

O-heterocyclic compounds

2,3-Benzofuran
6-Methylbenzofuran
3-/5-Benzofuran
4-Methylbenzofuran
3,6-Dimethylbenzofuran
4,5-Dimethylbenzofuran
4,6-Dimethylbenzofuran
α-Naphthofuran
β-Naphthofuran
Dibenzofuran
1-Methyldibenzofuran
2-Methyldibenzofuran
2,3,5,6-Dibenzbenzofuran
1,9-Benzoxanthene

Phenols

Phenol
o-Cresol
m-Cresol
p-Cresol
2,4-Xylenol
2,6-Xylenol
2,5-Xylenol
2,3-Xylenol
3,5-Xylenol
3,4-Xylenol
m-Ethylphenol
p-Ethylphenol
2,4,5-Trimethylphenol
3-Ethyl-5-methylphenol
2,3,5-Trimethylphenol
7-Hydroxybenzofuran
4-Hydroxyindene
3,4,5-Trimethylphenol
2,3,5,6-Tetramethylphenol
5-Hydroxyindane
4-Hydroxyindane
2-Hydroxybiphenyl
1-Naphthol
2-Naphthol
4-Hydroxyphenol
2-Hydroxydibenzofuran
2-Hydroxyfluorene
Hydroxyanthracene
2-Hydroxyphenanthrene

N-heterocyclic compounds

2,3,4,6-Tetramethylpyridine
Quinoline
Isoquinoline
2-Methylquinoline
8-Methylquinoline
3-Methylisoquinoline
Indole
1-Methylisoquinoline
2,8-Dimethylquinoline
7-Methylquinoline
6-Methylquinoline
3-Methylquinoline
1,3-Dimethylisoquinoline
5-Methylquinoline
4-Methylquinoline
5-/7-Methylquinoline
3-Methylindole
6-Methylisoquinoline
7-Methylindole
4-Methylindole
5-Methylindole
5,8-Dimethylquinoline
2-methylindole
2,4,6-Trimethylquinoline
Hydroacridine
Acridine
Phenanthridine
Carbazole
2-Methylcarbazole
3-Methylcarbazole

Table 3 (continued)

Amines	2,5-Xylidine
Aniline	3,5-Xylidine
o-Toluidine	2,3-Xylidine
m-Toluidine	1-Naphthylamine
p-Toluidine	2-Naphthylamine
2,4-Xylidine	

their physical properties. A typical example of a technical specification of creosote for wood impregnation is given in Table 2. However, already in 1945 as many as 348 chemical compounds (organic and inorganic) with boiling points from −252.5 °C (hydrogen) to 519–520 °C (picene and ammonium chloride) had been shown to be formed by the carbonization of coal [80].

Nestler has reviewed the early work on the physical and chemical characterization of of wood preserving creosote [67, 68] and compiled from [80] about 160 compounds present in creosote with a distillation range of 175–450 °C, Table 3. In the first quantitative investigations it was estimated that about twenty compounds constitutes the major part of creosote, and Lorenz and Gjovik [56] for example, concluded that 18 compounds make up about 90% of the total product, Table 4. All these compounds are aromatic hydrocarbons and their nitrogen and oxygen heterocyclic analogues. The structure of a number of specific organic compounds, representing different groups of chemicals, typical for creosote are given in Fig. 1.

Table 4. Major components in creosote [56]

Compound	Percentage	Boiling point (°C)
Naphthalene	3.0	218
2-Methylnaphthalene	1.2	241
1-Methylnaphthalene	0.9	245
Biphenyl	0.8	256
Dimethylnaphthalenes	2.0	268
Acenaphthene	9.0	279
Dibenzofuran	5.0	287
Fluorene	10.0	293–295
Methylfluorenes	3.0	318
Phenanthrene	21.0	340
Anthracene	2.0	340
Carbazole	2.0	355
Methylphenanthrenes	3.0	354–355
Methylanthracenes	4.0	360
Fluoranthene	10.0	382
Pyrene	8.5	393
Benzofluorenes	2.0	413
Chrysene	3.0	448
Total	90.4	

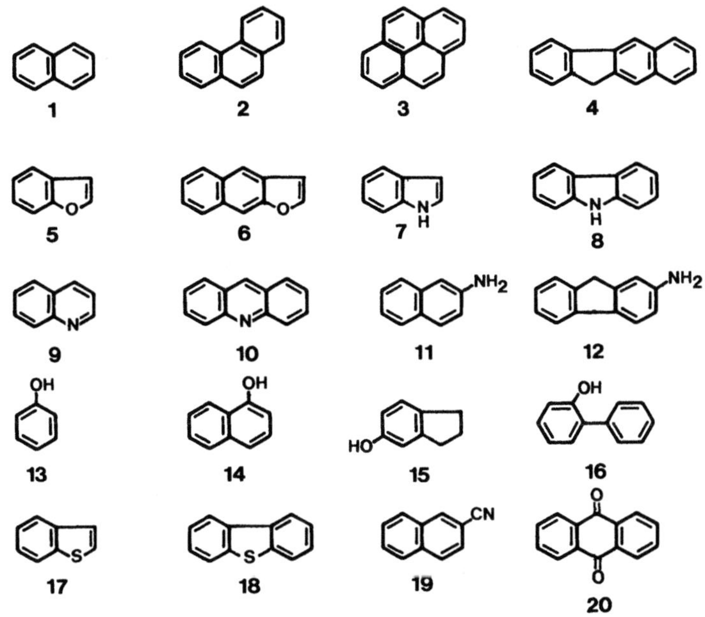

Fig. 1. Structures of some creosote components representing the basic structural features of important groups of compounds:

Aromatic hydrocarbons

1. Naphthalene
2. Phenanthrene
3. Pyrene
4. 2,3-Benzfluorene

Aromatic oxygen heterocyclics

5. 2,3-Benzofuran
6. 2,3-Naphthofuran

Secondary nitrogen heterocyclics

7. Indole
8. Carbazole

Tertiary nitrogen heterocyclics

9. Quinoline
10. Acridine

Aromatic amines

11. 2-Naphthylamine
12. 2-Aminofluorene

Phenols

13. Phenol
14. 1-Naphthol
15. 5-Indanol
16. 2-Phenylphenol

Sulphur heterocyclics

17. 2,3-Benzothiophene
18. Dibenzothiophene

Misc compounds

19. 1-Cyanonaphthalene
20. Anthraquinone

Since the 1970's more advanced chromatographic methods have been developed and used for the analysis of creosote and related oil products (e.g. [10, 58, 71, 73, 85, 87]). Until today a large number of publications describe the separation and identification of specific components in such products. The development of separation techniques has also allowed the detailed studies of specific subgroups of structurally related compounds.

Two of the most versatile separation methods are those described by Later and coworkers [52] for solvent refined coal, and by Hertz et al. [41] for shale oil.

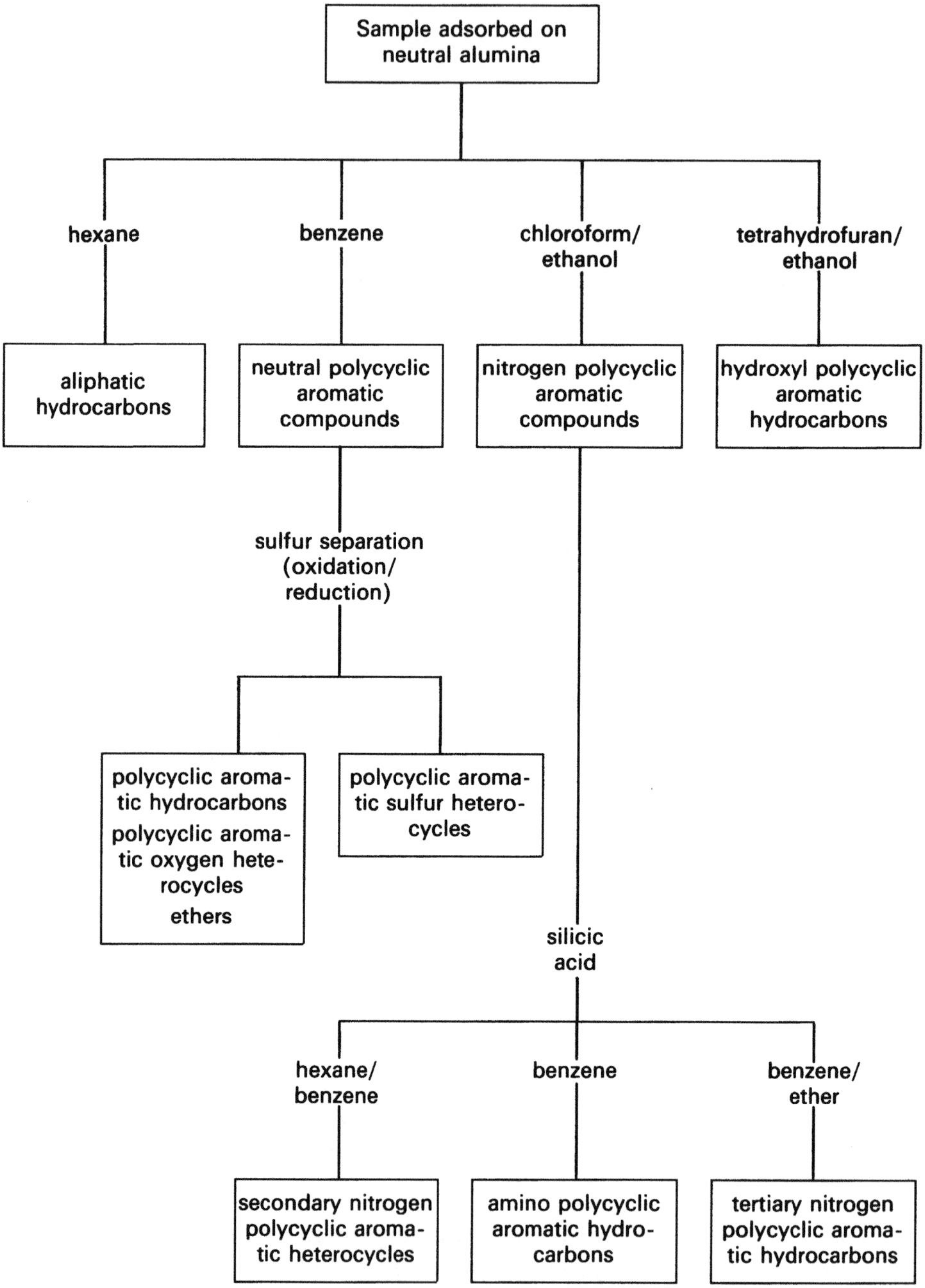

Fig. 2. Chemical class separation for synthetic fuel products (after [52])

Table 5. Compounds identified in the aluminium oxide subfractions from a solvent refined coal liquid heavy distillate (bp 260–450 °C), separation scheme as in Fig. 2 [52]

Aliphatic hydrocarbon fraction

C_{16}–C_{19} *n*-alkanes

Neutral polycyclic aromatic hydrocarbon fraction

C_1-Naphthalene
C_2-Naphthalene
C_3-Naphthalene
Biphenyl
Acenaphthene

Dibenzofuran
Fluorene
C_1-Biphenyl/acenaphthene (2 isomers)
C_4-Naphthalene/C_1-Fluorene
C_1-Dibenzofuran

C_2-Biphenyl
C_1-Fluorene/C_2-Biphenyl
C_5-Naphthalene/C_2-Fluorene
Dibenzothiophene
Tetrahydrophenanthrene

C_3-Biphenyl
Phenanthrene
C_3-Biphenyl/C_1-Dibenzothiophene
1-Phenylnaphthalene
C_1-Phenanthrene

2-Phenylnaphthalene
C_2-Phenanthrene
Fluoranthene
Pyrene
Benzo[*a*]fluorene

Benzo[*b*]fluorene
C_1-Pyrene/C_1-Fluoranthene
C_2-Pyrene/C_2-Fluoranthene
Benz[*a*]anthracene
Chrysene

C_1-Benz[*a*]anthracene/C_1-Chrysene
Benzofluoranthene
Benzopyrene

Nitrogen polycyclic aromatic fraction

C_1-Indole
C_2-Indole
C_3-Indole
C_2-Quinoline
C_4-Quinoline
C_2-Diphenyl ether
Tetrahydrobenzoquinoline
Phenanthrene
Benzoquinoline (2 isomers)
C_1-Tetrahydrobenzoquinoline (2 isomers)

Tetrahydrocarbazole/benzoquinoline
Carbazole
C_1-Benzoquinoline (5 isomers)
C_1-Tetrahydrocarbazole
C_1-Carbazole (2 isomers)

C_2-Carbazole (3 isomers)
C_3-Benzoquinoline (2 isomers)
C_3-Carbazole
Azafluoranthene
Azapyrene

C_4-Carbazole
C_1-Azapyrene/C_1-Azafluoranthene/C_5-Carbazole
C_2-Azapyrene/C_2-Azafluoranthene
Tetrahydrobenzocarbazole
Naphthoquinoline

Benzocarbazole
C_1-benzocarbazole

Hydroxyl polycyclic aromatic hydrocarbon fraction

C_1-Indanol
C_2-Indanol
C_3-Indanol
C_4-Indanol
C_5-Indanol

Hydroxybiphenyl (3 isomers)
C_1-Hydroxybiphenyl (5 isomers)
C_2-Hydroxybiphenyl (4 isomers)
C_3-Hydroxybiphenyl (4 isomers)
C_4-Hydroxybiphenyl (2 isomers)

C_5-Hydroxybiphenyl
Hydroxyfluorene (4 isomers)
C_1-Hydroxyfluorene (2 isomers)
C_2-Hydroxyfluorene
C_3-Hydroxyfluorene

Naphthylphenol (3 isomers)
C_1-Naphthylphenol (2 isomers)

Table 6. Nitrogen polycyclic aromatic compounds identified in the silicic acid subfractions from a solvent refined coal liquid heavy distillate (bp 260–450 °C), separation scheme as in Fig. 2 [52]

Secondary nitrogen polycyclic aromatic heterocycles fraction

C_2-Indole
C_3-Indole
C_4-Indole
C_1-Diphenyl ether
C_2-Diphenyl ether

C_3-Diphenyl ether
Phenanthrene
Tetrahydrocarbazole
Carbazole
Carbazole isomers (2)

C_1-Carbazole
C_2-Carbazole
C_3-Carbazole
C_4-Carbazole
C_5-Carbazole

Benzocarbazole (2 isomers)
C_1-Benzocarbazole
C_2-Benzocarbazole

Amino polycyclic aromatic hydrocarbons fraction

Naphthylamine
Aminobiphenyl
C_1-Aminobiphenyl
C_2-Aminobiphenyl
Aminofluorene

Benzo[*h*]quinoline
Carbazole
C_1-Benzoquinoline
C_2-Benzoquinoline
C_3-Benzoquinoline
C_4-Benzoquinoline
Aminophenanthrene/anthracene
C_1-Aminophenanthrene
Tetrahydronaphthoquinoline (2 isomers)
Aminopyrene/fluoranthene

Naphthoquinoline (2 isomers)
C_1-Aminopyrene/fluoranthene
Benzanthrone
C_1-Benzanthrone

Tertiary nitrogen polycyclic aromatic heterocycles fraction

C_2-Quinoline
C_3-Quinoline
C_4-Quinoline
C_5-Quinoline
Tetrahydroquinoline

Acridine
Phenanthridine/benzoquinoline
C_1-Tetrahydroquinoline
C_1-Benzoquinoline
C_2-Benzoquinoline

C_3-Benzoquinoline
Azafluoranthene
Azapyrene
C_1-Azapyrene/fluoranthene
Naphthoquinoline

Table 7. Compounds identified in the polycyclic aromatic sulfur heterocycles fraction of a solvent refined coal liquid heavy distillate (260–450 °C), separation scheme as in Fig. 2 [52]

Fluorene	Anthracene
C_1-Fluorene	Fluoranthene
Tetrahydrodibenzothiophene	Pyrene
Dibenzothiophene	Phenanthro[4,5-*bcd*]thiophene
C_1-Dibenzothiophene	C_1-Phenanthro[4,5-*bcd*]thiophene
C_2-Dibenzothiophene	Benzonaphthothiophene
C_3-Dibenzothiophene	C_1-Benzonaphthothiophene
Phenanthrene	C_2-Benzonaphthothiophene

The former procedure, used by us [94] and Jentoft [47] in studies of creosote pollution, is described schematically in Fig. 2, and in Tables 5–7 the individual components identified by Later and coworkers [52], in the different fractions of a solvent refined coal oil, are given. Although this example is not from creosote, the types of compounds found are similar to, or identical with, those found in wood impregnating oil.

A similar separation method, based on a combination of chromatographic methods and precipitation of the basic components as salts, has been used by Parees and Kamzelski [74] on coal derived liquids. Basically, the same type of compounds as those in Tables 5–7 were identified.

Thus, in the investigations mentioned so far, a large number of compounds have been identified or indicated in coal oil products. Of different reasons many of these components have been suspected to cause specific environmental and health effects. A number of special studies aimed at the identification of specific groups of compounds have therefore been undertaken. The most important groups of compounds will thus be treated individually below, with reference to some relevant publications.

Neutral Components

As mentioned above, the most investigated and well-known neutral components of coal tar liquids are the PAHs. The simplest group of neutral components are though the alkanes which, however, probably does not have any significance for

Table 8. Heterocyclic sulfur compounds identified in coal-tar distillates [12, 13, 50, 52, 55, 106]

Benzo[*b*]thiophene
2,3-Dihydrobenzo[*b*]thiophene
Methyldihydrobenzo[*b*]thiophenes (5 isomers)
C_2-Dihydrobenzo[*b*]thiophene (5 isomers)
3-Methylbenzo[*b*]thiophene
5-Ethylbenzo[*b*]thiophene
C_2-Benzo[*b*]thiophene (8 isomers)
Tetrahydrodibenzothiophene
Dibenzothiophene and/or naphtho[1,2-*b*]thiophenes (3 isomers)
Naphtho[1,2-*b*]thiophene
Methyldibenzothiophenes (7 isomers)
C_2-Dibenzothiophenes (8 isomers)
Phenanthro[4,5-*bcd*]thiophenes (2 isomers)
Methylphenanthro[4,5-*bcd*]thiophene
Benzo[*b*]naphtho[2,1-*d*]thiophene/Benzo[*b*]naphtho[2,3-*d*]thiophene/ Phenanthro[1,2-*b*]thiophene/Phenanthro[4,3-*b*]thiophene (6 isomers)
Methylbenzo[*b*]naphtho[2,1-*d*]thiophenes (7 isomers)
C_2-Benzo[*b*]naphtho[2,1-*d*]thiophenes (4 isomers)
Benzophenanthro[4,5-*bcd*]thiophenes (7 isomers)
Methylbenzophenanthro[4,5-*bcd*]thiophenes (3 isomers)
Dinaphthothiophenes (4 isomers)
Perylo[1,12-*bcd*]thiophenes (4 isomers)

the biological effects of the total liquids. Other important neutral components are the oxygen-, sulfur-, and some nitrogen-containing heterocyclic compounds. The latter group is treated together with the basic nitrogen components below.

The sulfur-containing components in oils have a specific interest as a source for sulfur oxides during combustion (acidification of the environment). Furthermore, sulfur analogues of PAH have lately been indicated as carcinogens and mutagens [55], and have been shown to be generally more toxic against *Daphnia magna* than the corresponding PAHs [26]. Some sulfur heterocycles also bioconcentrate in *Daphnia* to a greater extent than the PAHs [26] and show a strong tendency for accumulation and retention in short-necked clams and in eels [72].

The content of sulfur heterocycles (PASH) in fish and its relation to the levels in the sediment have been discussed by Vassilaros et al. [102] in terms of their physicochemical characteristics (aqueous solubility, octanol/water partition and volatility). In fish caught near a coking plant outfall several PASH could be detected (dibenzothiophene and alkylated dibenzothiophenes, naphtol[2,3-*b*]thiophene, phenanthro[4,5-*bcd*]thiophene) but with another composition profile than that found in the sediments.

Specific isolation techniques for organic sulfur compounds from coal tar liquids, as well as from other oils, have been developed by several groups [12, 13, 50, 52, 55, 106]. Examples of specific sulfur compounds identified are given in Table 8.

Acidic Components

Early specifications of creosote demanded an acid content, that is, mostly phenols, of as much as 10%. However, during later years the phenol content is kept low, often below 3%, one reason being that a high phenol content causes

Table 9. Phenols identified in a solvent refined coal middle distillate (bp 180–392 °C) [105]

2,6-Dimethylphenol	3-Isopropylphenol
Phenol	2,3,5-Trimethylphenol
2-Methylphenol	4-*n*-Propylphenol
2,5-Dimethylphenol	3-*n*-Propylphenol
4-Methylphenol	4-*sec*-Butylphenol
2,4-Dimethylphenol	4-Indanol
3-Methylphenol	5-Indanol
2-Isopropylphenol	6-Methyl-4-indanol
2,3-Dimethylphenol	7-Methyl-4-indanol
2-*n*-Propylphenol	5,6,7,8-Tetrahydro-1-naphthol
3,5,-Dimethylphenol	7-Methyl-5-indanol
4-Ethylphenol	2-Phenylphenol
3-Ethylphenol	5,6,7,8-Tetrahydro-2-naphthol
3,4-Dimethylphenol	1-Naphthol
4-Isopropylphenol	2-Naphthol
	3-Phenylphenol
	4-Phenylphenol

an undesirable "perspiration" of impregnated wood that makes it sticky. Literature data on phenol content in coal tar oils generally indicate rather high levels, from 4–6% [87], to 12–14% [31, 52], and even as high as 20% [85].

Several methods for the separation and isolation of phenols from coal tar distillates have been developed [84, 105], but the phenols are often obtained in a separate fraction also in the more general separation schemes mentioned above. A number of phenols representative for coal tar distillates are given in Table 9.

Phenols often have a high acute toxicity against aquatic organisms and a likewise high water solubility. Their leaching from e.g. creosote might therefore cause acute environmental effects. Therefore several investigations have been devoted specifically to the effects and behaviour of these compounds, see below.

Carboxylic acids do not seem to be present in significant amounts in coal tar distillates. However, Nestler [67] claimed the presence of benzoic acid in creosote and some higher aliphatic acids (palmitic and stearic) have been identified in high-boiling coal tar oils [71].

Basic Components, Including Non-Basic N-Heterocyclics

Under this heading belong mainly secondary and tertiary heterocyclic compounds (azaarenes) and aromatic amines. The different groups of nitrogen-containing compounds are those that have obtained the greatest interest lately, not only as components of coal tar distillates, but also as components of other types

Table 10. Basic nitrogen compounds identified in some coal liquefaction products [14]

Basic fraction of coal extract in recycle solvent

Tetrahydroquinoline	3,4-Benzoquinoline
Methyltetrahydroquinolines (4 isomers)	5,6-Benzoquinoline
C_2-tetrahydroquinolines (6 isomers)	Methylbenzoquinolines (8 isomers)
C_3-tetrahydroquinolines (5 isomers)	C_2-Benzoquinolines (14 isomers)
C_4-tetrahydroquinoline	C_3-Benzoquinolines (8 isomers)
Isoquinoline	C_3-Azafluorenes (3 isomers)
Quinoline	C_4-Azafluorenes (4 isomers)
2-Methylquinoline	Azafluoranthene/pyrene
4-Methylquinoline	Methylazafluoranthene/pyrene or azabenzofluorene (4 isomers)
Methylquinolines (4 isomers)	C_2-azafluoranthene/pyrene or methylazabenzofluorene (2 isomers)
C_2-Quinolines (10 isomers)	C_2-Alkylazacyclopenteno[*def*]phenanthrene
C_3-Quinolines (11 isomers)	C_3-Alkylazacyclopenteno[*def*]phenanthrene or tetrahydrodibenzoquinolines (5 isomers)
C_4-Quinolines (12 isomers)	3,4-Benzacridine
C_5-Quinolines (6 isomers)	Dibenzoquinoline
C_6-Quinoline	Diphenylamines or aminobiphenyls (3 isomers)
Octahydrobenzoquinolines (2 isomers)	Methyldiphenylamine (12 isomers)
Methyloctahydrobenzoquinoline	C_2-Diphenylamines (14 isomers)
Tetrahydrobenzoquinoline/ methyldiphenylamine	C_3-Diphenylamines (14 isomers)
7,8-Benzoquinoline	
2,3-Benzoquinoline	

Table 10 (continued)

Basic fraction of light ends product

C_2-Pyridines (3 isomers)	Aniline
C_3-Pyridines (8 isomers)	o-Toluidine
C_4-Pyridines (9 isomers)	m-Toluidine
C_5-Pyridines (7 isomers)	o-Ethylaniline
C_6-Pyridines (7 isomers)	m-Ethylaniline
C_7-Pyridines (8 isomers)	p-Ethylaniline
Azaperhydroacenaphthene	2,4-Dimethylaniline
Methylazaperhydroacenaphthenes (2 isomers)	2,6-Dimethylaniline
Methyltetrahydroquinolines (2 isomers)	2,5-Dimethylaniline
C_2-Tetrahydroquinolines (9 isomers)	3,5-Dimethylaniline
C_3-Tetrahydroquinolines (5 isomers)	3,4-Dimethylaniline
C_2-Quinoline	2,3-Dimethylaniline
	C_3-Anilines (7 isomers)
	C_4-Anilines (7 isomers)
	C_5-Anilines (6 isomers)

Basic fraction of LSE digester condensate

2-Picoline	Aniline
3-Picoline	o-Toluidine
4-Picoline	m-Toluidine
2,6-Lutidine	Methylaniline
2,4-Lutidine	o-Ethylaniline
2,5-Lutidine	m-Ethylaniline
2,3-Lutidine	p-Ethylaniline
3,4-Lutidine	2,4-Dimethylaniline
2-Ethylpyridine	2,6-Dimethylaniline
4-Ethylpyridine	2,5-Dimethylaniline
C_2-Pyridine	3,5-Dimethylaniline
C_3-Pyridines (10 isomers)	2,3-Dimethylaniline
C_4-Pyridines (11 isomers)	C_2-Aniline
C_5-Pyridines (8 isomers)	C_3-Anilines (3 isomers)
Dihydroindole or azaindane (2 isomers)	2,4,6-Trimethylaniline
Tetrahydroquinoline or methylhydroindole or methylazaindane (3 isomers)	C_4-Anilines (2 isomers)
Methyltetrahydroquinoline or C_2-dihydroindole or C_2-azaindane (3 isomers)	C_5-Anilines (3 isomers)
C_3-Dihydroindole or C_3-azaindane	C_6-Aniline
1,2,3,4-Tetrahydroquinoline	Quinoline
Methyltetrahydroquinolines (7 isomers)	Isoquinoline
C_2-Tetrahydroquinolines (16 isomers)	2-Methylquinoline
C_3-Tetrahydroquinolines (11 isomers)	4-Methylquinoline
C_4-Tetrahydroquinolines (3 isomers)	Methylquinolines (4 isomers)
C_5-Tetrahydroquinoline	C_2-Quinolines (6 isomers)
Octahydrobenzoquinoline	2,6-Dimethylquinoline
Diphenylamine or aminobiphenyl or aminoacenaphthene	2,7-Dimethylquinoline
	C_3-Quinolines (5 isomers)
	C_4-Quinolines (2 isomers)

Table 11. Amino polycyclic aromatic compounds identified in a heavy distillate (bp 260–450 °C) from a solvent refined coal [53]

Aminoindan	Aminophenanthridine/aminoacridine
C_1-Aminoindan	C_1-Aminophenanthridine/acridine
1-Aminonaphthalene	C_2-Aminophenanthridine/acridine
2-Aminonaphthalene	C_3-Aminophenanthridine/acridine
C_1-Aminonaphthalene	Aminophenylnaphthalene
C_2-Aminonaphthalene	C_1-Aminophenylnaphthalene
C_3-Aminonaphthalene	C_2-Aminophenylnaphthalene
Aminoquinoline	C_3-Aminophenylnaphthalene
C_1-Aminoquinoline	C_4-Aminophenylnaphthalene
C_2-Aminoquinoline	Aminofluoranthene/aminopyrene (2 isomers)
C_3-Aminoquinoline	C_1-Aminofluoranthene/pyrene
C_4-Aminoquinoline	C_2-Aminofluoranthene/pyrene
3-Aminobiphenyl	C_3-Aminofluoranthene/pyrene
4-Aminobiphenyl	C_4-Aminofluoranthene/pyrene
Aminoacenaphthylene/aminobiphenyl	Aminochrysene/aminobenzanthracene
C_1-Aminoacenaphthylene/aminobiphenyl	C_1-Aminochrysene/benzanthracene
C_2-Aminoacenaphthylene/aminobiphenyl	C_2-Aminochrysene/benzanthracene
C_3-Aminoacenaphthylene/aminobiphenyl	C_3-Aminochrysene/benzanthracene
C_4-Aminoacenaphthylene/aminobiphenyl	C_4-Aminochrysene/benzanthracene
Aminofluorene	Aminobenzo[*ghi*]fluoranthene/aminocyclopentapyrene
C_1-Aminofluorene	C_1-Aminobenzo[*ghi*]fluoranthene/cyclopentapyrene
C_2-Aminofluorene	C_2-Aminobenzo[*ghi*]fluoranthene/cyclopentapyrene
C_3-Aminofluorene	C_3-Aminobenzo[*ghi*]fluoranthene/cyclopentapyrene
C_4-Aminofluorene	C_4-Aminobenzo[*ghi*]fluoranthene/cyclopentapyrene
Aminocarbazole	Aminobenzopyrene/aminoperylene/aminobenzofluoranthene
C_1-Aminocarbazole	C_1-Aminobenzopyrene/perylene/benzofluoranthene
C_2-Aminocarbazole	C_2-Aminobenzopyrene/perylene/benzofluoranthene
C_3-Aminocarbazole	C_3-Aminobenzopyrene/perylene/benzofluoranthene
Aminoanthracene/aminophenanthrene (3 isomers)	C_4-Aminobenzopyrene/perylene/benzofluoranthene
C_1-Aminoanthracene/phenanthrene	Aminodibenzopyrene/aminodibenzofluoranthene
C_2-Aminoanthracene/phenanthrene	C_1-Aminobenzopyrene/dibenzofluoranthene
C_2-Aminoanthracene/phenanthrene	
C_4-Aminoanthracene/phenanthrene	
C_5-Aminoanthracene/phenanthrene	

of oils. Among these compound groups are found several with well-known biological effects, such as mutagenicity and carcinogenicity.

The nitrogen-containing components of coal tar liquids can be separated from other components by several of the group separation techniques described above (e.g. Fig. 2). In addition, several specific methods have been developed for the different subgroups of nitrogen-containing compounds [14, 15, 28, 39, 53, 86]. The results of some relevant investigations of coal oils are summarized in Tables 10 and 11.

The knowledge of the biological effects of azaarenes, which in many cases are analogous to those of the PAH-analogues, have resulted in some specific investigations of such compounds in environmental media. Thus, the separation and

Table 12. Heterocyclic nitrogen compounds and aromatic amines identified in a wood preservation wastewater [1] and in coal tar contaminated ground water [76]

Compound	Wastewater mg/kg	Ground water µg/l
2-Methylpyridine	–	41
Aniline	–	705
4-Methylaniline	–	647
3-Methylaniline	–	297
Quinoline	260	–
Isoquinolone	69	–
2-Methylquinoline	55	21
8-Methylquinoline	11	–
Methylazanaphthalene	95	–
7-Methylquinoline	38	–
Methylazanaphthalenes (2 isomers)	47	–
2,6-/2,7-Dimethylquinoline	21	–
C_2-Azanaphthalenes (3 isomers)	66	–
Methylvinylazanaphthalenes (2 isomers)	14	–
C_3-Azanaphthalene	12	–
4-Azafluorene	16	13
7,8-Benzoquinoline	53	7
Acridine	55	106
5,6-Benzoquinoline/phenanthridine	71	2
Methylbenzoazanaphthalenes (4 isomers)	350	–
Vinylbenzoazanaphthalene	3.0	–
Azafluoranthenes/azapyrenes (3 isomers)	54	–
Methylazafluoranthenes/azapyrenes (2 isomers)	4.4	–
Dibenzoazanaphthalenes (2 isomers)	5.2	–

analysis of complex mixtures of azaarenes have been studied [62, 88], and their behaviour during wastewater treatment have been investigated [92]. In Table 12 are given levels of azaarenes and aromatic amines identified in a wood preservative wastewater [1] and in contaminated ground water [76]. In addition to the substances given in Table 12 Pereira and coworkers [76] identified further a number of organic bases in the ground water and the oil-tar phase of the contaminated aquifer in St. Louis Park (see above). The compounds identified are given in Table 13.

Miscellaneous Components

A few minor groups of compounds present in creosote and other coal tar oils should be mentioned besides the more important ones above. Among early identified groups are the aromatic nitriles which seem to be relatively abundant in modern creosote products, including heteroaromatic nitriles (see below). Ketones and quinones are also present and are probably oxidation products of alkylated aromatics and aromatics of the anthracene and phenanthrene type (see also below on chlorination of coal tar leachates).

Table 13. Organic bases identified in the aqueous and oily-tar phases of an aquifer contaminated by coal tar wastes [76]

Aqueous phase	Oil-tar phase
C_1-Pyridines (3 isomers)	Azaacenaphthylenes (2 isomers)
C_2-Pyridines (2 isomers)	Azafluorene
Aniline	Aminoacenaphthene/aminobiphenyl
C_1-Anilines (3 isomers)	C_1-Aminoacenaphthene/aminobiphenyls (3 isomers)
C_2-Anilines (6 isomers)	C_2-Aminoacenaphthene/aminobiphenyl
C_3-Anilines (6 isomers)	Benzoquinoline/acridine/phenanthridines (8 isomers)
1,2,3,4-Tetrahydroquinoline	C_1-Benzoquinoline/acridine/phenanthridines (8 isomers)
C_1-1,2,3,4-Tetrahydroquinoline	C_2-Benzoquinoline/acridine/phenanthridines (5 isomers)
Quinoline	Azacyclo[*def*]phenanthrene
C_1-Quinolines (2 isomers)	Azafluoranthene/azapyrenes or anthracene/phenanthrene nitriles (3 isomers)
C_2-Quinolines (5 isomers)	
Isoquinoline	Azabenzofluorene/naphthoindoles (2 isomers)
Phenylpyridines (2 isomers)	C_1-Azabenzofluorene/naphthoindole
1-Naphthylamine	Benzacridine/azabenzoanthracenes or phenanthrenes/azachrysene/azatriphenylenes (3 isomers)
2-Naphthylamine	
Aminobiphenyls (2 isomers)	C_1-Benzacridine/azabenzoanthracenes or phenanthrenes/azachrysene/azatriphenylenes (7 isomers)
C_1-Aminobiphenyl	
C_1-Aminoacenaphthene	C_2-Benzacridine/azabenzoanthracenes or phenanthrenes/azachrysene/azatriphenylenes
4-Azafluorene	
1(2H)-Isoquinoline or quinoline	Azabenzofluoranthenes (6 isomers)
C_1-2(1H)-Quinoline (3 iosmers)	C_1-Azabenzofluoranthenes (4 isomers)
Benzo[*h*]quinoline	Azabenzoperylene
Acridine	Dibenzacridines (3 isomers)
Benzo[*f*]quinoline	C_1-Dibenzacridine
Benzoquinoline isomer	Azadibenzopyrenes
C_1-Benzoquinoline isomer	
Anthracene or phenanthrene nitrile, azafluoranthene or azapyrene	

Elemental sulfur have been indicated in creosote while the presence of other inorganic compounds probably differs with coal source and water content. Finally, the recent detailed studies aiming at the understanding of the biological activity of coal tar products have resulted in the identification of some rather special groups of compounds: Heterocyclic compounds containing *both* nitrogen and sulfur [13], hydroxyaromatic thiophenes [69], aminothiophenes [70], and hydroaromatic compounds [108].

Water Soluble Components

Another way to classify the components of coal tar oils might be based on the physical properties of special importance for environmental behaviour, such as water solubility and partition coefficients (see below). It has for example been concluded by Giddings and coworkers that the water soluble fraction of coal liquids are several orders of magnitude more toxic than those from petroleum [35].

Table 14. Compounds identified in coal tar leachate samples before and after chlorination [3]

Compound	Sample[a]	Compound	Sample[a]
Aromatic hydrocarbons		$C_{14}H_9N$	BN
Naphthalene	BN, C	Azafluoranthene	BN
Biphenyl	BN, C	Azapyrene	BN
Acenaphthene	BN	4*H*-Benzo[*def*]carbazole	BN
Fluorene	BN, C	Benzo[*a*]carbazole	BN
Phenanthrene, Anthracene	BN, C	Benz[*c*]acridine (2 isomers)	BN
4*H*-Cyclopenta[*def*]phenanthrene	BN	*Sulfur heterocycles*	
Fluoranthene	BN, C	Benzothiophene	BN
Pyrene	BN	Dibenzothiophene	BN
Benzofluorene	BN		
Benzofluoranthene	BN	*Oxygen-substituted aromatic compounds*	
Benzo[*c*]phenanthrene	BN	Dibenzofuran	BN, C
Benz[*a*]anthracene	BN	Fluorenone	C
Chrysene, Triphenylene	BN	$C_{13}H_{10}O$ (Xanthene isomer)	C
Alkylated aromatic compounds		Xanthene	C
2-Methylnaphthalene	BN	Anthrone (several isomers)	C
1-Methylnaphthalene	BN	Xanthone	C
Dimethylnaphthalene	BN	Anthraquinone	BN, C
Methylfluorene	BN	$C_{14}H_8O_2$	BN, C
Methylcarbazole	BN		
Methyldibenzofuran	BN	*Halogenated compounds*	
Methylphenanthrene/anthracene	BN	Chloronaphthalene	C
Methylfluoranthene/pyrene	BN	Chlorofluorene	C
Azaarenes		Chlorodibenzofuran	C
		Chlorophenanthrene	C
Quinoline, Isoquinoline	BN	Bromonaphthalene	C
$C_{11}H_7N$	BN, C	Bromofluorene	C
Carbazole	BN	Bromodibenzofuran	C
Acridine	BN		
Phenanthridine	BN		

[a] BN = basic-neutral fraction of unchlorinated leachate water
C = chlorinated leachate water, 50 mg free chlorine/l

Thus, in the biological testing of today's creosote products we have included tests of creosote-saturated water phases (see below).

Earlier investigations of the water solubility of creosote components have generally involved studies of the dissolution of PAH components and other aromatics from impregnated wood. Ingram and coworkers [46], for example, established that the more low-molecular aromatics naphthalene, anthracene, acenaphthene, dibenzofuran, fluorene and 2-methylnaphthalene, constituted the major part of the components found in the water phase. More recently samples of leachate from a commercial coal tar have been analyzed for PAHs [3]. The investigation was undertaken in order to establish the potential for contamination of potable water from coating material, derived from coal tar, used in public water

Table 15. Components identified by gas chromatography/mass spectrometry in the water phase after equilibration with a creosote product. The corresponding reconstructed ion chromatogram is given in Fig. 3

Scan no.	Compound[a]
182	Phenol
237	C_1-Phenol
261	C_1-Phenol
315	C_2-Phenol
334	C_2-Phenol
352	C_2-Phenol
356	C_2-Phenol
387	Quinoline/Isoquinoline
408	C_3-Phenol
421	Indanone
434	1H-Indole
443	C_1-Quinoline/isoquinoline
453	Indanol
471	Indanol
478	C_1 Quinoline/isoquinoline
512	C_1-Indole
574	Acenaphthene/Biphenyl
586	Cyanonaphthalene
598	Dibenzofuran
601	Phenylphenol
610	Naphthol
645	9H-Fluorene
675	Diphenylamine/Aminobiphenyl

[a] Structures given as suggested by comparison with EI-spectra stored in the mass spectrometer computer

supply systems. The investigation, which also included the analysis of chlorinated leachate, is summarized in Table 14.

For biological testing of the water soluble fraction of Swedish creosote products we stirred a two-phase system of creosote product and distilled water (1:5, v/v) carefully at room temperature and in the dark. Extraction after acidification and gas chromatographic analysis showed that already after 24 h a chromatographic pattern was obtained that was not altered after additional equilibration. The major components in the water phase were identified by mass spectrometry and the total amount of organic material dissolved in the water phase was estimated gravimetrically to about 0.2 g/l. The results given in Table 15 and Fig. 3, show that the majority of the extracted components are phenols and certain relatively water-soluble weak bases (azaarenes and amines). The biological tests are discussed below.

These studies of the water-soluble constituents of coal tar oils indicate that also other components than PAHs might be valuable tracers of creosote contamination of the environment.

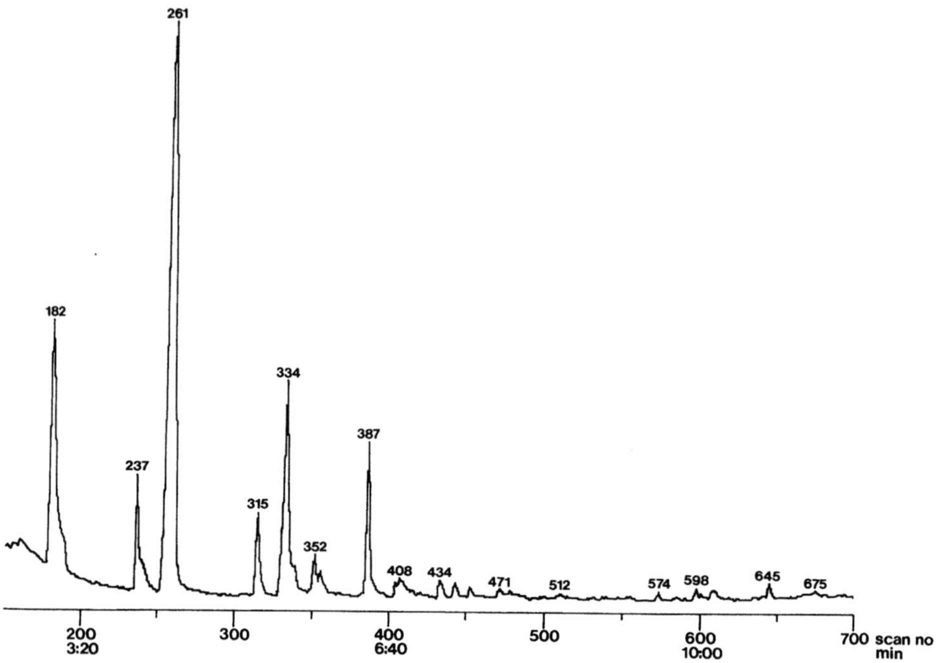

Fig. 3. Gas chromatographic-mass spectrometric total ion current chromatogram of an extract from a water phase equilibrated against creosote. Components identified are given in Table 15

Chemical Composition of Modern Creosote Products

In a Swedish project on the re-evaluation of the health and environmental safety of wood impregnating creosote, five products marketed in Sweden have been analyzed chemically and tested in standard biological tests [94, 97, 98]. All products were of foreign origin and are denoted A to E below. (On the basis of the similar labelling it could be suspected that product A and E were identical.) The intention was to investigate if today's creosote products differ from older ones in safety aspects and/or if any of the investigated product were safer than the other.

Initially all products were analyzed by gas chromatography with flame ionization detector (GC-FID). All products gave qualitatively and quantitatively very similar chromatograms with a first significant peak arising from naphthalene, Fig. 4A. Three products, however, B, C, and D, showed the presence of a number of components with shorter retention times than naphthalene. By mass spectrometric analysis these components were shown to be phenols (see below).

For further quantitative work on the environmental contamination from creosote residues nine single components have been used by us, marked in Fig. 4. These components constitute about 60% of the total creosote products, which should be compared with a figure of 62% as determined by Lorenz and Gjovik [56]. Thus, on a rough scale it seems as if the investigated modern products are similar to older ones.

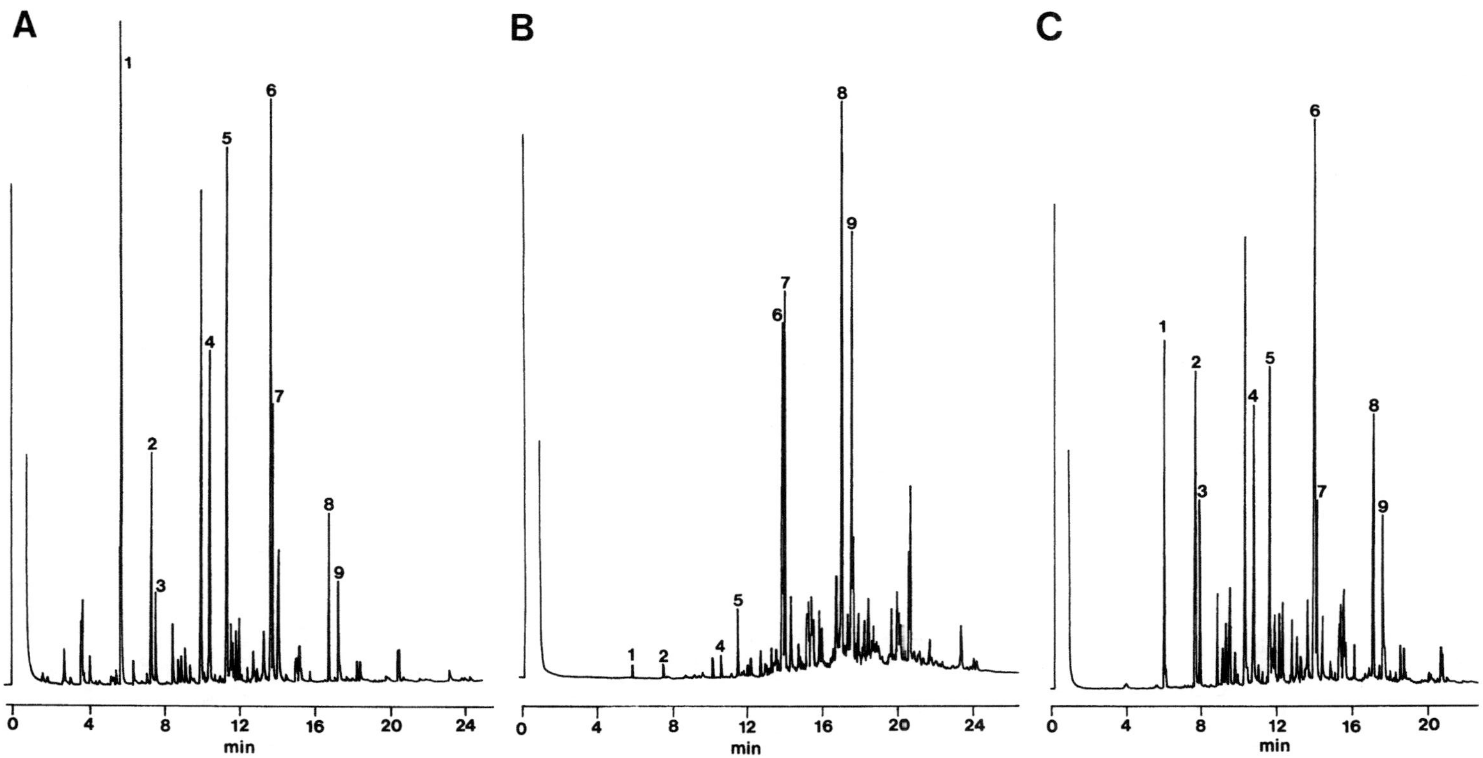

Fig. 4. Gas chromatogram of (A) a technical creosote product (Sweden 1984) and of creosote residues extracted from (B) the surface layers, and (C) at 50 cm deep at an old wood treatment facility, closed since 20 years [95]

1 Naphthalene
2 2-Methylnaphthalene
3 1-Methylnaphthalene
4 Dibenzofuran
5 Fluorene
6 Phenanthrene
7 Anthracene
8 Fluoranthene
9 Pyrene

Table 16. Components indicated by gas chromatography-mass spectrometry-computer in a Swedish creosote product (1984). Suggested structures based on comparisons of spectra with those stored in the computer. Scan numbers refer to Fig. 5

Scan No.	Compound	Scan No.	Compound
Aromatic hydrocarbons		754	C_1-Fluorene
172	C_3-Benzene	765	C_1-Fluorene
195	C_3-Benzene	807	Phenanthrene
227	Indan	811	Anthracene
234	Indene	875	C_1-Phenanthrene
268	C_1-Indan	879	C_1-Phenanthrene
313	C_1-Indan	889	4H-Cyclopenta[*def*]phenanthrene
322	C_1-Indene	923	Phenylnaphthalene
327	C_1-Indene	975	Fluoranthene
357	Naphthalene	1104	Pyrene
365	C_2-Indan	*Phenols*	
422	C_2-Indene	188	Phenol
447	C_1-Naphthalene	245	C_1-Phenol
461	C_1-Naphthalene	263	C_1-Phenol
517	Biphenyl	287	C_2-Phenol
527	C_2-Naphthalene		
536	C_2-Naphthalene	*O-heterocyclic compounds*	
547	C_2-Naphthalene	198	Benzofuran
562	C_2-Naphthalene	621	Dibenzofuran
569	Biphenylene	696	C_1-Dibenzofuran/9H-Xanthene
573	C_2-Naphthalene	705	C_1-Dibenzofuran/9H-Xanthene
598	Acenaphthene		
631	C_3-Naphthalene	*N-heterocyclic compounds*	
644	C_3-Naphthalene	399	Quinoline/Isoquinoline
656	C_3-Naphthalene	843	9H-Carbazole
671	Fluorene		
681	C_1-Biphenyl/Benzindan	*S-Heterocyclic compounds*	
687	C_1-Biphenyl/Benzindan	360	Benzothiophene
735	Dihydrophenanthrene	785	Dibenzothiophene
740	C_6-Benzene (?)		
749	C_1-Fluorene		

More detailed investigations were conducted by gas chromatography-mass spectrometry (GC-MS) and components were identified or indicated by comparison of mass spectra (electron impact) with those in the library of the computer in the mass spectrometer. The results from the analysis of product D are given in Table 16 and Fig. 5. In total 52 substances were indicated but only general structures could be given for many isomeric components. All compounds indicated are identical to, or closely related to, compounds already described as coal tar components.

In two creosote products (A and E) crystalline solids were present which after dissolution in acetone were investigated by GC-MS. It was shown that the crystals mostly consisted of phenanthrene and anthracene with impurities from

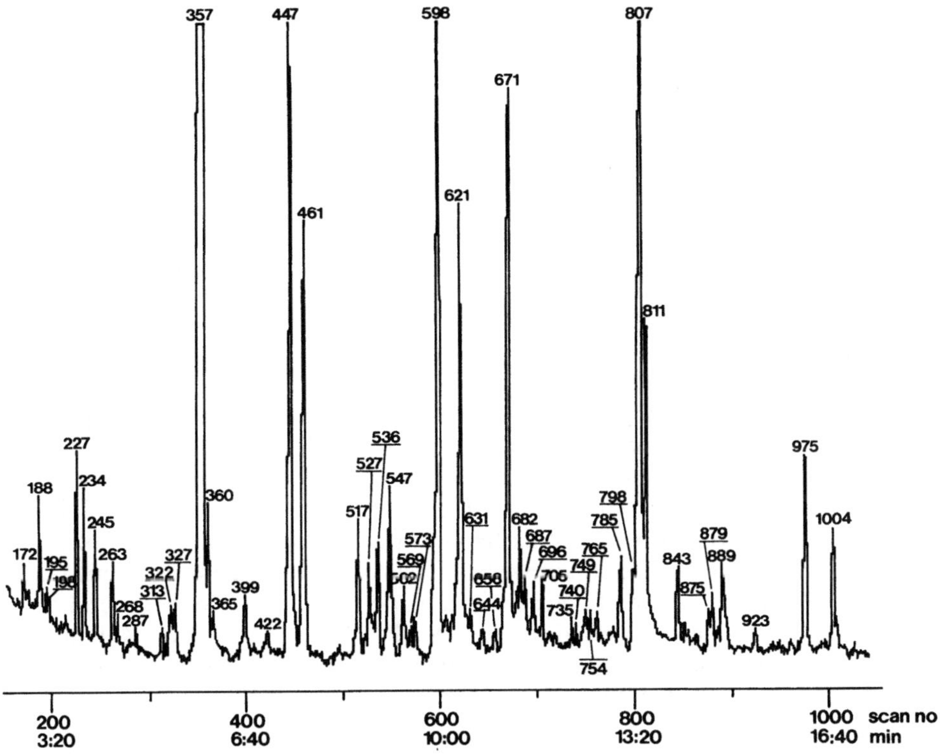

Fig. 5. Gas chromatographic-mass spectrometric total ion current chromatogram of a technical creosote product (Sweden 1984). Components identified are given in Table 16

Table 17. Nitrogen compounds and phenols indicated in a Swedish creosote product after fractionation as in Fig. 2

Quinoline/isoquinoline	Phenol
1H-Indole	C_1-Phenol (2 isomers)
C_1-Quinoline/isoquinoline (6 isomers)	C_2-Phenol (7 isomers)
C_2-Quinoline/isoquinoline (5 isomers)	C_3-Phenol (7 isomers)
C_3-Quinoline/isoquinoline (2 isomers)	C_4-Phenol
C_3-1H-Indole	Indanol (2 isomers)
Benzoquinoline/Phenanthridine/Acridine	Naphthol
9H-Carbazole	C_1-Naphthol
C_1-Benzoquinoline	
C_1-9H-Carbazole	
C_1-Benzacridine	
Aminobiphenyl	
Cyanoquinoline/isoquinoline (2 isomers)	
Cyanobiphenyl	
Cyanoanthracene (3 isomers)	
Cyano-9H-fluorene	

the other major aromatic components of creosote: naphthalene and alkyl derivatives, biphenyl, acenaphthene, dibenzofuran, fluorene, dibenzothiophene, and carbazole.

Fractionation of the products according to Fig. 2 (upper part) gave some additional information about the composition of the products. Analysis by GC-MS analogously to the above procedure revealed, again, the presence of a number of components of types earlier identified as constituents of coal tar oils, Table 17. The indication of several cyano derivatives is, however, somewhat surprising. Elemental sulfur was also indicated in the nitrogen compound fraction.

Biological Effects

In handbooks on toxicological properties of chemicals against man (e.g. [83]), it is generally stated that coal tar creosote is an irritant if inhaled or ingested. It is also concluded that "creosote is a recognized carcinogen of skin, forearm, scrotum, face, neck, and penis, end an experimental carcinogen of the lungs". Poisonings caused by creosote, on the other hand, is treated together with phenols, and phenolic products [21]. The carcinogenic effects of coal tar products is one main cause for the re-evaluation of creosote as a wood preservative. Background data in this respect has been reviewed by the US EPA [99] and by IARC [45], and in the following emphasis is put on other biological effects of importance for the environmental effects of creosote.

Mutagenicity

Creosote has been shown to be mutagenic to several strains of *Salmonella typhimurium* in the Ames' test, Fig. 6 [11]. It was suggested that the mutagenicity was

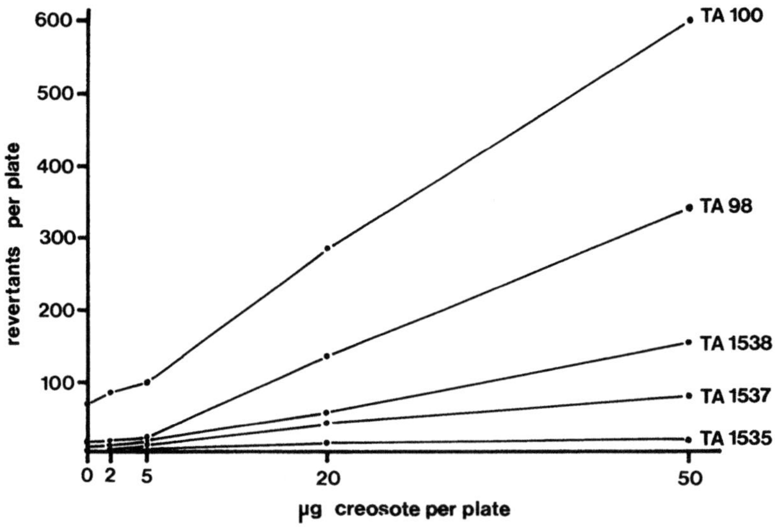

Fig. 6. Revertant colonies per plate of 5 *Salmonella typhimurium* strains as a function of creosote dose. Liver 9,000 g supernatant (S9 mix) added in all experiments (from [11])

caused mainly by the PAH components. However, the testing of different fractions of coal and shale derived liquids have lately indicated that the most potent mutagens in this type of products are aromatic amines and certain azaarenes [38, 42, 107]. The more specific separation and identification methods developed for oil products, as described above, have given increased evidence for this [39 and references cited therein].

Hydrogenation (hydrotreatment) of coal liquefaction products decrease their mutagenicity in the Ames' *Salmonella* assay as shown by Wilson and coworkers [107]. Fractions of solvent-refined coal was tested and the basic fraction, containing mostly aromatic amines, was the most mutagenic one. After hydrotreatment and subsequent fractionation of the oil no mutagenicity was observed in any fraction and no aromatic amines could be detected. Hydrotreatment of oil products is considered to generally reduce the concentration of nitrogen-, oxygen-, and sulfur compounds and is commonly used in petroleum refineries to upgrade crude oils for further refining.

Toxicity Against Aquatic Organisms

In spite of the large-scale use of creosote during a long period of time, there are surprisingly few reports on the effects of this product against aquatic organisms. On the other hand, there is an increasing literature on the environmental effects of modern coal derived oils of synfuel type and the consequences of the release of such oils to freshwater systems have been reviewed by Giddings and coworkers [35].

The results of four relatively late investigations of creosote are given in Tables 18–21. As evident from these tables creosote is highly toxic against many aquatic organisms, with 95h-LC50 values often below 1 mg/l.

Borthwick and Patrick [9] found in their investigations that the toxicity of creosote against estuarine organisms decreased if the test solutions were exposed to sunlight. About 50% reduction of toxicity to *Mysidopsis* was observed after

Table 18. Acute toxicity of marine grade creosote against some estuarine organisms at 20–21.2 $^0/_{00}$ salinity [9]

Species	Type of test, °C (solvent used)[a]	96h-LC50, mg/l (95% confidence interval)
Crassostrea virginica (Eastern oyster)	Flow-through, 21.4 (acetone)	0.71 (0.41–1.01)
Mysidopsis bahia	Static, 25.5 (triethylene glycol)	0.018 (0.015–0.021)
Panaeus duorarum (Pink schrimp)	Flow-through, 24.2 (acetone)	0.24 (0.18–0.34)
Cyprinodon variegatus (Sheepshead minnow)	Static, 25.0 (triethylene glycol)	0.72 (0.66–0.79)
Cyprinodon variegatus	Flow-through, 24.0 (acetone)	3.5 (2.9–4.2)

[a] Solvent used for administration of creosote to the water

Table 19. Toxicity of creosote against two crustaceans [60]

Species	Test temperature, °C	96h–LC50, mg/l
Homarus americanus, larvae	20	0.02
Homarus americanus, adult	10	1.76
Crangon septemspinosa	10	0.13
Crangon septemspinosa	20	0.11

Table 20. Acute toxicity of creosote products against some fish species in static tests [104]

Fish species	Type of sample	24h–LC50, mg/l (95% confidence interval)	96h–LC50, mg/l (95% confidence interval)	NOEL, mg
Lepomis machrochirus	Creosote-coal tar	3.72 (2.76–5.02)	0.99 (0.83–1.19)	0.75
Salmo gairdneri	Creosote-coal tar	4.42 (3.63–5.39)	0.88 (0.75–1.02)	0.49
Salmo gairdneri	Marine creosote	2.16 (1.60–2.90)	>0.56<0.75	0.32
Carassius auratus	Marine creosote	3.51 (3.34–3.69)	2.62 (2.30–2.97)	0.25

Table 21. Acute toxicity of water-soluble fractions of oil, creosote and single components against *Daphnia pulex* [32]

Test sample	48h–LC50, %WSF[a] (95% confidence interval)	48h–LC20, mg/l	48h–LC30, mg/l
Naphthalene	57.52 (3.40 mg/l) (46.02–76.79)	0.28–0.38	0.51–0.68
Phenanthrene	≫100 (1.14 mg/l)	0.096–0.13	0.31–0.41
No 2 Fuel oil	34.10 (24.94–41.78)	5.6[b]	10.0[b]
Creosote	2.91 (2.52–3.36)	1.0[b]	1.8[b]

[a] WSF = water soluble fraction

[b] Percentage WSF

3–5 days in the sun. Chemical analysis by gas chromatography indicated a half-life of a similar magnitude for the components analyzed, i.e. aromatic compounds and PAHs.

In addition to the acute effects, Geiger and Buikema [32, 33] also observed a number of sublethal effects of creosote on *Daphnia pulex*. Thus, a decrease of growth rates, reduced number of broods, disturbances of molting and an increase in abortion rates were observed. Also phenanthrene, but not naphthalene, caused these effects.

Hydrotreatment seems to reduce the toxicity of coal liquefaction products to aquatic animals, analogously to what has been found for the mutagenicity (see above). Cada and Kenna [16] tested the water soluble fraction of a pilot coal liquid blend after different grades of hydrotreatment. The raw oil had a 96h-LC50 value corresponding to 0.17% admixture of the water soluble fraction to the test water, while the high-severity hydrotreated oil gave only 5% mortality in 100% pure water soluble fraction.

The results of our toxicity tests of the five abovementioned creosote products A–E are recorded in Table 22. Tests with Bleak (*Alburnus alburnus, Pisces*) and *Nitocra spinipes* (*Crustaceae*) were performed by standard methods using acetone solutions [5, 6]. In the Microtox tests the water soluble fractions of the creosote products were tested by the standard technique in 100% aqueous medium [4]. In addition the products were tested by adding acetone solutions to the Microtox system, a modified test procedure developed in our laboratory [96]. In the *Nitocra* and Microtox tests two extracts from creosote polluted sediments were also investigated.

The cold water fish species tested by us seem to be less sensitive to creosote than the warm water species earlier tested, Tables 18 and 20. Creosote also seems to be less toxic to *Nitocra* than to earlier tested crustaceans, Tables 19 and 21.

The Microtox results are difficult to interprete. Both the tests with the water soluble fractions and the tests with acetone solutions of whole creosote give the same degree of toxicity when calculated on total weight of organic material added. But, as discussed above, the chemical composition of the two test solutions are evidently not identical and the toxicity must thus most probably be caused by different components.

There are no indications that there is a difference in toxicity between the five creosote products tested. The only major chemical difference established by us, the higher phenol contents of products B, C, and D, does not seem to influence the toxicity, not even in the tests with the water soluble fractions.

The two extracts from creosote polluted sediments also showed a high toxicity against *Nitocra* and in the Microtox test. The exact time of the event of pollution of the sediments is not known to us, but it can be assumed with reasonable certainty to be in the order of 10–20 years ago. Toxicity values given in Table 22 are calculated on amount of extracted material which was estimated gravimetrically. Gas chromatograms of the extracts are given in Fig. 7 together with that of a creosote standard. As not all extracted material could be redissolved in acetone the estimated toxicity most probably is even higher based on pure creosote residues. It can thus be concluded that even after long periods of environmental exposure the toxicity of creosote residues still is considerable.

It has not been within the scope of this review to cover the aquatic toxicity of the single components of creosote. However, a few data for some well known constituents are given in Table 23 for comparative reasons. This material indicate that only few single components show an acute toxicity of the same magnitude as the complex creosote products, and in those cases only against a few species. The toxicity of creosote therefore seems to be the results of synergistic effects or the effect of a few specific components of very high toxicity. The wide variety of

Table 22. Toxicity of Swedish creosote products (1984) and extracts from creosote-polluted sediments against two aquatic organisms and in the Microtox™ assay [94, 97]

Sample	Tests with acetone solutions (500 mg acetone/l test medium)				Tests with water soluble fraction[a]			
	Bleak *Alburnus alburnus*	*Nitocra spinipes*	Microtox™		Microtox™			
	96h–LC50, mg/l (95% confidence interval)	96h–LC50, mg/l (95% confidence interval)	5min–EC50, mg/l	15min–EC50, mg/l	5min–EC50		15min–EC50	
					%	mg/l	%	mg/l
Creosote A	10.52 (7.01–13.01)	0.89 (0.82–0.99)	0.38	0.43	0.18	0.36	0.20	0.40
B	10.57 (8.97–12.74)	1.34 (0.94–1.40)	0.41	0.53	0.15	0.30	0.18	0.36
C	11.60 (9.83–13.44)	0.96 (0.82–1.23)	0.41	0.56	0.21	0.42	0.23	0.46
D	9.09 (7.59–10.17)	1.56 (1.35–1.83)	0.52	0.63	0.07	0.14	0.07	0.14
E	7.93 (5.38–9.54)	0.76 (0.67–0.85)	0.40	0.49	0.21	0.42	0.24	0.48
Sediment 1[b]	–	0.51 (0.42–0.57)	0.81	0.88	–	–	–	–
2[b]	–	0.55 (0.42–0.72)	0.27	0.36	–	–	–	–

[a] Values calculated as percentage admixture of aqueous phase to test system and as total amount of water soluble material added, assuming a total of 200 mg/l creosote components

[b] Values based on extractable organic matter, see text. Gas chromatograms of extracts are given in Fig. 7

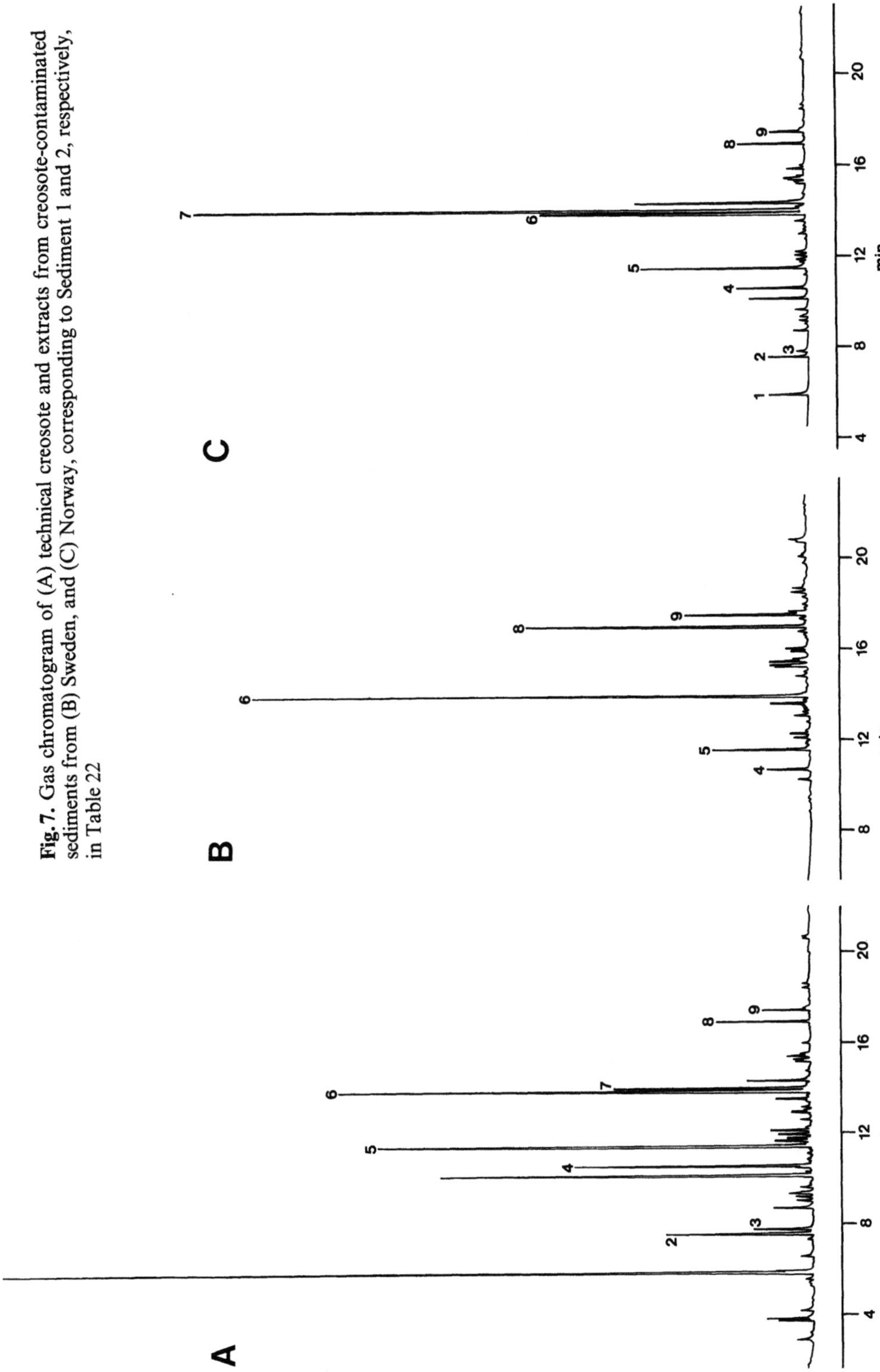

Fig. 7. Gas chromatogram of (A) technical creosote and extracts from creosote-contaminated sediments from (B) Sweden, and (C) Norway, corresponding to Sediment 1 and 2, respectively, in Table 22

Table 23. Acute toxicity against aquatic organisms of selected components of creosote and other coal tar oils

Compound	Test species	Type of test	Toxicity, mg/l	Ref.
Naphthalene	*Daphnia magna* (Crustaceae)	48h–LC50 (20 °C, static)	24.1	[75]
	Pimephales promelas (Pisces)	96h–LC50 (25 °C, flow-through)	6.08 (5.74–6.44)	[43]
1-Methylnaphthalene	*Nitocra spinipes* (Crustaceae)	96h–LC50 (21 °C, static)	13 (3–56)	[6]
Anthracene	*Daphnia magna*	48h–LC50 (20 °C, static)	Saturated sol. non-toxic	[75]
Phenanthrene	*Daphnia magna*	48h–LC50 (20 °C, static)	1.0	[75]
Acenaphthene	*Cyprinodon variegatus* (Pisces)	96h–LC50 (25 °C, static)	2.2 (1.7–2.8)	[40]
Fluoranthene	*Cyprinodon variegatus*	96h–LC50 (25 °C, static)	> 560	[40]
Dibenzofuran	*Cyprinodon variegatus*	96h–LC50 (25 °C, static)	1.8 (1.0–3.2)	[40]
Diphenyl ether	*Cyprinodon variegatus*	96h–LC50 (25 °C, static)	2.4 (1.8–3.2)	[40]
Benzo[*b*]thiophene	*Daphnia magna*	48h–LC50 (19 °C, static)	63.7 (48.3–103)	[26]
Dibenzothiophene	*Daphnia magna*	48h–LC50 (19 °C, static)	0.466 (0.324–0.708)	[26]
Phenol	*Salmo gairdneri* (Pisces)	96h–LC50 (14 °C, static)	8.9	[19]
	Pimephales promelas	96h–LC50 (14 °C, static)	67.5	[19]
	Pimephales promelas	96h–LC50 (25 °C, static)	24.9	[19]
	Pimephales promelas	96h–LC50 (25 °C, flow-through)	28 (23–34)	[77]
	Pimephales promelas	48h–LC50 (25 °C, static)	8.3 (7.1–9.7)	[77]
	Daphnia magna	48h–LC50 (20 °C, static)	30.1	[75]
	MICROTOX	5min–EC50	28	[25]
		5min–EC50	22	[78]
		5min–EC50	25	[54]
ortho-Cresol	*Leuciscus idus* (Pisces)	48h–LC50 (20 °C, static)	18	[48]
	Salmo gairdneri	48h–LC50 (20 °C, static)	13	[90]
	Salmo gairdneri	96h–LC50 (14 °C, static)	8.4	[19]
	Poecilia reticulata (Pisces)	48h–LC50 (24 °C, static)	38	[51]
	Pimephales promelas	48h–LC50 (20 °C, static)	34	[90]
	Pimephales promelas	96h–LC50 (14 °C, static)	18.2	[19]

Table 23 (continued)

Compound	Test species	Type of test	Toxicity, mg/l	Ref.
	Daphnia magna	48h–LC50 (20 °C, static)	5.0	[75]
	Daphnia magna	48h–LC50 (19 °C, static)	9.5	[90]
	MICROTOX	15min–EC50	15.4 (8.4–28.3)	[20]
		5min–EC50	32	[54]
meta-Cresol	*Salmo gairdneri*	96h–LC50 (14 °C, static)	8.9	[19]
	Pimephales promelas	96h–LC50 (14 °C, static)	55.9	[19]
	Daphnia magna	48h–LC50 (20 °C, static)	18.8	[75]
	MICROTOX	5min–EC50	8.2	[54]
para-Cresol	*Salmo gairdneri*	96h–LC50 (14 °C, static)	7.9	[19]
	Pimephales promelas	96h–LC50 (14 °C, static)	28.6	[19]
	Daphnia magna	48h–LC50 (20 °C, static)	1.4	[75]
	MICROTOX	5min–EC50	1.3	[54]
2,4-Dimethylphenol	*Pimephales promelas*	96h–LC50 (25 °C, flow-through)	17 (16–18)	[77]
	Pimephales promelas	48h–LC50 (25 °C, static)	9.5 (8.4–11)	[77]
2,4,6-Trimethylphenol	*Daphnia magna*	48h–LC50 (20 °C, static)	0.3	[75]
2-Biphenylol	*Pimephales promelas*	96h–LC50 (25 °C, flow-through)	5.99 (5.70–6.30)	[43]
1-Naphthol	*Pimephales promelas*	96h–LC50 (25 °C, flow-through)	4.24 (4.12–4.37)	[43]
	MICROTOX	5min–EC50	5.66	[17]
		15min–EC50	3.8	[24]
Pyridine	*Leuciscus idus*	48h–LC50 (20 °C, static)	196	[48]
	Salmo gairdneri	48h–LC50 (20 °C, static)	560	[90]
	Poecilia reticulata	48h–LC50 (24 °C, static)	1 390	[51]
	Pimephales promelas	48h–LC50 (20 °C, static)	115	[90]
	Daphnia magna	48h–LC50 (19 °C, static)	1 080	[90]
	MICROTOX	15min–EC50	2 120 (1,335–3,427)	[20]
Trimethylpyridine	*Daphnia magna*	48h–LC50 (20 °C, static)	47.5	[75]
Quinoline	*Daphnia magna*	48h–LC50 (20 °C, static)	28.5	[75]

Table 23 (continued)

Compound	Test species	Type of test	Toxicity, mg/l	Ref.
Isoquinoline	*Daphnia magna*	48h–LC50 (20 °C, static)	4.1	[75]
2-Methylquinoline	*Daphnia magna*	48h–LC50 (20 °C, static)	27.3	[75]
2,6-Dimethylquinoline	*Daphnia magna*	48h–LC50 (20 °C, static)	34.1	[75]
Acridine	*Daphnia magna*	48h–LC50 (20 °C, static)	2.3	[75]
		NOEC (20 °C, static)	0.4	[75]
Aniline	*Leuciscus idus*	48h–LC50 (20 °C, static)	49	[48]
	Salmo gairdneri	48h–LC50 (15 °C, static)	43	[90]
	Poecilia reticulata	48h–LC50 (14 °C, static)	100	[51]
	Pimephales promelas	48h–LC50 (20 °C, static)	65	[90]
	Daphnia magna	48h–LC50 (19 °C, static)	0.64	[90]

chemical structures present in creosote also indicate the intervention of a large number of mechanisms of toxicity.

Many of the references given in Table 23 contain further valuable information in connection with testing and environmental effects of specific chemicals related to coal tar oils.

The acute toxicity of the water soluble fractions (WSF) of coal- and shale oils and petroleum against two fresh water algae have been investigated by Giddings and Washington [34]. The coal based oils were toxic to both species at concentrations around 1% WSF in the test media. Hydrotreatment reduced the toxicity somewhat against one of the species, *Microcystis aeruginosa*. Shale oils and petroleum products were in general less toxic than the coal products and some petroleum oils were non-toxic even when tested as 100% WSFs.

Toxicity Against Mammals and Birds

In Registry of Toxic Effects of Chemical Substances [79] the oral toxicity of coal tar creosote to the rat, as LD50, is recorded to 725 mg/kg and the oral LDL0 values for dog, cat, and rabbit is given to 600 mg/kg for all species. This is, in perspective of the aqueous toxicity data, surprisingly high values. Similarly high values are recorded for the toxicity against bird species. Webb [104] have reported 8d-LC50 values of 1,261 (744–2,139) mg/kg against Bobwhite quail (*Colinus virginianus*) and 10,388 (1,177–91,712) mg/kg against Mallard duck (*Anas platyrhynchus*). The no effect levels were estimated to 215 and 2,150 mg/kg, respectively.

Toxicity Against Higher Plants

An aquatic vascular plant, *Spirodela polyrrhiza*, has been tested with water soluble fractions of coal distillates after different degrees of hydrotreatment [49]. The raw oils stopped growth completely at a level of 5% WSF while the same level of WSF from the high-severity treated oil reduced growth by 32%

We are not aware of any other data on the toxicity of creosote against higher plants. Therefore we tested the five abovementioned new Swedish products against *Allium cepa* according to well-known techniques [29, 98]. The experiments were performed using acetone solutions and the root length was recorded after

Table 24. Effect of five creosote products on growth of roots, measures as root length, of *Allium cepa*. Concentrations causing 10 and 50% reduction of root lengths during 4 days exposure in the presence of 0.5% acetone [98]

Sample	96h–EC10 mg/l	96h–EC50 mg/l
Creosote A	2.3	19
B	3.5	34
C	0.8	20
D	3.3	18
E[a]	3.4	20
	2.9	24

[a] Two experiment series

four days growth in different concentrations of creosote. The results are given in Table 24 and indicate that product B might be slightly less toxic than the other. In comparison with known herbicidal organic compounds, such as phenoxyacetic acids, the toxicity of creosote against *Allium* is low. Lower threshold levels for growth retardation caused by phenoxyacetic acids and esters are about 0.1 mg/l [30].

Bioaccumulation and Persistence

With knowledge about the structures and the physical/chemical properties of the components of creosote it is evident that several substances can be suspected to bioaccumulate in organisms when released into the environment. Such assumptions have also been tested in bioaccumulation experiments with some components, mostly though without direct connection to studies of the creosote complex. A few examples have been collected in Table 25.

Partition studies of coal tar constituents have been performed by Rostad and colleagues [82] in connection with a case of ground water contamination in the USA. After identification of a large number of components in the oil/tar and the aqueous phase (Table 1), the partition of 31 components of different chemical classes were investigated in the oil/water system and compared to their octanol/water partition behaviour. It was concluded that the oil/water partition coef-

Table 25. Water solubility, partition coefficient octanol/water (P) and bioaccumulation factors (BCF) in fish for some creosote components[a]

Compound	Water solubility, mg/l	log P	log BCF	Species
Phenol	67,000	1.46		
p-Cresol	24,000	2.51		
2,4-Dimethylphenol	17,000	2.42	2.18	*Lepomis macrochirus*
1-Naphthol	700	2.98		
Diphenylamine		3.42	1.48	*Pimephales promelas*
Quinoline	60,000	2.03	~0.9	*Pimephales promelas*
Acridine		3.40	~2	*Pimephales promelas*
Benz[*a*]acridine			~2	*Pimephales promelas*
Dibenz[*a,h*]acridine			~2	*Pimephales promelas*
Dibenzofuran	10	4.12	3.13	*Pimephales promelas*
Diphenyl ether	21	4.21	2.33	*Salmo gairdneri*
Dibenzothiophene		4.38		
Biphenyl	7.0	3.76	2.64	*Salmo gairdneri*
Naphthalene	30	3.37	2.63	*Pimephales promelas*
2-Methylnaphthalene	25	4.11	2.61	*Lepomis macrochirus*
Acenaphthene	3.9	3.92	2.59	*Lepomis macrochirus*
Fluorene	1.7	4.38	3.11	*Pimephales promelas*
Phenanthrene	1.0	4.46	3.42	*Pimephales promelas*
2-Methylphenanthrene	0.3	4.86	3.48	*Pimephales promelas*

[a] Data collected from [18, 59, 61, 66, 91, 103]

ficients of the compounds studied were reasonably comparable with their respective octanol/water partition coefficients, Table 26.

Bioaccumulation experiments with native creosote have been performed with lobsters in connection with toxicity tests [60]. Creosote levels in the tissues (hepatopancreas) were determined by fluorescence measurements, which is mainly sensitive to polycyclic aromatic compounds, including heterocyclic components. Creosote levels of about 15,000, 24,000, and 10,000 μg/g fat were estimated in the hepatopancreas after 100–150 h exposition to 2.5, 1.3, and 0.3 mg creosote per liter, respectively. At the higher dose levels the lobsters died shortly after the experiments.

The interest in azaarenes have resulted in bioaccumulation experiments also with such components [91]. The bioaccumulation of quinoline, isoquinoline, acridine, benz[*a*]acridine and dibenz[*a,h*]acridine was investigated in the Fathead minnow, results are included in Table 25. Theoretical calculations of the bioaccumulation factors (BCF) based on the octanol/water partition coefficients according to Veith and coworkers [103] gave a reasonable agreement with observed values for quinoline and acridine, but gave too high values for the other compounds. Experiments with ^{14}C-labelled dibenzacridine showed that metabolites accumulated to a higher level (BCF about 400) than the parent compound. The BCF of dibenzacridine in *Daphnia pulex,* on the other hand, was estimated to 3,500.

The persistence of creosote in the environment is clearly evident from the many cases of creosote pollution reported. We have, for example, found creosote

Table 26. Coal-tar/water (K_{tw}) and octanol/water (P) partition coefficients for selected compounds identified in connection with studies of coal-tar contaminated ground-water [82]

Compound	log K_{tw}[a]	log P[b]
Polycyclic aromatic compounds		
Indene	3.68	2.92
Naphthalene	4.00	3.37
2-Methylnaphthalene	4.76	3.97
Acenaphthene	5.07	3.92
9H-Fluorene	4.52	4.38
1-Methylfluorene	4.80	n. a.
Anthracene	4.76	4.45
2-Methylanthracene	5.07	5.07
Benz[*a*]anthracene	5.70	5.61
Dibenz[*a,h*]anthracene	5.80	6.50
Nitrogen heterocycles		
Quinoline	4.20	2.03
2-Methylquinoline	n. d.	2.23
3-Methylindole	3.02	2.60
Acridine	3.36	3.40
Carbazole	4.01	3.29
Benz[*c*]acridine	5.32	4.45
Dibenzo[*a,i*]carbazole	5.20	6.40
Dibenz[*a,j*]acridine	6.75	n. a.
Sulfur heterocycles		
Benzo[*b*]thiophene	3.70	3.12
3-Methylbenzothiophene	4.70	n. a.
Dibenzothiophene	5.45	4.38
Benzo[*b*]naphtho[2,1-*d*]thiophene	5.78	n. a.
Benzo[*b*]naphtho[1,2-*d*]thiophene	4.45	n. a.
Oxygen heterocycles		
Benzofuran	2.96	2.67
2-Methylbenzofuran	5.05	n. a.
Dibenzofuran	4.74	4.12
Benzo[*b*]naphtho[2,3-*d*]furan	5.08	n. a.

[a] n. d. = not determined
[b] n. a. = not available

residues which show gas chromatographic patterns virtually identical with native creosote in soil at wood treatment plants that has been closed for more than 20 years. In surface layers the more volatile constituents have evaporated, but are still distinguishable in the lower layers, Fig. 4. Similar redistributions can be observed in aquatic sediments where the same components have decreased relatively, in these cases due to the higher water solubility, Fig. 7.

The degradation of phenols from creosote waste waters have been studied by Roberts and Hedrick [81]. Isolates of the species *Pseudomonas* and *Alcaligenes* reduced phenol content in batch cultures, but only one type of phenol was totally degraded by single strains. However, the isolates were inhibited if the creosote levels were not diluted before addition to the media.

In the ground water contamination incident discussed above [44, 76], it was found that the phenolic components of the coal tar disappeared faster from the pollution source than could be accounted for by the dilution gradient [27]. As sorption of the phenols on aquifer sediments is negligible as compared to another marker used, naphthalene, microbial degradation of the phenols was suggested. It was concluded by laboratory experiments that anaerobic bacteria in the ground water were degrading the phenols to carbon dioxide and methane. Evidence for anaerobic degradation of naphtalene was not obtained either in the field or in the laboratory.

The distribution of two ^{14}C-labelled creosote components, phenanthrene and acenaphthene, mixed with native creosote, in a terrestrial microcosm has been studied by Gile and coworkers [37]. The substances were administered as impregnated wood poles in a system containing soil, ryegrass, invertebrates, and a pregnant Gray-tailed vole as only vertebrate. The highest levels of radioactivity was found in the animals living in and on the soil, with a concentration of phenantherene and acenaphthene in the vole of 7.2 and 32 ppm, respectively (whole body weight). The mass balance of ^{14}C in the system is given in Table 27. In the same study the fate of pentachlorophenol (included in Table 27), dieldrin and tributyltin oxide was also investigated.

The fate of five radiolabelled constituents of coal and other coal-conversion products – anthracene, fluorene, carbazole, dibenzofuran, and dibenzothiophene – in a model ecosystem has been investigated by Lu and coworkers [57]. In the terrestrial ecosystems the tritium labelled components were applied to Sorghum leaves whereafter they were dispersed by consumption by salt marsh caterpillars (*Estigmene acrea*) into water containing plankton, algae (*Oedogonium cardiacum*), *Daphnia magna*, and snail (*Physa* sp.). Subsequently mosquito larvae

Table 27. ^{14}C Mass balance after distribution of the labelled creosote components phenanthrene and anthracene to a terrestrial model ecosystem. The compounds were mixed with native creosote and administered as impregnated wood posts. Comparison is made with another wood impregnating agent, pentachlorophenol [37]

Compartment	% of chemical applied		
	Phenanthrene	Acenaphthene	Pentachlorophenol
Wood post	95	93.5	96
Water	0.0	0.0	0.0
Soil	2.7	4.3	3.6
Plants	0.1	0.04	0.06
Animals[a]	0.8	1.2	0.24
Air	1.4	1.0	0.1

[a] Earthworms (*Lumbricus* sp.), pill bugs (*Armadillarium* and *Porcellia* sp.), mealworm larvae (*Tenebrio molitor*), gray crickets (*Acheta domesticus*), garden snails (*Helix pomata*) and a female gray-tailed vole (*Microtus canicaudus*)

Table 28. Partition behaviour (ecological magnification) and concentration of some tritium labelled creosote components in the biota of a terrestrial model ecosystem [57]

Compound	P[a]	Conc. of parent compound in biota/ conc. of parent compound in water				% of parent compound of total extractable activity in species			
		Fish[b]	Mosquito[c]	Snail[d]	Alga[e]	Fish[b]	Mosquito[c]	Snail[d]	Alga[e]
Anthracene	8,135	1,029	631	2,714	670	48.3	49.5	22.8	41.0
Fluorene	1,345	1,400	3,330	76,680	4,020	9.1	36.3	37.3	5.8
Carbazole	2,560	125	112	134	49	29.3	26.8	12.3	10.1
Dibenzofuran	1,527	947	2,094	2,858	82	48.4	56.3	30.9	9.3
Dibenzothiophene	3,120	45	356	1,505	138	9.7	24.9	33.2	39.3

[a] Partition coefficient octanol/water
[b] *Gambusia affinis*
[c] *Culex pipiens*
[d] *Physa* sp.
[e] *Oedogonium cardiacum*

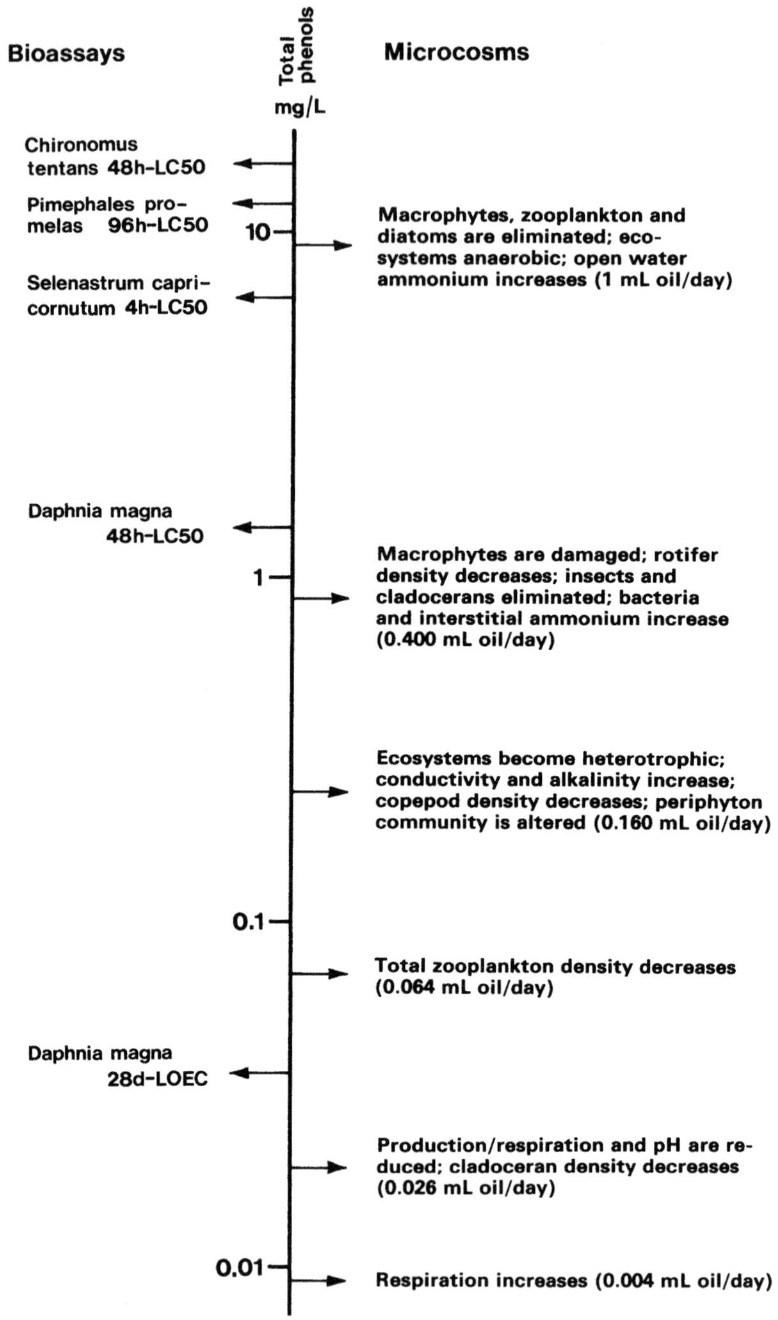

Fig. 8. Toxicity spectrum for a coal oil (unrefined middle distillate, b.p. 196–268 °C) comparing results from single-species bioassays and tests with 64 l freshwater microcosms. For bioassays, the scale represents the concentration of phenols in dilutions of a 1:8 water-soluble fraction. Phenol levels in the microcosms are expressed as the average concentration during 56 days of oil treatment (after [31])

(*Culex pipiens*) and fish (*Gambusia affinis*) were added to complete the food chain.

The supposed environmental properties of the chemicals were expressed quantitatively as ecological magnification (EM = concentration of parent compound in the organism/concentration of parent compound in the water), Table 28. The behaviour of the compounds in this respect was well correlated to their respective octanol/water partition coefficients. Also the degradation of the compounds were followed and several metabolites were identified. Major biodegradation routes were as follows: Anthracene and fluorene were biologically converted to hydroxy and keto analogues. Carbazole was N-methylated and N-acetylated. Dibenzothiophene was oxidized to the sulfoxide and sulfone while dibenzofuran was relatively inert to biodegradation.

Extensive microecosystem studies relevant in this context have been performed at the Oak Ridge National Laboratory by the group of Giddings, Franco, and Southworth [31, 36, 93]. The coal oil studied was an unrefined middle distillate, b.p. 196–268 °C, containing about 12% phenols, and it was established that 95% of the water soluble components of the oil were phenols.

In an initial study aquatic microcosms of 55 l containing mainly aquatic plants (*Elodea canadensis*) were used to investigate the behaviour of the phenols [93]. The phenols that dissolved from the oil into the water accumulated with sequential additions and attained a maximum after 28 days, whereafter degradation rates exceeded input rates. Significant differences in degradation rates were observed between different alkylphenol isomers, but no compound seemed refractory. After discontinuation of oil addition the phenols dissappeared within several weeks. The major process for removing the phenols from the systems was considered microbial degradation and other processes were judged negligible in comparison.

Further studies were made using 67 l fresh water microcosms which were treated for 8 weeks with oil in amounts of 0.003–7 ml per week which correspond to approximately 0.01 to 10 mg/l dissolved phenols [31]. At the highest dose level the systems were severely damaged with elimination of macrophytes (*Elodea canadensis, Potamogeton* sp.), zooplankton (e.g. *Simocephalus vetulus, Cyclops vernalis, C. varicans, Eucyclops agilis*), and insects (*Chironomus tentans*). The microcosms also became anaerobic and did not recover within 5 months after the treatment period. At lower dosages temporary effects on ecosystem metabolism, water chemistry, and community structure were observed.

When the studies were extended to 15 m^3 outdoor ponds [37] qualitatively and quantitatively similar results were obtained and an exposure level of approximately 3% of the 48h-LC50 for *Daphnia magna,* 1.7 mg/l, was concluded safe for this model ecosystem. On the basis of these studies a "toxicity spectrum" of the coal tar oil was constructed and is reproduced in Fig. 8.

Conclusions

On the basis of the present knowledge about the health and environmental effects of coal tar creosote and its constitutents, the use of this type of product has been restricted in at least two countries. Both in the USA and Sweden the registration of creosote as a pesticide has been changed. Handling of creosote will be allowed only for specifically trained personell and sale to the general public will be prohibited. Uses for other purposes than wood treatment will also not be allowed.

Acknowledgements

We are indebted to Mr Ingvar Björklund, Miss Margareta Hansson, Mrs Kerstin Nylund, Miss Birgitta Samuelsson, Mrs Siv Ullstrand and Miss Ulla Wideqvist for technical assistance with biological tests and analytical investigations.

References

1. Adams, J., Giam, C.-S.: Polynuclear azaarenes in wood preservative wastewater. Environ. Sci. Technol. *18*, 391 (1984)
2. Agnew, R.W.: Removal and treatment of contaminated river bottom muds: Field demonstration. U.S. EPA Municipal Environmental Research Laboratory, Cincinatti, OH, EPA-600/s2-84-006 (1984)
3. Alben, K.: Gas chromatographic-mass spectrometric analysis of chlorination effects on commercial coal-tar leachate. Anal. Chem. *52*, 1825 (1980)
4. Beckman Instruments, Inc., Carlsbad, Calif.: Microtox™ System Operating Manual, Beckman Instructions 015-555879 (1982)
5. Bengtsson, B.-E.: The harpacticoid *Nitocra spinipes* (Crustacea) as a test organism in brackish water toxicological bioassays. INSERM *106*, 421 (1981)
6. Bengtsson, B.-E., Tarkpea, M.: The acute toxicity of some substances carried by ships. Mar. Pollut. Bull. *14*, 213 (1983)
7. Black, J.J.: Movement and identification of a cresote-derived PAH-complex below a river pollution point source. Arch. Environ. Contam. Toxicol. *11*, 161 (1982)
8. Black, J.J., Hart, T.F., Evans, E.: HPLC studies of PAH pollution in a Michigan trout stream, in: Polynuclear Aromatic Hydrocarbons: Chemical Analysis and Biological Fate (ed.) Cooke, M., Dennis, A.J., p. 343, Columbus, OH, Batelle Press, 1980
9. Borthwick, P.W., Patrick, J.M., Jr.: Use of aquatic toxicology and quantitative chemistry to estimate environmental deactivation of marine-grade creosote in seawater. Environ. Toxicol. Chem. *1*, 281 (1982)
10. Borwitsky, P.W., Schomburg, G.: Separation and identification of polynuclear aromatic compounds in coal tar by using glass capillary chromatography including combined gas chromatography-mass spectrometry. J. Chromatog. *170*, 99 (1979)
11. Bos, R.P., Hulshof, C.T.J., Theuws, J.L.G., Henderson, P.T.: Mutagenicity of creosote in the Salmonella/microsome assay. Mutat. Res. *119*, 21 (1983)
12. Burchill, P., Herod, A.A., Pritchard, E.: Identification of sulphur heterocycles in coal tar and pitch. J. Chromatogr. *242*, 51 (1982)
13. Burchill, P., Herod, A.A., Pritchard, E.: Determination of nitrogen-sulphur mixed heteroatomic compounds and sulphur heterocycles in an anthracene oil. J. Chromatogr. *242*, 65 (1982)
14. Burchill, P., Herod, A.A., Pritchard, E.: Estimation of basic nitrogen compounds in some coal liquefaction products. J. Chromatogr. *246*, 271 (1982)
15. Burchill, P., Herod, A.A., Mahon, J.P., Pritchard, E.: Comparison of methods for the isolation of basic nitrogen compounds from coal. J. Chromatogr. *265*, 223 (1983)
16. Cada, G.F., Kenna, M.: Effetiveness of hydrotreatment in reducing the toxicity of a coal liquefaction product to juvenile Channel catfish. Bull. Environ. Contam. Toxicol. *34*, 746 (1985)

17. Curtis, C., Lima, A., Lozano, S.J., Veith, G.D.: Evaluation of a bacterial bioluminescence bioassay as a method for predicting acute toxicity of organic chemicals to fish, in: Aquatic Toxicology and Hazard Assessment: Fifth Conference (eds.) Pearson, J.G., Foster, R.B., Bishop, W.E., p. 170, American Society for Testing and Materials, *ASTM STP 766,* 1982
18. Davies, R.P., Dobbs, A.J.: The prediction of bioconcentration in fish. Water Res. *18,* 1253 (1984)
19. DeGraeve, G.M., Geiger, D.L., Meyer, J.S., Bergman, H.L.: Acute and embryo-larval toxicity of phenolic compounds to aquatic biota. Arch. Environ. Contam. Toxicol. *9,*, 557 (1980)
20. De Zwart, D., Slooff, W.: The Microtox as an alternative assay in the acute toxicity assessment of water pollutants. Aquat. Toxicol. *4,* 129 (1983)
21. Dreisbach, R.H.: Hanbook of poisoning, prevention, diagnosis, and treatment, p 363, Los Altos, Calif., Lange Medical Publications 1980
22. Dunn, B.P., Stich, H.F.: Monitoring procedures for chemical carcinogens in coastal waters. J. Fish. Res. Board Can. *33,* 2040 (1976)
23. Dunn, B.P., Fee, J.: Polycyclic hydrocarbon carcinogens in commercial seafoods. J. Fish. Res. Board Can. *36,* 1469 (1979)
24. Dutka, B.J., Kwan, K.K.: Comparison of three microbial toxicity screening tests with the Microtox test. Bull. Environ. Contam. Toxicol. *27,* 753 (1981)
25. Dutka, B.J., Nyholm, N., Petersen, J.: Comparison of several microbiological toxicity screening tests. Water Res. *17,* 1363 (1983)
26. Eastmond, D.A., Booth, G.M., Lee, M.L.: Toxicity, accumulation, and elimination of polycyclic aromatic sulfur heterocycles in *Daphnia magna.* Arch. Environ. Contam. Toxicol. *13,* 105 (1984)
27. Ehrlich, G.G., Goerlitz, D.F., Godsy, E.M., Hult, M.F.: Degradation of phenolic contaminants in ground water by anaerobic bacteria: St. Louis Park, Minnesota. Groundwater *20,* 703 (1982)
28. Felice, L.J.: Determination of alkyl anilines and alkyl pyridines in solvent refined coal distillates and aqueous extracts by gas chromatography/mass spectrometry. Anal. Chem. *54,* 869 (1982)
29. Fiskesjö, G.: The Allium test as a standard in environmental monitoring. Hereditas *102,* 99 (1985)
30. Fiskesjö, G., Lassen, C., Renberg, L.: Chlorinated phenoxyacetic acids and chlorophenols in the modified *Allium* test. Chem.-Biol. Interactions *34,* 333 (1981)
31. Franco, P.J., Giddings, J.M., Herbes, S.E., Hook, L.A., Newbold, J.D., Roy, W.K., Southworth, G.R., Stewart, A.J.: Effects of chronic exposure to coal-derived oil on freshwater ecosystems: I. Microcosms. Environ. Toxicol. Chem. *3,* 447 (1984)
32. Geiger, J.G., Buikema, A.L., Jr.: Oxygen consumption and filtering rate of *Daphnia pulex* after exposure to water-soluble fractions of naphthalene, phenanthrene, No 2 fuel oil and coal-tar creosote. Bull. Environ. Contam. Toxicol. *27,* 783 (1981)
33. Geiger, J.G., Buikema, A.L., Jr.: Hydrocarbons depress growth and reproduction of *Daphnia pulex* (Cladocera). Can. J. Fish. Aquat. Sci. *30,* 830 (1982)
34. Giddings, J.M., Washington, J.N.: Coal-liquefaction products, shale oil and petroleum. Acute toxicity to freshwater algae. Environ. Sci. Technol. *15,* 106 (1981)
35. Giddings, J.M., Herbes, S.E., Gehrs, C.W.: Coal liquefaction products. Environ. Sci. Technol. *19,* 14 (1985)
36. Giddings, J.M., Franco, P.J., Cushman, R.M., Hook, L.A., Southworth, G.R., Stewart, A.J.: Effects of chronic exposure to coal-derived oil on freshwater ecosystems: II. Experimental ponds. Environ. Toxicol. Chem. *3,* 465 (1984)
37. Gile, J.D., Collins, J.C., Gillett, J.W.: Fate and impact of wood preservatives in a terrestrial microcosm. J. Agric. Food Chem. *30,* 295 (1982)
38. Guerin, M.R., Ho, C.-H., Rao, T.K., Clark, B.R., Epler, J.L.: Polycyclic aromatic primary amines as determinant mutagens in petroleum substitutes. Environ. Res. *23,* 42 (1980)
39. Haugen, D.A., Peak, M.J., Suhrbier, K.M., Stamoudis, V.S.: Isolation of mutagenic aromatic amines from a coal conversion oil by cation exchange chromatography. Anal. Chem. *54,* 32 (1982)

40. Heitmuller, P.T., Hollister, T.A., Parrish, P.R.: Acute toxicity of 54 industrial chemicals to Sheepshead minnows. (*Cyprinodon variegatus*). Bull. Environ. Contam. Toxicol. *27*, 596 (1981)
41. Hertz, H.S., Brown, J.M., Chester, S.N., Guenther, F.R., Hilpert, L.R., May, W.E., Parris, R.M., Wise, S.A.: Determination of individual organic compounds in shale oil. Anal. Chem. *52*, 1650 (1980)
42. Ho, C.-H., Clark, B.R., Guerin, M.R., Barkenbus, B.D., Rao, T.K., Epler, J.L.: Analytical and biological analyses of test materials from the synthetic fuel technologies IV. Studies of chemical structure-mutagenic activity relationships of aromatic nitrogen compounds relevant for synfuels. Mutat. Res. *85*, 335 (1981)
43. Holcombe, G.W., Phipps, G.L., Knuth, M.L., Felhaber, T.: The acute toxicity of selected substituted phenols, benzenes, and benzoic acid esters to Fathead minnows (*Pimephales promelas*). Environ. Pollut. Ser. A **35**, 367 (1984)
44. Hult, M.F., Schoenberg, M.E.: Preliminary evaluation of ground-water contamination of coal-tar derivatives, St. Louis Park area, Minnesota. U.S. Geological Survey Water-supply Paper *2211*, 53 (1984)
45. IARC: IARC Monographs on the Evaluation of Carcinogenic Risk of Chemicals to Humans, Vol 35, Polynuclear Aromatic Compounds, Part 4, Bitumens, Coal-Tars and Derived Products, Shale-oils and Soots, Lyon 1985
46. Ingram, L.L., Jr., McGinnis, G.D., Gjovik, L.R., Roberson, G.: Migration of creosote and its components from treated piling sections in a marine environment. Proc. Ann. Meet. Amer. Wood-Press. Assoc. *78*, 120 (1982)
47. Jentoft, A.: Kreosotforurensningar i Tröndelag. Prövetaking og karakterisering av polysykliske aromatiske forbindelser (Creosote pollution in Tröndelag county. Sampling and characterization of polyaromatic compounds), Universitetet i Trondheim, Kjemisk Institutt NDMT 1982 (in Norwegian)
48. Juhnke, I., Lüdemann, D.: Ergebnisse der Untersuchung von 200 chemischen Verbindungen auf akute Fischtoxicität mit dem Goldorfentest. Z. Wasser-Abwasser Forsch. *11*, 161 (1978)
49. King, J.M., Coley, K.S.: Toxicity of coal-distillates to *Spirodela polyrrhiza*. Bull. Environ. Contam. Toxicol. *33*, 220 (1984)
50. Kong, R.C., Lee, M.L., Tominaga, Y., Pratap, R., Iwao, M., Castle, R.N., Wise, S.A.: Capillary column gas chromatographic resolution of isomeric polycyclic aromatic sulfur heterocycles in a coal liquid. J. Chromatogr. Sci. *20*, 502 (1982)
51. Kuhn, R., Canton, J.H.: Vergleichende hydrobiologisch-toxikologische Befunde an Mikroorganismen und Makroorganismen biologischer Spektren, in: Reinhaltung des Wassers (eds.) Aurand, K., Spaander, J., p. 58, Berlin, Erich Schmidt Verlag, 1979
52. Later, D.W., Lee, M.L., Bartle, K.D., Kong, R.C., Vassilaros, D.L.: Chemical class separation and characterization of organic compounds in synthetic fuels. Anal. Chem. *53*, 1612 (1981)
53. Later, D.W., Lee, M.L., Wilson, B.W.: Selective detection of amino polycyclic aromatic compounds in solvent refined coal. Anal. Chem. *54*, 117 (1982)
54. Lebsack, M.E., Andersson, A.D., DeGraeve, G.M., Bergman, H.L.: Comparison of bacterial luminescence and fish bioassay results for fossil-fuel process waters and phenolic constituents, in: Aquatic Toxicology and Hazard Assessment: Fourth Conference (eds.) Branson, D.R., Dickson, K.L., p. 348, American Society for Testing and Materials, *ASTM STP 737*, 1981
55. Lee, M.L., Willey, C., Castle, R.N., White, C.M.: Separation and identification of sulfur heterocycles in coal-derived products, in: Polynuclear Aromatic Hydrocarbons: Chemistry and Biological Effects (eds.) Björseth, A., Dennis, A.J., p. 58, Columbus, OH, Battelle Press, 1980
56. Lorenz, L.F., Gjovik, L.R.: Analyzing creosote by gas chromatography. Relationship to creosote specifications. Proc. Amer. Wood-Preservers Assoc. *68*, 32 (1972)
57. Lu, P.-Y., Metcalf, R.L., Carlson, E.M.: Environmental fate of five radiolabeled coal conversion by-products evaluated in a laboratory model ecosystem. Environ. Health Perspect. *24*, 201 (1978)

58. Lucke, R.B., Later, D.W., Wright, C.W., Chess, E.K., Weimer, W.C.: Integrated, multiple-stage chromatographic method for the separation and identification of polycyclic aromatic hydrocarbons in complex coal liquids. Anal. Chem. *57,* 633 (1985)
59. Mackay, D.: Correlation of bioaccumulation factors. Environ. Sci. Technol. *16,* 274 (1982)
60. McLeese, D.W., Metcalfe, C.D.: Toxicity of creosote to larval and adult lobsters and *Crangon* and its accumulation in lobster hepatopancreas. Bull. Environ. Contam. Toxicol. *22,* 796 (1979)
61. Melancon, M.J., Jr., Lech, J.J.: Uptake, biotransformation, disposition, and elimination of 2-methylnaphthalene in several fish species, in: Aquatic Toxicology, (eds.) Marking, L.L., Kimerle, R.A. p. 15, American Society for Testing and Materials, *ASTM STP 667,* 1979
62. Merli, F., Novotny, M., Lee, M.L.: Fractionation and gas chromatographic analysis of azaarenes in complex mixtures. J. Chromatogr. *199,* 371 (1980)
63. Merrill, E.G., Wade, T.L.: Carbonized coal products as a source of aromatic hydrocarbons to sediments from a highly industrialized estuary. Environ. Sci. Technol. *19,* 597 (1985)
64. National Academy of Sciences: The International Mussel Watch. The National Research Council, NAC, Washington, DC, 1980
65. National Research Council Canada: Polycyclic aromatic hydrocarbons in the aquatic environment: Formation, sources, fate, and effects on aquatic biota. National Research Council Canada Publication No NRCC 18981, 1983
66. Neely, W.G., Branson, D.R., Blau, G.E.: The use of partition coefficient to measure the bioconcentration potential of organic chemicals in fish. Environ. Sci. Technol. *8,* 1113 (1974)
67. Nestler, F.H.M.: The characterization of wood-preserving creosote by physical and chemical methods of analysis. US Department of Agriculture, USDA Forest Service Research Paper FPL 195, 1974
68. Nestler, F.H.M.: Characterization of wood-preserving coal-tar creosote by gas liquid chromatography. Anal. Chem. *46,* 46 (1974)
69. Nishioka, M., Lee, M.L., Kudo, H., Muchiri, D.R., Baldwin, L.J., Pakray, S., Stuart, J.G., Castle, R.N.: Determination of hydroxylated thiophenic compounds in a coal liquid. Anal. Chem. *57,* 1327 (1985)
70. Nishioka, M., Campbell, R.M., West, W.R., Smith, P.A., Booth, G.M., Lee, M.L., Kudo, H., Castle, R. N.: Determination of aminodibenzothiophenes in a coal liquid. Anal. Chem. *57,* 1868 (1985)
71. Novotny, M., Strand, J.W., Smith, S.L., Wiesler, D., Schwende, F.J.: Compositional studies of coal tar by capillary gas chromatography/mass spectrometry. Fuel *60,* 213 (1981)
72. Ogata, M., Miyake, Y, Fujisawa, K., Kira, S., Yoshida, Y.: Accumulation and dissipation of organosulfur compounds in short-necked clam and eel. Bull. Environ. Contam. Toxicol. *25,* 130 (1980)
73. Osuský, A.: Zur qualitativ-gaschromatographischen Untersuchung von Imprägnieröl STIA. Holz als Roh- und Werkstoff *33,* 308 (1975)
74. Parees, D.M., Kamzelski, A.Z.: Characterization of coal-derived liquids using fused silica capillary column GC-MS. J. Chromatogr. Sci. *20,* 441 (1982)
75. Parkhurst, B.R., Bradshaw, A.S., Forte, J.L., Wright, G.P.: The chronic toxicity to *Daphnia magna* of acridine, a representative azaarene present in synthetic fossil fuel products and wastewaters. Environ. Pollut. Ser. A *24,* 21 (1981)
76. Pereira, W.E., Rostadt, C.E., Garbarino, J.R., Hult, M.F.: Groundwater contamination by organic bases derived from coal-tar wastes. Environ. Toxicol. Chem. *2,* 283 (1983)
77. Phipps, G.L., Holcombe, G.W., Fiandt, J.T.: Acute toxicity of phenol and substituted phenols to the Fathead minnow. Bull. Environ. Contam. Toxicol. *26,* 585 (1981)
78. Qureshi, A.A., Flood, K.W., Thompson, S.R., Janhurst, S.M., Inniss, C.S., Rokosh, D.A.: Comparison of a luminescent bacterial test with other bioassays for determining toxicity of pure compounds and complex effluents, in: Aquatic Toxicology and Hazard Assessment: Fifth Conference (eds.) Pearson, J.G., Foster, R.B., Bishop, W.E., p. 179, American Society for Testing and Materials, *ASTM STP 766,* 1982
79. Registry of Toxic Effects of Chemical Substances, 1979 Edition, Volume 1 (eds.) Lewis Sr, R.J., Tatken, R.L., p. 441, US Department of Health and Human Services, 1980

80. Rhodes, E.O., Woolridge, S.E.: The chemical nature of coal tar, in: Chemistry of Coal Utilization, p. 1357, John Wiley and Sonc, Inc., New York, 1945
81. Roberts, S.J., Hedrick, H.G.: Degradation of creosote phenolic compounds by selected bacterial isolates. Dev. Ind. Microbiol. *20,* 471 (1979)
82. Rostad, C.E., Pereira, W.E., Hult, M.F.: Partitioning studies of coal-tar constituents in a two-phase contaminated ground-water system. Chemosphere *14,* 1023 (1985)
83. Sax, N.J.: Dangerous Properties of Industrial Materials 5th Edition, p. 520, Van Nostrand Reinhold Co, New York, 1979
84. Schabron, J.F., Hurtubise, R.J., Silver, H.F.: Chromatographic and spectrometric methods for the separation, characterization, and identification of alkylphenols in coal-derived solvents. Anal. Chem. *51,* 1426 (1979)
85. Schiller, J.E., Mathiason, D.R.: Separation method for coal-derived solids and heavy liquids. Anal. Chem. *49,* 1225 (1977)
86. Schmitter, J.-M., Ignatiadis, I., Arpino, P., Guichon, G.: Selective isolation of nitrogen bases from petroleum. Anal. Chem. *55,* 1685 (1983)
87. Schultz, R.V., Jorgenson, J.W., Maskarinec, M.P., Novotny, M., Todd, L.J.: Characterization of polynuclear aromatic and aliphatic hydrocarbon fractions of solvent-refined coal by glass capillary gas chromatography/mass spectrometry. Fuel *58,* 783 (1979)
88. Shinohara, R., Kido, A., Okamoto, Y., Takeshita, R.: Determination of trace azaarenes in water by gas chromatography-mass spectrometry. J. Chromatogr. *256,* 81 (1983)
89. Sirota, G.R., Uthe, J.F.: Polynuclear aromatic hydrocarbon contamination in marine shellfish, in: Polynuclear Aromatic Hydrocarbons: Chemical Analysis and Biological Fate (eds.) Cooke, M., Dennis, J., p. 329, Battelle Press, Columbus, OH, 1981
90. Sloof, W., Canton, J.H., Hermens, J.L.M.: Comparison of the susceptibility of 22 freshwater species to 15 chemical compounds. I. (Sub)acute tests. Aquatic Toxicol. *4,* 113 (1983)
91. Southworth, G.R., Keffer, C.C., Beauchamp, J.J.: Potential and predicted bioconcentration. A comparison of observed and predicted bioconcentration of azaarenes in the Fathead minnow (*Pimephales promelas*). Environ. Sci. Technol. *14,* 1529 (1980)
92. Southworth, G.R., Keffer, J.L.: Mobilization of azaarenes from wastewater treatment plant biosludge. Bull. Environ. Contam. Toxicol. *32,* 445 (1984)
93. Southworth, G.R., Herbes, S.E., Franco, P.J., Giddings, J.M.: Persistence of phenols in aquatic microcosms recieving chronic inputs of coal-derived oil. Water Air Soil Pollut. *24,* 283 (1985)
94. Sundröm, G. (ed.): Kreosotmedlens miljöeffekter. Litteraturgenomgång samt kemiska och biologiska test av svenska kreosotprodukter (Environmental effects of creosote products. Literature survey and chemical and biological tets of Swedish products). Mimeographed report in Swedish to the Products Control Division, National Environmental Protection Board, Contract No. 2-83/84 (1985)
95. Sundström, G., Nylund, K.: Förekomst av kreosotprodukter i jordprover från en nedlagd träimpregneringsanläggning (Creosote residues in the soil of an abandoned wood treatment facility). Mimeographed report in Swedish, Rapport *85-03* (1985)
96. Tarkpea, M., Hansson, M., Samuelsson, B.: Comparison of the Microtox test with the 96h-LC50 test for the harpacticoid *Nitocra spinipes*. Ecotoxicol. Environ. Safety (in press, 1985)
97. Tarkpea, M., Larsson, Å, Hansson, M.-B., Samuelsson, B.: Brackish Water Toxicology Section, Emission and Product Control Laboratory, National Environmental Protection Board, Studsvik, S-611 82 Nyköping, Sweden. Data generated for [94]
98. Ullstrand, S., Björklund, I.: Biology Section, Emission and Product Control Laboratory, National Environmental Protection Board, Box 1302, S-171 25 Solna, Sweden. Data generated for [94]
99. US Environmental Protection Agency: Creosote. Special Review Position Document 2/3, Washington DC, Aug. 1984
100. US Environmental Protection Agency: Wood Preservation Pesticides: Creosote, Pentachlorophenol, Inorganic Arsenicals. Position Document 4, Washington DC, July 1984
101. US Environmental Protection Agency: Removal of contaminated river bottom muds: Field demonstration, NTIS PB 84-128 022, 1984 (Hansen, C.A., Sanders, R.G.: Removal of harzardous material spills from bottoms of flowing river bodies, US EPA Cincinatti EPA-600/2-81-137, 1981)

102. Vassilaros, D.L., Eastmond, D.A., West, W.R., Booth, G.M., Lee, M.L.: Determination and bioconcentration of polycyclic aromatic sulfur in heterocycles in aquatic biota, in: Polynuclear Aromatic Hydrocarbons: Physical and Biological Chemistry (eds.) Cooke, M., Dennis, J., Fisher, G.L., p. 845, Battelle Press. Columbus, OH, 1981
103. Veith, G.D., DeFoe, D.L., Bergstedt, B.V.: Measuring and estimating the bioconcentration factor of chemicals in fish. J. Fish. Res. Board Can. *36,* 1040 (1979)
104. Webb, D.A.: Some environmental aspects of creosote. Proc. Ann. Meet. Amer. Wood-Preservers Assoc. *71,* 176 (1975)
105. White, C.M., Li, N.C.: Determination of phenols in a coal liquefaction product by gas chromatography and combined gas chromatography/mass spectrometry. Anal. Chem. *54,* 1570 (1982)
106. Willey, C., Iwao, M., Castle, R.N., Lee, M.L.: Determination of sulphur heterocycles in coal liquids and shale oils. Anal. Chem. *53,* 400 (1981)
107. Wilson, B.W., Peterson, M.R., Pelroy, R.A., Cresto, J.T.: *In-vitro* assay for mutagenic activity and gas chromatographic-mass spectral analysis of coal liquefaction material and the products resulting from its hydrogenation. Fuel *60,* 289 (1981)
108. Wozniak, T.J., Hites, R.A.: Analysis of hydroaromatics in coal-derived synthetic fuels: Applications to pilot-plant samples. Anal. Chem. *57,* 1320 (1985)
109. Zitko, V.: Aromatic hydrocarbons in aquatic fauna. Bull. Environ. Contam. Toxicol. *14,* 621 (1975)

Elemental Phosphorus

R. F. Addison

Marine Ecology Laboratory
Bedford Institute of Oceanography
Dartmouth, N. S., B2Y 4A2, Canada

Summary

P_{el} is manufactured throughout the world on a relatively large scale, for eventual oxidation mainly to phosphoric acid and phosphates, although a small amount is used as P_{el} for poisons and munitions. The nature of its chemical processing is such that releases of P_{el} to the environment are likely to occur at the site of manufacture, use, or (occasionally) during shipment.

P_{el} is acutely toxic to all organisms examined so far. The lethal dose to man is around 1 $mg \cdot kg^{-1}$ body weight.

Introduction

Elemental phosphorus (P_{el}) occurs as several allotropes, of which the most important commercially is the α-(white) form. This is the "white" or "yellow" phosphorus which is manufactured by reduction of phosphate ore and used for the preparation of various P(III) or P(V) compounds, mostly phosphoric acid and the phosphates. P_{el} has caused at least one major water pollution event, and the abundance of "grey literature" reports about its environmental behaviour suggests that there is some concern that it may be involved in other instances of pollution. Although P_{el} is fairly reactive, it is stable enough in dilute suspension to cause problems of acute and chronic toxicity to aquatic organisms. Furthermore, it is accumulated by aquatic biota and may survive in their adipose tissue for several months. Although P_{el} is readily oxidised in air, it can be protected by solution in fat or other matrices and be kept stable for several months or more as is illustrated by its use as a rodent poison, or (formerly) in matches.

This review summarises information about the manufacture, use, analysis, environmental behaviour, and toxicology of P_{el}. Much of this information has emerged from studies of the Long Harbour, Newfoundland, pollution event of 1969–1970; detailed descriptions of this have been published previously [1, 2].

Manufacture and Use

P_{el} is manufactured by coke (C) reduction of phosphate ore, usually nowadays in an electric furnace with silica as a flux. The product P_{el} is distilled off and condensed under water. Silicates, ferrophosphorus and fluorides (if the starting ore is fluorapatite) are formed as by products. In some plants, the condensed P_{el} is immediately re-oxidised to P_2O_5 and hydrated to form phosphoric acid; in other cases, P_{el} may be stored or shipped for subsequent oxidation or other uses.

P_{el} is a "heavy" chemical. World capacity for its production in 1965 was just under 900,000 metric tonnes, but by 1980 this had grown to about 1,350,000 metric tonnes [3, 4] (Table 1). Between 1965 and 1980, US production capacity for P_{el} dropped from about 63% to about 40% of world capacity; however, U.S.S.R. production capacity increased from 11% to 33% of world capacity over that period. This increase in P_{el} manufacture in the U.S.S.R. may explain the growing Russian literature on P_{el}. US demand for P_{el} is expected to exceed capacity by the mid-1990's [4] although this projection depends very much on demand for detergent phosphates [5].

To put these production figures in perspective, world production of the polychlorinated biphenyls was around 50,000 tonnes/yr in 1970 [6]; US production of the insecticide DDT was around 64,000 tonnes in 1965 [7].

Most of the P_{el} manufactured in the US is re-oxidized to phosphoric acid and phosphates. This apparently roundabout process (reduction of phosphate ore to P_{el} followed by its re-oxidation to phosphate) presumably offers advantages in ef-

Table 1. Capacity, production and use of elemental phosphorus in 1965 and 1980 (compiled from Refs. [3] and [4])

Variable	1965	1980
World production capacity (tonnes/yr)	8.9×10^5	13.5×10^5
U.S. Production (tonnes/yr)	5.2×10^5	4.2×10^5 (approx.)
% U.S. Production re-oxidized to phosphoric acid and phosphates	75	70 (approx.)
% P_{el} converted to P(III) and P(V) compounds other than phosphates	8	16 (approx.)
% P_{el} production used as P_{el}	2	3 (approx.)

ficiency of shipping (it is cheaper to ship P than, say, H_3PO_4) and of product purity. Phosphates prepared by this route represented about a quarter of all the phosphates manufactured in the US in 1965 [3]. However, during the 1970's the demand for phosphates as detergent builders declined due to their contributing to the eutrophication of inland waters, and so by 1980 a larger fraction of the P_{el} produced was used for the synthesis of products other than detergent phosphates. A relatively small fraction of P_{el} is used unchanged, as a component of rodent or roach poisons or of munitions; this might represent about 2% of all P_{el} manufactured in the US – i.e., approx. 10,000 tonnes.

Analysis of P_{el}

This topic is reviewed elsewhere [8] and in this section, only a summary and update are given.

Approaches to the analysis of P_{el} fall into three groups:

(a) isolation of P_{el} by distillation, organic extraction or complexation as a Group IB phosphide, followed by its oxidation to phosphate which is determined spectrophotometrically;

(b) isolation of P_{el} by gas-liquid chromatography (GLC) followed by its determination in a P-specific detector;

c) several specialised methods, including neutron activation of P_{el} isolated by any procedure followed by radiometric determination, or spectrophotometric methods designed for reaction mixtures.

Various methods from the first of these groups were used almost exclusively until the early 1970's; the commonest sample types were air (for occupational hygiene studies) and biological tissues (in forensic toxicology). However, the GLC approach which we introduced in 1970 [9] has tended to replace other methods, as it is more sensitive, specific and convenient. Table 2 summarises comparisons of minimum detectable limits by various analytical methods as reported in the original literature, or inferred from it [8]. It is clear that the GLC approach is usually about one order of magnitude more sensitive than the classical methods.

Table 2. Comparison of the sensitivities of various approaches to P_{el} analysis

Sample type	Analytical approach	Detection limit	References
Air	Collection in xylene, oxid'n and spectrophotometry	$6 \times 10^{-5}\ g \cdot m^{-3}$	[11]
	Collection in xylene, GLC	$4 \times 10^{-7}\ g \cdot m^{-3}$	[12]
Water	Organic extraction, oxid'n and spectrophotometry	10^{-6}–$10^{-7}\ g \cdot ml^{-1}$	[8]
	Organic extraction, GLC	$10^{-10}\ g \cdot ml^{-1}$	[9]
Biological samples	Distillation, complexation, oxid'n, spectrophotometry	$5 \times 10^{-8}\ g \cdot g^{-1}$	[8, 43]
	Distillation, complexation, oxid'n, neutron activation	$5 \times 10^{-10}\ g \cdot g^{-1}$	[10]
	Organic extraction, GLC	$10^{-10}\ g \cdot g^{-1}$	[8]

In addition, the GLC approach allows distinction between P_{el} and related industrially important materials such as phosphine, PH_3 (a major component of alkaline process waters from P_{el} manufacture and which is not separated from P_{el} by classical analytical methods [9]) and the GLC approach usually involves less sample manipulation than some more elaborate methods such as the neutron activation method of Krishnan and Gupta [10].

P_{el} Analysis in Air

This is usually based on the collection of airborne P_{el} in organic solvents such as the xylene impingers used by Rushing [11]. Bohl and Kaelble [12] have revised Rushing's original method by using the same impinger/collection system but carrying out the P_{el} estimation by GLC. A more recent paper [13] describes a personal sampler for airborne P_{el}: this is based on adsorption of P_{el} on a Tenax-GC column which is eluted with xylene and analysed by GLC. Either of these analytical approaches [12, 13] allows the determination of P_{el} concentrations in air well below the O.S.H.A. standard of $10^{-4} g \cdot m^{-3}$; both have been proposed as standard monitoring methods [14].

P_{el} Analysis in Water

The recent studies have been based on organic extraction followed by GLC. Maddock and Taylor [15] used the sample: extractant ratios we recommended [9] but they substituted a modified flame ionisation detector (F.I.D.) for the final determination of P_{el}; this detector has approximately the same sensitivity as has the flame photometric detector (F.P.D.). Bentley et al. [16] used a much larger sample: extractant ratio (100–200 : 2, v : v) which allowed a lower detection limit (Table 1, loc.cit.) of about $5 \times 10^{-11} g \cdot ml^{-1}$.

P_{el} Analysis in Sediments

Apart from our previous description of P_{el} analysis in sediments [9] the only report on this subject has been that of Pearson et al. [17] who used identical methods.

P_{el} Analysis in Biota

Other than our own description of P_{el} analyses in biota [9], the only reports have been those of Maddock and Taylor [15] and of Bentley et al. [16]; both these groups used methods identical to ours.

Physical and Chemical Properties of P_{el}

P_{el} is a white waxy solid, m.p. 44.1 °C. It consists of P_4 tetrahedra. The commercial product is usually slightly discoloured by traces of red phosphorus which itself occurs in at least six different crystalline forms, and to which white phosphorus slowly interconverts. P_{el} boils at 280.5 °C. It is virtually insoluble in water, its solubility being reported as $3 \times 10^{-6} g \cdot ml^{-1}$ [18]. This concentration has been produced experimentally (e.g., [16]) for toxicity studies. However, the term "solubility" as applied to materials such as P_{el} is usually an operational definition; it refers to passage of the preparation through a filter of some selected pore size such as 0.45 µm. Whether aqueous P_{el} in preparations which pass through such filters is in pure solution or in colloidal form is not yet clear, although the difference between "dissolved" and "collodial" P_{el} has been invoked to explain variable aquatic toxicity data [15, 19].

Although P_{el} is only slightly soluble in water, it is readily soluble in organic solvents [4] and (by inference) in lipids. If its aqueous solubility is taken as $3 \times 10^{-6} g \cdot ml^{-1}$, its octanol-water partition coefficient ($K_{O/W}$) and "bioconcentration factor" (BCF) can be predicted from equations derived by Mackay [20]. $Log_{10} K_{O/W}$ is then calculated to be 4.08, and the BCF to be approximately 600. This latter figure seems to be in the range of those produced in experimental bioaccumulation studies, as we shall see below.

The vapour pressure-temperature relationship of P_{el} is shown by van Wazer [4]; at normal environmental temperatures, P_{el} should have a V.P. below 0.1 torr. This is appreciable, and would explain the presence of P_{el} largely in the vapour form, as opposed to the particulate form, in P_{el} manufacturing plants [12].

The accurate determination of many of the physical and chemical properties of P_{el} is confounded by the problem of its chemical reactivity. Although P_{el} is sufficiently stable to survive for weeks or longer in organic solvents [9, 15, 16], P_{el} in water decays rapidly. This is apparently due to reaction with oxygen dissolved in water [15], but reaction rates vary with pH and temperature. At least in its early stages, the reaction is apparently first order with respect to P_{el} with half lives ranging from less than an hour to a few hours [21, 22]. The products of oxidation are complex; P(V) in the form of phosphate (PO_4^{3-}) seems to be the final product [22] but Lai and Rosenblatt [23] using $^{32}P_{el}$ have identified various P(III) products by TLC; these products included hypophosphite ($H_2PO_3^{2-}$) and phosphite (HPO_3^{2-}), both of which were converted to phosphate.

Toxicology of P_{el}

P_{el} is toxic to a wide variety of organisms, but the mechanism of its action is not understood in any detail. Most studies of its toxicology have been undertaken in response to:

(a) acute poisoning (often suicidal) or burns in humans;
(b) occupational exposure in P_{el} manufacturing or processing industries;
(c) experimental studies on laboratory organisms, often in response to pollution incidents.

Table 3 summarises its acute toxicity to various organisms.

Effects on Humans and Other Mammals

The literature on this subject up until the mid 1970's has been reviewed by Fletcher [24]; this section summarises and updates that review.

Table 3. Summary of selected acute toxicity values of P_{el} to various biota (compiled from Ref. [24] unless otherwise stated)

Organism	Lethal Dose ($mg \cdot kg^{-1}$)	LC50 ($\mu g \cdot l^{-1}$)		References
		96 h	48 h	
Mammals:				
Man	≈ 1			
Mouse	16			
Rat	7			
Freshwater fish				
Salmon (*Salmo salar*)		16		
Trout (*S. gairdneri*)		22		
Bluegill (*Lepomis macrochirus*)		6–44		[19, 24]
Marine fish				
Cod (*Gadus morhua*)		2.8, 6.5		[15, 24]
Herring (*Clupea harengus*)		3.7		[21]
Freshwater invertebrates				
Shrimp (*Paleomonetes* sp.)		32		
Water flea (*Daphnia magna*)			30	
Marine invertebrates				
Lobster (*Homarus americanus*)		≈200		[21]

The lethal dose of P_{el} to man is around 1 $mg \cdot kg^{-1}$ body weight. Death follows from 12–72 h, depending on dose, and may be due to circulatory, renal or hepatic failure or to a combination of these. P_{el} appears to act as a cellular poison. There is no specific treatment for P_{el} poisoning: the strategy is usually to remove as much P_{el} as possible from the victim's digestive tract as quickly as possible, and this may be followed by treatment with dilute $KMnO_4$ to oxidise any remaining P_{el}. A more recent approach is to attempt to adsorb any P_{el} remaining in the tract to activated charcoal [25].

P_{el} burns, caused either by munitions or manufacturing accidents, are painful and slow to heal. They may lead to renal malfunction, as has been shown in experimental studies in rabbits. Emergency treatment involves immersing the injury in water or in dilute Cu^{2+} solution.

Chronic P_{el} poisoning had been reported since the mid-nineteenth century in workers occupationally exposed to P_{el} fumes during the manufacture of safety matches. The classical symptoms were "phossy jaw" (necrosis of the lower jaw) and this could be accompanied by increased fragility of the long bones. Although these symptoms of P_{el} poisoning have seldom been seen since the mid-1930's, a report [26] in the Russian literature suggests that more subtle symptoms of occupational exposure may be found: these include incipient liver and bone disorders.

In former times, P_{el} was recommended [27] "with doubtful success" in the treatment of rickets, osteomalacia, as a nervine tonic and as an aphrodisiac (maximum daily dose 3 mg); this is an indication of the robustness of our grandfathers, if of nothing else.

Experimental studies of the toxicity of P_{el} to laboratory animals have shed relatively little light on the mechanism of P_{el} toxicity. Apart from changes to bone tissue, the effects of P_{el} include fatty degeneration of the liver (with resulting changes in circulatory lipid and carbohydrate levels [28–31], effects on enzyme systems [30–35], and effects on the brain and nervous system, kidney necrosis, depressed liver protein synthesis, increase in the smooth endoplasmic reticulum, and electrocardiographic changes (summarized in [24]). P_{el} therefore seems to act as a non-specific cellular poison, but its mode of action remains unknown.

Effects on Aquatic Organisms

Most work on the aquatic toxicology of P_{el} has been done in response to pollution incidents such as that at Long Harbour, Newfoundland, or the chronic discharges from a munitions plant in Arkansas. In addition, the existence of several reports in the Russian literature suggests that P_{el} pollution is at least anticipated there.

96 hr LC 50's of P_{el} to aquatic organisms are usually below 100 $\mu g \cdot l^{-1}$, and often are below 10 $\mu g \cdot l^{-1}$. If an incipient lethal level exists, it must be well below 1 $\mu g \cdot l^{-1}$ for many species. Even brief laboratory exposures to P_{el} may result in later death and this is consistent with field observations of migrating herring showing signs of P_{el} poisoning many miles from the zone of potential exposure.

As in mammals, P_{el} appears to act on various tissues and organs in aquatic organisms; these include gills and liver and nervous system and bone and scale effects. Hemolysis occurs readily in herring and in some other marine species – it was this feature that gave rise to the name "the red herring problem" – but hemolysis seems not to be a general response.

The dynamics of P_{el} in fish have been studied fairly extensively. P_{el} is lipid-soluble and accumulates in lipid-rich tissues. A projected bioconcentration factor from experimental studies in salmon and cod is in the range 10^3–10^4, which – given the errors involved in these measurements and predictions – is not too different from the calculated BCF shown above. Steady state concentrations in various tissues seem to be attained fairly rapidly, in a matter of hours, and depuration rates are also rapid; first order half lives may also be in the range of hours. Aquatic invertebrates are also poisoned by P_{el}. Both marine and fresh-waters forms are sensitive, with LC 50's in the same range as those for fish [16, 21, 24, 36].

As is the case in mammals, the mechanism of P_{el} toxicity to aquatic biota is not understood. However, enough data exist to state that the mechanism is almost certainly not directly related to the "oxygen demand" of P_{el} during oxidation. Thus, a realistic steady state blood concentration in cod exposed to 8 $\mu g \cdot l^{-1}$ P_{el} (approximately the LC 50) is about 1.2 $\mu g \cdot ml^{-1}$. Assuming complete oxidation to P(V) as phosphate, this would require about 1.6 $\mu g \cdot ml^{-1}$ O_2, which represents only about 1% of the O_2 capacity of most fish blood [37].

P_{el} as a Pollutant

The western literature contains references to several reports which suggest that P_{el} has either caused, or has been expected to cause, pollution problems.

The event best documented in the western literature was the Long Harbour, Newfoundland, spill. In that event, P_{el} in waste waters was discharged over about a four month period from a plant with an annual production capacity of about 40,000 tonnes. It is still not clear how much P_{el} was actually discharged (a rate of about 1,500 lb/day, equivalent to a total of 90 tonnes, is quoted [38]) but analysis of sediments (the expected "sink" for P_{el}) confirmed the presence of appreciable amounts of P_{el} [39, 40]. These discharges into Placentia Bay coincided with the migration of herring, and resulted in mass herring mortalities, and also kills of several other species. After various studies by government and other agencies, the plant operators dredged up the contaminated sediments and installed various effluent containment and treatment systems which appear to have effectively eliminated discharges.

Another occurrence of P_{el} pollution is described in a series of reports describing the use and discharge of P_{el} at the Pine Bluff, Arkansas, munitions plant. Here, P_{el} (used in smoke bombs and incendiary devices) has been discharged to a holding pond; the impact of P_{el} on aquatic biota and its bioaccumulation has been described [17]. Although sediment P_{el} levels were lower in general than in Long Harbour, Nfld., they were high enough to cause major changes in species distribution and diversity.

In addition to these fairly detailed descriptions of pollution by P_{el}, there are (as noted above) in the Russian literature various reports describing the effects of P_{el} on fish and other aquatic biota, and even setting limits for P_{el} concentrations in potable water [41]. Taken together, these reports suggest that discharges of P_{el} either have occurred, or are expected; given the growth of the P_{el} manufacturing industry in Russia, the most probable source of these discharges would be production plants.

The shipping of P_{el} has led to at least one potential spill, when a tractor-trailer carrying P_{el} caught fire. The report of this incident [42] describes containment methods, and eventual disposal of damaged P_{el} drums by detonation.

Conclusions

P_{el} is manufactured throughout the world on a relatively large scale, for eventual oxidation mainly to phosphoric acid and phosphates, although a small amount is used as P_{el} for poisons and munitions. The nature of its chemical processing is such that releases of P_{el} to the environment are likely to occur at the site of manufacture, use, or (occasionally) during shipment.

P_{el} is acutely toxic to all organisms examined so far. The lethal dose to man is around 1 $mg \cdot kg^{-1}$ body weight. The 96 hr LC 50 to aquatic biota may be as low as 10 $\mu g \cdot l^{-1}$ for some species; no incipient lethal level has been described. The mechanism of toxicity has not been established; however, P_{el} appears to act as a non-specific cellular poison, affecting bone structure, liver, kidney, heart and brain function, and, in fish, gills. Although the classical symptoms of P_{el} poisoning (phossy jaw, etc.) are no longer observed, there is some evidence that occupational exposure may lead to liver and kidney malfunction. In this context, it is curious that no direct measurements of P_{el} accumulation have been attempted in occupationally exposed workers: given the ease with which P_{el} accumulates in experimentally exposed aquatic biota, it would be surprising if P_{el} was not present in occupationally exposed workers.

Once released to the environment, P_{el} oxidises, eventually to phosphate. The rate and mechanism of oxidation probably depend on the environmental reservoir in which P_{el} is found, and on conditions: P_{el} oxidises fairly rapidly in water (though it is stable enough to maintain acutely toxic concentrations) but in anoxic sediments it is fairly long lived, and it is stable for several months in adipose tissues. Dredging has been used successfully to remove some contaminated sediment from the waters of Placentia Bay, the site of one major P_{el} spill.

References

1. Jangaard, P.M. (ed): Effects of elemental phosphorus on marine life. Atlantic Regional Office, Research and Development, Fisheries Research Board of Canada, Halifax, N.S., Canada, Circular No. 2 (1972)
2. Idler, D.R., Fletcher, G.L., Addison, R.F.: Effects of yellow phosphorus in the Canadian environment. National Research Council of Canada, Associate Committee on Scientific Criteria for Environmental Quality, Publication No. NRCC 17587, Ottawa, Canada 1981
3. Van Wazer, J.R.: Phosphorus and the phosphides. In: Kirk-Othmer Encyclopedia of Chemical Technology, 2nd Ed., Vol. 15, Interscience, N.Y. 1968
4. Van Wazer, J.R.: Phosphorus and the phosphides. In: Kirk-Othmer Encyclopedia of Chemical Technology, 3rd Ed. Vol. 17, Interscience, N.Y. 1982
5. Arthur D. Little, Inc: Environmental considerations of selected energy conserving manufacturing process options. Vol. 13. Phosphorus/phosphoric acid industry report. Available as NTIS PB 264 279 (1976)
6. Addison, R.F.: Env. Sci. Technol. *17*, 486A (1983)
7. Anon.: Cleaning our environment: the chemical basis for action. American Chemical Society, Washington D.C. 1969
8. Addison, R.F.: Methods of analysis of elemental phosphorus. In: Idler, D.R., Fletcher, G.L., Addison, R.F.: Effects of yellow phosphorus in the Canadian environment. National Research Council of Canada, Associate Committee on Scientific Criteria for Environmental Quality, Publication No. NRCC 17587, Ottawa, Canada 1981
9. Addison, R.F., Ackman, R.G.: J. Chromatog. *47*, 421 (1970)
10. Krishnan, S.S., Gupta, R.C.: Anal. Chem. *42*, 557 (1970)
11. Rushing, D.E.: Amer. Ind. Hyg. Assoc. J. *23*, 383 (1962)
12. Bohl, C.D., Kaelble, E.F.: Amer. Ind. Hyg. Assoc. J. *34*, 306 (1973)
13. Dillon, H.K., Barrett, W.J., Eller, P.M.: Amer. Ind. Hyg. Assoc. J., *39*, 608 (1978)
14. Anon.: NIOSH Manual of Analytical Methods, 2nd Ed., Vol. 1. US Dept. Health, Education and Welfare, Cincinnati, Ohio 1977
15. Maddock, B.G., Taylor, D.: Water Res. *10*, 289 (1976)
16. Bentley, R.E., Dean, J.W., Hollister, T.A., Leblanc, G.A., Sauter, S.: Laboratory evaluation of the toxicity of elemental phosphorus (P_4) to aquatic organisms. Available as NTIS AD A01 785 (1978)
17. Pearson, J.G., Bender, E.S., Taormina, D.H., Manuel, K.L., Robinson, P.F., Asaki, A.E.: Effects of elemental phosphorus on the biota of Yellow Lake, Pine Bluff Arsenal, Arkansas, March 1974 – January 1975. Available as NTIS AD A035 925 (1976)
18. Weast, R.C. (Ed).: Handbook of Chemistry and Physics, 4th Ed., CRC Press, Cleveland Ohio 1966
19. Isom. B.G.: J. Water Poll. Cont. Fed. *32*, 1312 (1960)
20. Mackay, D.: Env. Sci. Technol. *16*, 274 (1982)
21. Zitko, V.: Aiken, D.E., Tibbo, S.N., Besch, K.W.T., Anderson, J.M.: J. Fish Res. Bd. Canada *27*, 21 (1970)

22. Addison, R.F.: Analysis of elemental phosphorus and some of its compounds by gas-liquid chromatography. In Proc. International Symposium on the Identification and Measurement of Environmental Pollutants, Ottawa, Ont. Canada 1971
23. Lai, M.G., Rosenblatt, D.H.: Identification of transformation products of white phosphorus in water. Available as NTIS AD A041 068 (1977)
24. Fletcher, G.L.: Effects of yellow phosphorus on human health. In: Idler, D.R., Fletcher, G.L., Addison, R.F.: Effects of yellow phosphorus in the Canadian environment. National Research Council of Canada, Associate Committee for Scientific Criteria for Environmental Quality, Publication No. NRCC 17587, Ottawa 1981
25. Snodgrass, W.R., Doull, J.: Vet. Human Toxicol. *24* Suppl. 96 (1982)
26. Ozerova, V.V., Rusakova, G.S., Korenevskaya, S.P.: Chem. Abst. *76*, 17541q (1972)
27. Squire, P.W.: Pocket Companion to the British Pharmacopeia, J. and A. Churchill, London 1915
28. Jacqueson, A., Thevenin, M., Warnet, J.-M., Claude, J.-R., Truhaut, R.: Arch Toxicol. Suppl. *2*, 327 (1979)
29. Jacqueson, A., Thevenin, M., Warnet, J.-M., Bonnaud, G., Claude, J.R., Truhaut, R.: Chromatography Symp. Ser. *1*, 169 (1979)
30. Strelyukhina, N.A., Lukashev, A.A.: Chem. Abst. *93*, 231998t (1980)
31. Strelyukhina, N.A., Lukashev, A.A.: Chem. Abst. *93*, 231999u (1980)
32. Lukashev, A.A., Nekrasova, A.S.: Chem. Abst. *93*, 144085z (1980)
33. Strelyukhina, N.A., Lukashev, A.A., Nekrasova, A.S.: Chem. Abst. *92*, 192101c (1979)
34. Nekrasova, A.S., Lukashev, A.A., Strelyukhina, N.A.: Chem. Abst. *92*, 191202d (1979)
35. Ribotta, P.B., Parola, M., Barrera, G., Carasso, M.C., Bosia, B., Paradisi, L.: Soc. Ita. Biol. Sper. Boll. *58*, 1589 (1982)
36. Lakhnova, V.A.: Chem. Abst. *87*, 79279a (1977)
37. Prosser, C.L., Brown, F.A.: Comparative Animal Physiology, 2nd. Ed. W.B. Saunders Co., Philadelphia 1965
38. Idler, D.R.: Chemistry in Canada *21*, 16 (1969)
39. Ackman, R.G., Addison, R.F., Ke, P.J., Sipos, J.C.: Fish. Res. Bd. Canada Tech. Rept. no. 233, Halifax Laboratory, Halifax, N.S. Canada 1971
40. Addison, R.F., Zinck, M.E., Ackman, R.G., Chamut, P.S., Jamieson, A.: Fisheries. Res. Bd. Canada, Tech. Rept. no. 303, Fisheries Research Board of Canada, Ottawa 1972
41. Krasovskii, G.N., Shortanbaeva, M.A., Varshavskaya, S.P., Vasyukovich, L.Ya., Egorova, N.A.: Chem. Abst. *91*, 96328r (1979)
42. Lafornara, J.P., Massey, T.I.: Proc. National Conf. Control of Hazardous Material Spills, 58 (1980)
43. Curry, A.S., Rutter, E.R., Chin-Hua, L.: J. Pharm. Pharmacol. *10*, 635 (1958)

Molybdenum

Gordon A. Parker

Department of Chemistry, University of Toledo
Toledo, OH 43606, USA

Summary

Molybdenum is perhaps not as well known for its environmental effects as certain other metals; nevertheless, molybdenum plays an important role in the modern world. It is an alloying element in metallurgical alloys and an essential compound in various life processes. Its environmental impact has been widely studied by those with a specific interest in molybdenum and knowledge of these studies should be available to a wider audience. The presence of molybdenum upon the environment can, under certain circumstances, have a pronounced effect upon plant and animal life, including man. Earlier reviews of molybdenum in the environment are available [31, 83].

History

It was not until the mid eighteenth century that molybdenum, as MoS_2, was recognized as a distinct substance. Prior to that time it had been confused with graphite. Both have similar appearance and properties. The Swedish chemist,

Scheele, first prepared molybdenum oxide in 1778 but it remained until 1782 before Hjelm isolated metallic molybdenum by reduction of the oxide with carbon [169, 218].

Occurrence, Production, Uses

Occurrence

Molybdenum is present in most of the earth's crust to the extent of about 1.2 μg/g [66]. Its abundance is approximately the same as that of tungsten and about ten times less than for chromium. Chromium, molybdenum and tungsten are closely related in their chemical and physical properties. Widely scattered throughout the earth its principal ore is molybdenite, MoS_2, in which it is present to the extent of about 0.5%. Molybdenum deposits are located in the western United States and elsewhere about the world. Other ores of commercial importance are powellite, $CaMoO_4$, and wulfenite, $PbMoO_4$. Table 1 lists other mineral deposits containing molybdenum.

Trace amounts of molybdenum are present in many rock formations, especially igneous rocks. Selected examples of molybdenum distribution in various rocks and minerals are listed in Table 2. The molybdenum content of soils worldwide varies considerably. An average value of about 2 μg/g seems reasonable [86] although variation from near zero to approximately 20 μg/g Mo and higher are reported. Different soils exhibit different molybdenum content. In general, sandy soils are low in molybdenum while iron rich soils, through adsorption of molybdenum by hydrous iron oxides, contain greater amounts of molybdenum. Soils rich in organic matter too, through complexation with molybdenum, are high in their molybdenum content. Typical values for molybdenum are reported for sandy soil, a lateritic soil and loam as 0.4–1.0, 2.0–2.9, and 10.7–

Table 1. Selected molybdenum containing minerals [25, 95]

Mineral	Composition
Major sources of molybdenum	
Molybdenite	MoS_2
Powellite	$CaMoO_4$
Wulfenite	$PbMoO_4$
Minor sources of molybdenum	
Belonesite	$MgMoO_4$
Chillagite	$3PbWO_4 \cdot PbMoO_4$
Eosite	$3PbO \cdot V_2O_4 \cdot MoO_3$
Ferrimolybdite	$Fe_2(MoO_4)_3 \cdot 8H_2O$
Ilsemannite	$MoO_2 \cdot 4MoO_3$
Jordisite	$MoO_3 \cdot SO_3 \cdot 5H_2O$
Koechlinite	Bi_2MoO_6
Lindgrenite	$Cu_3(MoO_4)_2(OH)_2$
Pateraite	$CoMoO_4$

Table 2. Concentration of molybdenum in selected rocks and minerals

Location	Molybdenum concentration	Year reported	Reference
Molybdenite			
Caucasus, USSR	0.42–1.1%	1975	[61]
Greenland	0.11–0.18%	1961	[78]
Wolframite			
Cornwald, England	0.5%	1966	[49]
Uganda	0.07–0.2%	1957	[84]
Scheelite			
Colorado, USA	0.007%	1966	[49]
Uranium Ore			
Mexico	0.2–0.8%	1973	[104]
Granites	1.0 μg/g	1961	[208]
Sandstones	0.2 μg/g	1961	[208]
Shales	2.6 μg/g	1961	[208]
Carbonates	0.4 μg/g	1961	[208]
Phosphates	1–200 μg/g	1948	[162]
	6 μg/g	1981	[8]
Silicates	0.2–13 μg/g	1966	[28]

Table 3. Concentration of molybdenum in selected soils

Location	Molybdenum concentration, μg/g	Year reported	Reference
Australia	0.36– 4.4	1976	[96]
Ontario, Canada	1.65	1979	[56]
Cuba	1 – 18	1979	[146]
Scotland	0.8 – 1.5	1978	[209]
Rajastham, India	0.9 – 2.4	1974	[110]
USSR	0.1 –> 50	1974	[181]
USA	0.01– 17	1977	[106]

15.9 μg/g Mo respectively [163]. Basic soils, by formation of soluble molybdate ion, MoO_4^{2-}, are generally more rich in molybdenum than acid soils, where insoluable hydrous molybdenum oxide formation is favored. Addition of lime to soil favors higher molybdenum content both by increasing pH and through compound formation between Ca and Mo. Table 3 lists selected values for soils in various countries. A more comprehensive list is found in the work by Kabata-Pendias and Pendias [86].

Trace amounts of molybdenum, in the microgram per liter range, are found in most of the natural waters of the world. Concentration of molybdenum varies and is reported for sea water to fall within the range 10–12 μg/L [133] or 2–19 μg/L [22] depending upon the particular study. A mean value of 10 μg Mo/L is cited by one investigator worldwide in sea water [4]. Fresh waters are assumed to contain similar amounts. This does not imply that higher concentrations of molybdenum are absent in some natural waters. A survey of 100 cities in the United

Table 4. Concentration of molybdenum in selected natural waters

Location	Molybdenum concentration, μg/L	Year reported	Reference
Sea water			
Atlantic Ocean	7.3 –7.9	1980	[107]
Eastern Atlantic	7.5	1983	[201]
Western Atlantic	6.3–14.0	1959	[223]
North Atlantic	0.5– 1.0	1966	[30]
North Atlantic	12.8–13.2	1985	[210]
Eastern Pacific	8.8	1984	[99]
Western Pacific	1.5	1983	[135]
Indian Ocean	9.5–13.3	1966	[191]
Japan Sea	11.5	1985	[174]
Tokyo Bay	7.7	1969	[91]
Tokyo Bay	10 –13	1974	[109]
Tokyo Bay	9.3	1984	[200]
English Channel	12 –16	1966	[30]
Irish Sea	8.4	1968	[161]
Irish Sea	11.8	1985	[210]
Baltic Sea	4.9	1977	[142]
Baltic Sea	2.3	1980	[130]
Fresh water (lakes and rivers)			
Black Sea	7.0	1979	[143]
Lake Ontario	5.6	1969	[33]
River, Minnesota USA	65 – 7.0	1982	[55]
River, Salzburg Austria	0.5– 1.4	1975	[102]
Drinking water			
Mineral water, France	0 –36	1973	[149]
Vienna, Austria	0.2–13	1976	[103]
Seville, Spain	0.8– 1.0	1983	[201]

States, for example, found molybdenum values as high as 68 μg Mo/L although 96% of the samples tested contained less than 10 μg/L and the mean value reported was 1.4 μg/L [47]. As much as 6,900 μg Mo/L was found in certain rivers in the USSR [115] and 1,100 μg/L in the United States [58]. From 0.1–7.0 μg/g Mo has been reported in river sediments in Europe [158]. Pollution from industrial sites and other sources can, of course, contribute to abnormally high molybdenum levels in water and sediments. Table 4 lists selected values of molybdenum content of various water sources.

Production

The following values pertaining to the production and consumption of molybdenum are taken from the review by Bilhorn [15].

Commercial mining of molybdenum in the western world occurs primarily in the United States, Canada, and Chile. Production from these and other western countries totaled approximately 45 M kg in 1983. As shown in Table 5, this is less than in previous years and, as with other commodities, reflects the overall down-

Table 5. Production of molybdenum in Western countries [15]

Year	1977	1978	1979	1980	1981	1982	1983
Production, M kg	83	88	90	101	98	80	45

Table 6. World consumption of molybdenum [15]

Year	1977	1978	1979	1980	1981	1982	1983
Consumption, M kg	83	91	93	83	77	63	63

ward economic trend of the past few years worldwide. Price for molybdenum averaged about 9.00 $US/kg in 1983.

Production, of course, mirrors consumption and although less than in previous years molybdenum consumption is holding steady and expected to increase as world demand increases (see Table 6). Approximately 22% of the molybdenum consumed in 1983 was in the United States. Western Europe accounted for approximately 40% of world consumption. Eastern Europe 14%, Japan 18%, and about 6% was distributed about other manufacturing countries.

Uses

Molybdenum is used primarily in the manufacture of steels. Approximately 74% of the 1983 consumption served this purpose. Approximately 9% of the molybdenum used in 1983 went to the manufacture of other specialty alloys. The remaining molybdenum went to production of molybdenum metal (6%), molybdenum chemicals (10%), and other miscellaneous uses (1%) [15].

Molybdenum is added to steels in amounts ranging from a few tenths of a percent to several percent depending on the intended use of the finished product [7, 129]. In general, addition of molybdenum to steels improves the strength of the steel, especially at high temperatures. It also improves wear resistance, corrosion resistance, and metal fatigue. Molybdenum containing alloys are found in steam tubing and heat exchangers with coal and nuclear generating power plants. Its low thermal neutron absorption makes it particularly useful for the latter where metal components in the primary cooling system are directly exposed to high levels of radiation. Its wear and fatigue resistance makes it useful for gears in automotive assemblies. Its corrosion resistance is of importance in pipes and other components where sea water comes in contact with metal surfaces [44, 141]. Molybdenum's high electrical conductivity coupled with a coefficient of expansion similar to that of glass make it useful in glass metal seals for electronic tubes and valves. Because of its ability to withstand high temperatures, it is used in resistance heating elements. Various molybdenum containing alloys, ferrous and nonferrous, are employed in the production of engines for space crafts because of their ability to withstand high temperatures and still perform their specified functions satisfactorily. Molybdenum alloys are used in sampling tubes for sampling stack gases and particulates. Again, their high temperature strength is desirable for this application. In the laboratory, crucibles made from molybdenum and molybdenum containing alloys are useful because of their ability to withstand corrosion from most common laboratory acids and because of their high temperature strength.

Molybdenum trioxide is the starting material for the preparation of most molybdenum containing compounds. In addition, it, alone or in combination with other metal oxides, catalyzes the removal of sulfur from petroleum and coal [160]. Hydrodesulfurization is an important and necessary step if one is to remove sulfur from fossil fuels. Molybdenum compounds are also useful in catalyzing other selected organic reactions [73]. Molybdenum sulfide, MoS_2, both in appearance and properties is similar to graphite. It is used as a solid lubricant or added to lubricating oils [111]. Other molybdenum compounds and organomolybdenum compounds exhibit corrosion inhibiting properties [164]. Added to automobile radiator coolant, for example, they serve as a rust preventor. Molybdenum, as ammonium paramolybdate, is an analytical reagent for the determination of phosphorus, and other elements, as the phosphomolybdate heteropoly ion or as the reduced heteropoly ion, heteropoly blue, with its distinct color intensity.

Chemistry

Molybdenum is a silvery-white transition metal belonging to the chromium group (chromium, molybdenum, tungsten) of the periodic table. It is known for its high melting point and its ability to maintain its integrity at high temperatures. Table 7 lists pertinent physical characteristics. The metal is attacked by concentrated nitric acid or sulfuric acid and readily dissolves in aqua regia [$HCl:HNO_3$, 3:1 (v/v)] forming molybdenum(VI) compounds. The most common oxide, MoO_3, also dissolves in mineral acids and in strong caustic. A variety of binary molybdenum compounds exist with the general formula Mo_nX_r. X equals oxide, sulfide, nitride, carbide, phosphide, halide, etc. and n and r are integers. Oxy salts also are common with the general formula MoO_nX_r. Above pH 6 the predominate form of molybdenum(VI) is the molybdate ion, MoO_4^{2-}. In a manner analogous to the more familiar chromate ion, CrO_4^{2-}, it forms stable salts with alkaline metal ions and generally insoluble salts with alkaline earth and transition metal ions. Sodium molybdate dihydrate, $Na_2MoO_4 \cdot 2H_2O$, is a commercially available form of molybdenum. Unlike chromate ion, however, molybdate ion can undergo

Table 7. Physical properties of molybdenum

Symbol	Mo
Atomic number	42
Atomic weight	95.94
Melting point	2610 °C
Boiling point	5560 °C
Density	10.22 g/cm³ (20 °C)
Specific heat	0.25 J/g °C

extensive polymerization in solution depending upon both molybdenum concentration and pH. The following are typical [166, 207].

$$8H^+ + 7MoO_4^{2-} \rightarrow Mo_7O_{24}^{6-} + 4H_2O$$

$Mo_7O_{24}^{6-}$ paramolybdate ion predominates at pH ~4

$$12H^+ + 8MoO_4^{2-} \rightarrow Mo_8O_{26}^{4-} + 6H_2O$$

$Mo_8O_{26}^{4-}$ octamolybdate ion predominates at pH ~1–2

As the acidity increases further (lower pH) precipitation of hydrous molybdenum(VI) oxide, $MoO_3 \cdot xH_2O$, occurs. In strong acid solution, this precipitate dissolves yielding molybdenyl ion, MoO_2^{2+}, and higher cationic polymers. Each of the various isopoly ions is itself capable of protonization to varying degrees resulting in a large number of possible species [77, 152]. Ammonium paramolybdate, $(NH_4)_6Mo_7O_{24} \cdot 4H_2O$, is an alternate commercial form of molybdenum.

Perhaps better known than the isopoly molybdates are the heteropoly molybdates. Here, clusters of molybdenum and oxygen groups are found about a central metal atom, often phosphorus. The 12-molybdophosphate ion, $PMo_{12}O_{40}^{3-}$ is well known in the colorimetric method for determination of phosphorus [186]. A unique feature of heteropoly ions is their ability to form intensely colorful solutions in the presence of a mild reducing agent, the heteropoly blue color.

Lower oxidation states of molybdenum, V, IV, III, II exist. Like the more stable molybdenum(VI) they can polymerize in solution and frequently require the presence of a complexing agent if air oxidation to the more stable Mo(VI) is to be avoided. Molybdenum has the ability to form complex ions in each of its oxidation states. Many of these complexes occur with biologically active materials in living systems, plant, and animal. The importance of molybdenum to life and its role in the environment are intimately associated with its complexing ability in these systems. Details of these interactions are brought forth in subsequent sections of this report.

Of importance, too, is the ability of molybdenum to change from one oxidation state to another within inorganic ions in aqueous solution or as complex ions in (perhaps) living systems. A reduction potential, ε, versus pH diagram for se-

Table 8. Selected standard electrode potentials for molybdenum [76, 112]

										ε°, V
H_2MoO_4	+	$2H^+$	+	$1e^-$	→	MoO_2^+	+	$2H_2O$		+0.4
H_2MoO_4	+	$6H^+$	+	$3e^-$	→	Mo^{3+}	+	$4H_2O$		+0.20
MoO_4^{2-}	+	$8H^+$	+	$6e^-$	→	Mo^0	+	$4H_2O$		+0.15
MoO_4^{2-}	+	$4H_2O$	+	$6e^-$	→	Mo^0	+	$8OH^-$		−0.92
MoO_3	+	$6H^+$	+	$6e^-$	→	Mo^0	+	$3H_2O$		+0.06
MoO_3	+	$3H_2O$	+	$6e^-$	→	Mo^0	+	$6OH^-$		−0.77
MoO_2^+	+	$4H^+$	+	$2e^-$	→	Mo^{3+}	+	$2H_2O$		~0.0
MoO_2^+	+	$4H^+$	+	$5e^-$	→	Mo^0	+	$2H_2O$		−0.09
Mo^{3+}	+			$3e^-$	→	Mo				−0.20

lected molybdenum species is available in the literature along with a compilation of other electrochemical reactions of molybdenum [76]. Selected standard electrode potential values taken from this reference and elsewhere [112] are summarized in Table 8.

Analytical Methods

Widely scattered throughout the environment molybdenum determination requires careful experimental skill [46]. Few analytical procedures are unique for molybdenum in complex natural samples. Because of this, separations to remove interfering constituents or to isolate molybdenum from its matrix are generally necessary prior to actual determination. In general, low molybdenum concentrations require caution, too, in selection of a suitable method for determination. Trace analytical methods are required and they must be skillfully performed.

Sample Collection

As with any analytical procedure, the results are no better than the sample upon which the analysis is performed. Those wishing to determine molybdenum must acquire a sample representative of the larger body from which it is taken. Air samples should be collected from various localities within the region of study, at various times of the day and at various heights above the ground. Aquatic samples should be collected at various depths and at various distances from the shore. If some contributing factor is prevalent, for example, an industrial site suspected of discharging excessive molybdenum salts, distances from this source should be recorded. Collections should be spread over a period of time. Rain on any given day may enhance molybdenum content in a stream by passing excess soil run-off into the stream. On the other hand, the soil itself if tested following a heavy rain could be unduly deficient in soluble molybdenum salts. Information regarding soil sample location, depth, and type should accompany each sample. A region containing sandy soil will contain different amounts of molybdenum than is present in a rich fertile soil with a large percentage of organic components. Plant and animal samples should be labeled regarding the particular part of the body, be it leaf or limb, from which they were taken.

Samples once gathered should be properly handled and stored. In general, it is always better to analyze a sample as soon as possible following its collection. Containers in which samples are stored should be clean and free from any molybdenum contamination. Adsorption of ions on glass surfaces and/or leaching of metal ions from glass container walls are both possible and with trace amounts of component could contribute a significant fractional change in the concentration of component stored within the container. If it is desired to determine a particular oxidation state of molybdenum or a particular molybdenum complex rather than total molybdenum irregardless of form, then the sample should be analyzed immediately to avoid, for example, air oxidation of lower oxidation state molybdenum species. Total molybdenum, determined from stored samples, does not depend upon the oxidation state of Mo provided proper chemical treatment is performed to assure the correct oxidation state for molybdenum determined by the procedure selected. Solid samples can, generally, be safely stored for long periods of time although freezing of biological samples is recommended to retard putrefaction. Liquid samples should be acidified to prevent hydrolysis of heavy metal ions with resultant insoluble hydrous oxide formation [180]. Iron salts, for example, upon standing in neutral solutions form insoluble hydrous iron oxides which adsorb other ions upon their surfaces thus altering the trace metal content of the sample. Samples once gathered can be pooled if an overall average molybdenum content is desired or analyzed separately if molybdenum content of individual sites within a sample area is desired. Few samples can be analyzed directly upon collection and it is necessary to specify the pretreatment employed to render the sample suitable for analysis.

Dissolving

Rocks, minerals, and molybdenum ores are generally ground into a fine powder. There are specific, step-wise procedures for achieving uniform homogeneous laboratory size samples from a large heterogeneous rock or mineral sample. These include mechanical grinding to produce particles of uniform size followed by a cone and quarter procedure in which alternate quarters of the sectional sample are discarded, the remainder mixed, and the coning and quartering repeated until a suitable laboratory sample is obtained [220].

Liquid samples containing molybdenum in a soluble form require the least sample pretreatment. It may, however, be necessary to concentrate the molybdenum as described in a subsequent paragraph. Solid molybdenum containing samples require some treatment to render molybdenum into a soluble form. Solid samples, once obtained, are dissolved either with strong mineral acid [45, 92, 170] or by fusion. With fusion the finely ground solid sample is thoroughly mixed with flux, a solid chemical capable of reacting with the sample at elevated temperatures. The mixture is heated to a red-heat and upon cooling the residue taken up in mineral acid. Sodium carbonate [165] and sodium carbonate-potassium carbonate (1 : 1 w/w) [216] work well with various rock samples while sodium peroxide, Na_2O_2, [26] is best for molybdenum ores and concentrates.

Plant, animal, and soil samples require removal of organic matter before trace metals can be determined. Samples are dried to remove moisture and decomposed by either wet acid digestion using strong mineral acid [12, 97, 139, 168] or dry ashing at elevated temperatures, often in the presence of a suitable flux [29, 74]. Care must be exercised to avoid too high a temperature as molybdenum(VI) oxide is volatile above 550 °C and will be lost from the sample. More recent sample treatment involves use of a low temperature microwave ashing to remove organic material from trace mineral residues [13]. In addition to wet or dry digestion which treats total molybdenum content, soil samples are sometimes leached with water or suitable complexing agent to extract only available molybdenum [116, 119].

Air samples analyzed for their particular content are passed through a suitable filter. The filter is then destroyed by either wet or dry digestion while simultaneously rendering the particulates soluble.

Preliminary Treatment

Samples containing trace molybdenum are generally concentrated to enhance their molybdenum concentration. Methods for achieving this include co-precipitation resulting from adsorption upon the surface of another ion either initially present in the sample or added intentionally. Iron salts followed by addition of base [222] forming a hydrous iron oxide precipitate, manganese salts [30] forming MnO_2, thorium salts [98] forming $ThO_2 \cdot x\,H_2O$, and organic reagents [224] are among those used to concentrate trace molybdenum containing samples. Extraction of molybdenum or its complexes with a small amount of immiscible organic liquid into which it is preferentially dissolved also achieves concentration of molybdenum from a large sample volume into a smaller volume more suitable for subsequent treatment. Tributyl phosphate separates molybdenum from many other ions [42]. Various molybdenum complexes are extracted into a variety of organic solvents, among these are long chain amine containing ligands, for example, Aliquot 336 (tricaprylmethylammonium chloride) [213] and the liquid ion exchange resin Amberlite LA-2 [90].

Because, depending on pH and the presence of complexing agents, molybdenum can exist in solution as either an anion or cation species; ion exchange resins are useful in concentrating molybdenum while simultaneously removing potential interfering ions. In the presence of citrate ion at pH one, the negatively charged molybdenum citrate complex readily passes through a strong cation resin column while positive ions (Pb, Cu, Cr, Ni, Fe, V) are retained [100]. Molybdenum in hydrochloric acid solution is retained on an anion resin while cation species pass through. After thorough washing, the molybdenum itself is removed from the column by a small volume of dilute HCl solution [105].

Determination

Qualitatively the presence of molybdenum is determined by the orange-red color observed for molybdenum(V) in the presence of thiocyanate ion. Tin(II) solution is added to the acidified sample, reducing Mo(VI) to Mo(V), followed by addition of thiocyanate containing solution [159].

Precipitation. If present in sufficient amounts, for example in ores and minerals, molybdenum can be precipitated through formation of an insoluble compound upon addition of a suitable reagent, collected and weighed. Preliminary separation is necessary to remove possible interfering ions. Sulfide precipitation has traditionally been employed for molybdenum [198] but can be unsatisfactory because of co-precipitation of other ions and inconvenient because of the need to ignite the precipitate to the oxide before weighing. Precipitation of MoO_4^{2-} with lead(II) from weakly acidic solution is more convenient as the precipitate can be dried and weighed directly [53].

$$Pb^{2+} + MoO_4^{2-} \xrightarrow[\text{buffer}]{\text{acetate}} PbMoO_4(s)\,.$$

8-Hydroxyquinoline forms an insoluble material with molybdenum(VI), as molybdenyl ion, which can be collected and weighed [49].

$$MoO_2^{2+} + 2\,C_9H_7ON \xrightarrow{\text{pH 5}} MoO_2(C_9H_6ON)_2(s) + 2\,H^+\,.$$

Titration. Titrations utilizing molybdenum generally involve oxidation of molybdenum(V) with a sutiable oxidizing agent. Cerium(IV) is most commonly used [188]. EDTA titrant and similar complexones also react with molybdenum(VI) [156] and molybdenum(V) [94]. Certain ion-selective electrodes can be made responsive to molybdate ion [118, 185] or by following the concentration of another ion capable of reacting with molybdenum to determine its concentration indirectly [171, 199]. In general, these electrodes have been developed with pure solutions in the laboratory. Their applications to real world samples presently awaits further study.

Colorimetry. Traditionally colorimetric procedures have been employed for assay of environmental samples for molybdenum. Of these, the formation of an orange-red thiocyanate complex with molybdate ion in the presence of a mild reducing agent is, perhaps, best known. Measurement at 470 nm upon sample and standards allows, through the preparation of a calibration curve, quantitative determination of molybdenum in the parts per million range. The procedure is not entirely interference free and the literature should be consulted to determine proper steps for measurement in a particular sample type. The method is suitable for determining molybdenum in rocks [113], atmospheric particulates [197], salt brines [69], soil [67], animal [206], and plant [14, 29] materials. Improvements in the thiocyanate method for molybdenum have focused on addition of an auxiliary complexing agent to form a ternary species with the hope of improving both sensitivity and selectivity for molybdenum. These auxiliary ligands are generally high molecular weight quaternary amine salts, other electron rich ligands or surfactants [147, 148]. Addition of the dye crystal violet, for example, to the molybdenum thiocyanate procedure has been applied for determination of molybdenum in drinking water [81]. Other color forming reagents for determining molybdenum in environmental samples are listed in Table 9.

Atomic Absorption Spectroscopy. Trace amounts of molybdenum are conveniently determined by atomic absorption spectroscopy. This analytical technique has grown rapidly since its inception and is preferred by many for trace metal analysis at the parts per million to parts per billion level. Molybdenum, however, like other heavy transition metals does not as readily lend itself to this type of determination. Atomization of the components within a sample is necessary before

Table 9. Color forming reagents for determination of molybdenum in environmental samples

Reagent	Sample	Reference
Toluene-3,4-dithiol	Silicate rock	[92]
Toluene-3,4-dithiol	Ore	[84]
Toluene-3,4-dithiol	Soil	[187]
Toluene-3,4-dithiol	Liver	[139]
Phenylfluorone	Ore	[204]
Pyrogallol red + dimethyldioctadecylammonium ion	Natural water	[132]
Bromopyrogallol red	Wheat	[167]
Bromopyrogallol red + cetyltrimethylammonium ion	Sea water	[6]
Sodium thiosulfate	Beans	[71]

radiation from an external source can be absorbed by the individual atoms. Molybdenum and other heavy metals with their tendency to form refractory compounds in the flame are not easily atomized. Nitrous oxide-acetylene rather than air-acetylene is preferred as the flame fuel mixture because of its higher temperature. Flameless atomic absorption in which an electric furnace replaces the gaseous fuel mixture, because of its even higher temperature and better temperature control, is a better tool for molybdenum by this method although here, too, carbide formation, from the graphite rod within the furnace, can occur [134]. Pyrolytically coated graphite, commercially available, rather than uncoated graphite is better for molybdenum because of its higher sublimation point and greater resistance to oxidation [48, 215].

Other ions present with molybdenum effect the accuracy of an atomic absorption procedure. The presence of aluminum or ammonium salts in the sample enhance the atomic absorption signal and these salts are frequently added intentionally to sample and standards for increased sensitivity [41, 52]. It has been noted that if zirconium crucibles are used to contain plant samples for sodium cabonate, fusion prior to molybdenum determination by spectroscopic measurement decreased sensitivity results [27]. This is attributed to the presence of zirconium in the sample from the crucible with, perhaps, formation of refractory molybdenum carbide-zirconium carbide compounds when heated on a carbon electrode.

Atomic absorption spectroscopy is suitable for molybdenum in a variety of environmental samples including rocks [8, 28], sea water [205], potable water [63], and plant and animal samples [93].

Emission Spectroscopy. Emission spectroscopy has long been used in the study of environmental samples. Development of newer, more efficient, excitation sources has generated renewed interest in this technique both for qualitative and quantitative determination of metals, including molybdenum. Selected emission lines using conventional d.c. arc [1] and the newer inductively coupled plasma (ICP) source [17] are given for molybdenum as are procedures for determining molybdenum in rocks and minerals [145, 183], air borne particulates [192], sea water [125], soils [119, 203], animal tissue [136], and plants [51, 124].

Other Techniques. If available, x-ray fluorescence analysis and/or neutron activation analysis are extremely sensitive for analysis of environmental samples. Both have been used for determination of molybdenum [107, 123, 130, 202, 211].

Transport in the Environment

The natural distribution of molybdenum in the environment is affected by man's activity. Ores are often transported from one location to another for refining and ore tailings transported again to disposal sites. Fossil fuels are distributed to provide energy worldwide and their combustion products spread throughout the environment. Waste water from industrial processes and from municipalities are put back into the environment sometimes with preliminary treatment to remove harmful contaminants which may have accumulated and sometimes not. Molyb-

Table 10. Selected values for molybdenum in various fossik fuels and related materials

Fuel	Molybdenum concentration, μg/g	Year	Reference
Coal			
Lignite (Turkey)	80 –300	1984	[182]
Sub-bituminous (Western United States)	0.5 – 7	1976	[195]
Bituminous (Australia)	> 0.3 – 6	1977	[194]
Anthracite (Belgium)	1.2 – 2.4	1975	[16]
Fly Ash			
United States	6 – 40	1977	[59]
Australia	1 – 40	1963	[193]
Belgium	~34	1975	[16]
Petroleum			
Crude oil (Italy)	0.02– 10	1964	[36]
Shale oil (Central Asia)	12 – 25	1979	[10]

denum, along with other metals, nonmetals, organic components, etc., is a part of this distribution process. One finds, therefore, molybdenum appearing in the environment sometimes where it is not normally expected, sometimes where its presence may be intentional and beneficial, and sometimes where excessive molybdenum is harmful to those plants and animals, including man, with which it comes in contact.

Molybdenum is present in most fossil fuels. Table 10 lists typical amounts of molybdenum found in various fuels. Keep in mind that the values listed serve only as indicators of possible molybdenum levels. Insufficient data are available to accurately characterize all deposits of fossil fuels and, no doubt, wider variations occur in different regions of the world.

Molybdenum is also found in the discarded by-products from various industrial operations and municipal discharges. Greatest amounts, as expected, are observed for those industries in which molybdenum plays a dominate role. Streams located in the western United States, where molybdenum mining operations take place, have abnormally high molybdenum concentrations attributed to the mining operations. Industrial discharges can significantly increase molybdenum content in nearby waters. Tokyo Bay, for example, was tested to find 9.3 μg Mo/L except in a polluted area when the molybdenum content was 168 μg/L [200]. A similar observation is reported for the Black Sea where molybdenum concentration was 7 μg Mo/L except near a steam power plant where 600 μg Mo/L was measured [143]. Table 11 lists molybdenum content of selected mining, manufacturing, and municipal sources.

By-products from manufacturing, from burning fossil fuels and from municipal wastes contribute to excessive molybdenum presence in the atmosphere. Nikolaev estimates 4,000 ton per year of recycleable molybdenum in metal alloy manufacturing through ignition loss, mill scaling, flue dust, pickling, etc. [138]. Twice that amount, 8,000 ton Mo/yr is estimated as available through shale oil production [32]. Equally significant losses are cited for molybdenum during vari-

Table 11. Selected values for molybdenum in various manufacturing wastes

Description	Molybdenum concentration, μg/g	Year	Reference
Mining operations			
River near mining region, Colorado, USA	60	1982	[55]
Slag from ore refining	5–25	1970	[153]
Industrial operations			
Chloralkali brine	16–83	1981	[211]
Waste water from electrorefining	200	1979	[23]
Waste water from electroploting	1100	1981	[58]
Sewage sludge			
United Kingdom	< 50	1983	[43]
London	~ 9	1980	[189]
Switzerland	0.2–20	1979	[38]

ous mining and refining operations [177]. Industry is not unaware of these effects and efforts are being made to minimize them [5, 144]. Bag filters to remove particulate dust from factory stacks in mining [122] and steel-making operations [3, 9] are successful as is electrocoagulation and other measures [65] in molybdenum recovery. Waste water discharges from mining operations have been reduced in molybdenum content from 2–10 μg Mo/mL to >0.5 μg/mL through coprecipitation with hydrous iron oxides [39]. Expelling molybdenum in these waste products greatly enhances the presence of molybdenum in lakes, streams, soil, and eventually plants and animals.

Sometimes molybdenum is added to soils albeit not always intentionally as part of an effort to improve soil quality. Municipal sledge is sometimes substituted for fertilizer to upgrade crop production. Fly ash from coal fired operations is also used for this purpose. Molybdenum contamination is not unique because of these practices and possible deleterious effects of other metals present in sludge and ash are also of concern. One study of the effects of sludge from ten Canadian cities found only one instance where molybdenum content of plants grown upon the treated soil increased [217]. Other studies, however, gave different results. Addition of from 40–80 ton/ha of a sludge produced a raise in molybdenum levels in plants [79]. Fly ash in amounts of 112 ton/ha also produced increased molybdenum presence in forage crops grown on the treated soil [114].

The mere presence of molybdenum in soil does not, of course, preclude its availability for plant utilization. Available molybdenum content, that which can be taken up by plants, is often very different from total molybdenum content present in a soil. The availability of molybdenum for plant uptake is diminished in soils of low pH [19]. Other factors tending to retard molybdenum availability include soils with high iron content upon which molybdenum is strongly adsorbed, high organic content resulting in molybdenum tightly bound to the organic material, eroded and/or sandy soil from which molybdenum is easily washed away and soils containing calcium or other metal ions that form insoluble molybdenum compounds [175]. Attempts have been made to corrolate plant molybdenum content with available molybdenum in the soil upon which the plant was grown. These have met with only limited success [121]. Similar studies attempting to relate molybdenum content of natural waters with molybdenum presence in aquatic insects proved equally inconclusive [35].

The effects of molybdenum addition can be long lasting, as shown by one study in which up to 0.5 kg Mo/ha was added to soil. Subclover grown on the soil showed an increased molybdenum content (leaf) from 1.5 to >25 µg/g. A similar crop grown the following year exhibited 10 µg Mo/g (leaf) and in the third year 2 µg Mo/g (leaf) [150]. Some of these levels are beyond the toxic limit acceptable for molybdenum. More will be said regarding this later. A related study in which irrigation water high in molybdenum content was used in growing alfalfa projected after one year that in a three year period plant molybdenum levels would approach toxic levels [82].

Bowen has presented an interesting summary of the exchange of various elements, including molybdenum, within the earth's crust [18]. He estimates from natural sources (rock weathering, rain, dust, litter decay, etc.) approximately 0.48 mg Mo per square meter of soil surface per year added to uncultivated soil and approximately 0.61 mg Mo/m^2 yr added to cultivated soil; the additional input for cultivated soil coming from fertilizer and irrigation water. He goes on to

Table 12. Transfer of molybdenum within the environment

Loss of Mo from soils	from →	Weathering Plant uptake Mining	to →	Plants Natural waters Manufacturing
Loss of Mo from plants	from →	Animals Decay Harvesting	to →	Animals Soil Natural waters Man
Loss of Mo from natural waters	from →	Animals Plants Sedimentation	to →	Animals Plants Soil
Loss of Mo from animals	from →	Animals Decay Sewage	to →	Man Soil Natural waters
Gain of Mo for solis	from →	Weathering Plants Animals Manufacturing discharge		
Gain of Mo for plants	from →	Soil Natural waters		
Gain of Mo for natural waters	from →	Weathering Plants Animals Manufacturing discharge		
Gain of Mo for animals	from →	Plants Natural waters Manufacturing discharge		

estimate loss of molybdenum (drainage) for uncultivated soil of approximately 0.12 mg Mo/m^2 yr and for cultivated soil approximately 0.5 to 3 mg Mo/m^2 yr depending upon the extent of uptake by a particular plant crop.

Molybdenum is essential not only for plants but for animals, including man. Ruminants acquire molybdenum, primarily from forage crops, and man gains some of his nutrition from animal sources. The importance of this cycle to the overall agricultural economy of Australia during the 1940's has recently been recounted and serves as an interesting introduction to the importance of molybdenum in nutrition and health [131].

If pursued further, one would gain an even better understanding of the exchange of molybdenum between air, soil, water, plant, and animal. The general, subject of trace element transfer is treated extensively and the reader is referred elsewhere for an introduction to the topic [11]. Table 12 summarizes in general terms the transfer of molybdenum throughout the environment.

Enzymes

Molybdenum appears in several enzymes essential to both plant and animal life [21, 151, 196] and the study of these enzymes, especially their structure and their role in the various chemical reactions which they catalyze, is a current and active area of research. The conversion within plants of nitrogen, N_2, to ammonia, NH_3, depends upon the enzyme nitrogenase to catalyze the reaction [154, 190]. Nitrogen fixation, as it is called, is well known in certain plants (bean, clover, alfalfa, and other legumes). Nitrogenase contains molybdenum, iron, and sulfur. It is a two part enzyme, one fraction with molecular weight approximately 220,000 having a Mo:Fe:S ratio of 1:24:24 and the other with molecular weight approximately 56,000 having a Mo:Fe:S ratio of 1:4:4 [72]. The actual structure of this enzyme and the unique role of molybdenum within the enzyme are uncertain [137]. A second molybdenum containg enzyme, nitrate reductase, is also involved in the utilization of nitrogen by plants. In non-legumous plants, bacteria, fungi, etc. it aids in plant utilization of nitrogen by catalyzing conversion of nitrate ion, NO_3^- to nitrite ion, NO_2^-, as a first step in assimilation of nitrogen [140]. It is interesting to predict similar bonding of molybdenum in nitrate reductase and other molybdenum containing enzymes xanthine oxidase, xanthine dehydrogenase, aldehyde oxidase, and sulfite oxidase [70, 85, 184].

Xanthine oxidase and the related enzyme xanthine dehydrogenase catalyze the change of xanthine to uric acid, an important step in the conversion of purine based molecules. They are complex enzymes containing a molybdenum subunit, a flavin adeninedinucleotide portion and two iron-sulfur centers [20, 64]. They are usually isolated from cow's milk or chicken liver. Aldehyde oxidase is similar in composition to xanthine oxidase and catalyzes the conversion of aldehydes to carboxylic acids. Prime source for this enzyme is rabbit liver [196]. Sulfite oxidase catalyzes the reaction of sulfite ion, SO_3^{2-}, to sulfate ion, SO_4^{2-}, a final step in the degradation of sulfur containing amino acids. Present in animals and plants, it is generally isolated from cow's liver and contains two subunits of about equal molecular weight (approximately 60,000 units each); one with a molybdenum center and the other with a cytochrome unit [62].

Biological Effects and Toxicity

The importance of molybdenum to plants and animals lies in its ability, through enzyme action, to catalyze vital changes necessary for life processes. Completion of these processes is impaired if molybdenum deficiency persists. Fortunately, the amount of molybdenum necessary for plant and animal health is very small. This fact, however, creates another problem for molybdenum, which like copper, arsenic, and certain other chemical elements, is toxic to life forms in too great an excess. As pointed out by Manilla, one must be careful in using the word toxic for these substances. Toxicity, harmful effects from the presence of certain substances, is relative and does not imply complete absence of the substance under consideration [120].

Plants

Molybdenum is essential to plants for which nitrogen fixation is the means of acquiring nitrogen. One may generalize that molybdenum concentrations for plants, less than 0.2–0.3 μg (per gram of dried leaf), constitutes a deficiency while concentrations greater than 5–10 μg represents excessive, perhaps toxic, amounts [87]. The presence of excess molybdenum may not be harmful to plants although leaf chlorosis (yellowing) is known to occur in cabbage [88]. Molybdenum deficiency in soils is generally corrected by adding soluble molybdate, MoO_4^{2-}, usually mixed with fertilizer [176] or by soaking seeds prior to planting in a molybdate solution [172]. Care must be taken, however, to avoid excessive amounts of molybdenum, a distinct possibility, for example, if fly ash or sewage sludge, as described earlier, are used as a nutrient source. Suggested upper limit of molybdenum in fertilizer is 0.001% [221].

Of greatest concern in cases of excess molybdenum in plants is the effect it will have on animals grazing on the land where the plants are grown. It has been variously suggested that molybdenum present in excess of 2 μg Mo/g [128], 5 μg/g [89], or 10 μg/g [68] is toxic to animals.

Animals

Molybdenum in animals tends to concentrate in the liver [40] where increased enzyme activity is observed [108]. Concern is not so much for molybdenum deficiency as it is for excessive molybdenum being present. Fetal rat development is, for example, inhibited in the presence of excess molybdenum and other metals. Normal weight gain was less, internal hemorrhaging was observed, and skeletal ossification disrupted [219]. Gastrointestinal irrigation, diarrhea, weight loss, bone joint abnormalities, and other maladies have been attributed to molybdenum toxicity [117]. In cattle molybdenois, illnesses from the presence of excess molybdenum, is known as tearts and is characterized by severe diarrhea [57]. It causes metabolic disturbances in cattle and sheep and is of considerable concern to those engaged in raising these animals [50]. High molybdenum levels also interfere with the functions of sulfate [80, 173] and perhaps other elements [101] in animals. It is well established, for example, that excess molybdenum depletes the amount of copper in the body [128].

LD_{50} for rats (lethal dose for half of the injected subjects) is 125 mg/kg body weight as MoO_3 and 333 mg/kg as ammonium molybdates [24]. In ruminants chronic ingestion of between 20–100 mg Mo/kg body weight produces obvious symptoms of ill health [126].

Man

Blood molybdenum levels in man are reported as 0.015 μg/g [155], 0.005 μg/g [34], or other divergent values [34]. Urine content is listed as 34 μg/24 h [155], 300 μg/24 h [34], or other values [60, 206]. Molybdenum and other metals present in mother's milk are known to affect infants [37]. Liver and kidney malfunctions are reported in adults as a result of excess molybdenum [157]. Excess molybdenum also upsets the copper balance within the body [155]. Workers exposed to molybdenite or dust exhibit breathing disorders [179] and elevated uric acid excretion [124]. A limit of 5 mg/m^3 soluble molybdenum compounds and 15 mg/m^3 insoluble molybdenum compounds in air has been established in the United States [212]. One study of city air in 1964 found 10 fg/m^3 molybdenum, much less than the established limit [75]. Elevated molybdenum levels have been observed in the blood of cancer patients [2]. This does not imply that molybdenum is carcinogenic and may merely be a result of some circumstance caused by the cancer. In prospective of all that has been said regarding molybdenum toxicity, molybdenum is not considered as serious a concern regarding its harmful effects as are many other metals.

There is no established level of intake for molybdenum regarding health and nutrition at the present time. Obviously a trace amount is needed to maintain proper health, yet with excess molybdenum toxic effects are observed. Mertz [127] suggests 100 μg Mo per day as a possible value and Sittig reports that daily exposure of from 100–500 μg Mo occurs [178]. This exposure would satisfy the daily requirement. A list of the molybdenum content of a variety of common foods has recently been published [54].

Acknowledgements

The author thanks Mr. N. Parinandi for his discussions on the biological aspects of molybdenum chemistry and Mrs. S. Flick for typing the manuscript.

References

1. Addink, N.W.H.: Spectrochim. Acta *11*, 168 (1957)
2. Agrawal, Y.K.: Anal. Lett. *13B*, 357 (1980)
3. Alary, J., Bourbon, P., Esclassan, J., Lepert, J.-C., Vandaeh, J., Klein, F.: Water Air Soil Pollut. *20*, 137 (1983)
4. Allaway, W.H.: Adv. Agron. *20*, 235 (1968)
5. Andes, G.L.: Air Pollut. Control Assoc. 71st Ann. Meeting, Pittsburgh, PA, 1978; Pollut. Abstr. *10*, 79-01364 (1979)
6. Andreeva, I.Yu., Lebedeva, L.I., Kavelina, G.L.: Zh. Anal. Khim. *37*, 2202 (1982)
7. Archer, R.S., Briggs, J.Z., Loeb, C.M.: Molybdenum Steels, Irons, Alloys, New York, Climax Molybdenum Co. 1948

8. Barbooti, M.M., Jasin, F.: Talanta *28*, 359 (1981)
9. Barnard, P.G., Dressel, W.M., Fine, M.M.: US Bur. Mines Rept. 8218 (1981)
10. Bastova, S.M., Yurina, R.D., Vakhobova, R.U.: Zh. Anal. Khim. *34*, 935 (1979)
11. Baudo, R.: in Trace Element Specification in Surface Waters (ed.) Leppard, C.G., p. 275, New York, Plenum Press 1983
12. Benne, E.J., Linden, E.I.: J. Assoc. Off. Agric. Chem. *43*, 510 (1960)
13. Bentley, G.E., Markowitz, L., Meglen, R.R.: in Ultratrace Metal Analysis in Biological Science and Environment, Advances in Chemistry Series 172 (ed.) Risby, T.H., Washington, D.C., American Chemical Society 1979
14. Bergamin Filho, H., Medeiros, J.X., Reis, B.F., Zagatto, E.A.G.: Anal. Chim. Acta *101*, 9 (1978)
15. Bilhorn, W.W.: Mining J. Ann. Rev. 69 (1984)
16. Block, C., Dams, R.: Environ. Sci. Tech. *9*, 146 (1975)
17. Boumans, P.W.J.M.: Spectrochim. Acta *38B*, 742 (1983)
18. Bowen, H.J.M.: Environmental Chemistry of the Elements, London, Academic Press 1979
19. Bowie, S.H.U., Thornton, I.: Environmental Geochemistry and Health, p. 52, Dordrecht, Holland, D. Reidel Pub. Co. 1985
20. Bray, R.C.: in Proceedings of the Climax First International Conference on the Chemistry and Uses of Molybdenum (ed.) Mitchell, P.C.H., p. 216, Ann Arbor, MI, Climax Molybdenum Co. 1973
21. Bray, R.C., Swann, J.C.: Struct. Bond. *11*, 107 (1972)
22. Brewer, P.G.: in Chemical Oceanography (eds.) Riley, J.P., Skirrow, G., p. 415, New York, Academic Press 1975
23. Broekaert, J.A.C., Leis, F.: Anal. Chim. Acta *109*, 73 (1979)
24. Browing, E.C.: Toxicity of Industrial Metals, London, Butterworths 1969
25. Browning, P.E.: Introduction to the Rarer Elements, New York, John Wiley Pub. 1917
26. Budesinsky, B.W.: Analyst (London) *105*, 278 (1980)
27. Burridge, J.C., Hewitt, I.J.: Anal. Chim. Acta *154*, 30 (1983)
28. Butler, L.R.P., Mathews, P.M.: Anal. Chim. Acta *36*, 319 (1966)
29. Carel, A.B., Wimberley, J.W.: Anal. Lett. *15A*, 493 (1982)
30. Chan, K.M., Riley, J.P.: Anal. Chim. Acta *36*, 220 (1966)
31. Chappell, W.R., Petersen, K.K. (eds.): Molybdenum in the Environment, New York, Marcel Dekker 1976
32. Chappell, W.R., Runnells, D.D.: Eleventh Annual Conference on Trace Substances in Environmental Health Columbia, MO June 1977; Pollut. Abstr. *9*, 78-02956 (1978)
33. Chan, Y.K., Lum-Shue-Chan, K.: Anal. Chim. Acta *48*, 205 (1969)
34. Christian, G.D.: Anal. Lett. *12B*, 11 (1979)
35. Colborn, T.: in Aquatic Toxicology and Hazard Assessment Proceedings of the Fifth Annual Symposium Philadelphia, PA 1980, p. 316, Philadelphia, American Society for Testing Materials 1982
36. Colombo, U., Sironi, G.: Anal. Chem. *36*, 802 (1964)
37. Dang, H.S., Jaiswal, D.D., Wadhwani, C.N., Somasunderam, S., Dacosta, H.: Sci. Total Environ. *27*, 43 (1983)
38. Daniel, R.C., Hänni, E., Shariatmadari, H.: Mitt. Geb. Lebensmittelunters. Hyg. *70*, 49 (1979); Anal. Abstr. *37*, 3G3 (1979)
39. Dannenberg, R.O., Petersen, A.E., Altringer, P.B., Brooks, P.T.: US Bur. Mines Rept. 8686 (1982)
40. Daskalova, A.: in Proceedings of the 2nd International Workshop on Trace Element Analytical Chemistry in Medicine and Biology Neuherberg, FRG 1982, p. 89, New York, Walter de Gruyter 1982; Pollut. Abstr. *14*, 83-00925 (1983)
41. David, D.J.: Analyst (London) *93*, 79 (1968)
42. De Silva, M.E.M.: Analyst (London) *100*, 517 (1975)
43. Dept. Environ. Nat. Water Council [UK]: Methods Exam. Waters Assoc. Mater. 18 pp, (1983); Anal. Abstr. *46*, 2H66 (1984)
44. Derbyshev, A.S., Kabluchko, N.A.: Desalination *44*, 233 (1983)
45. Donaldson, E.M.: Talanta *27*, 79 (1980)

46. Dreesen, D.R.: Water Pollut. Cont. Fed. J. *51*, 2447 (1979)
47. Durfur, E.N.: US Geol. Sur. Circ. 1812 (1964)
48. Dymott, T.C., Wassall, M.P., Whiteside, P.J.: Analyst (London) *110*, 467 (1985)
49. Easton, A.J., Moss, A.A.: Miner, Mag. *35*, 995 (1966)
50. Ebens, R.J.: in Eleventh Annual Conference in Trace Substances in Environmental Health Columbia, MO 1977, p. 48, Columbia, MO, University of Missouri 1977; Pollut. Abstr. *9*, 78-02947 (1978)
51. Ecrement, F., Burelli, F.P.: Analusis *2*, 306 (1973)
52. Edgar, R.M.: At. Absorpt. Newsl. *14*, 68 (1975)
53. Elwell, W.T., Wood, D.F.: Analytical Chemistry of Molybdenum and Tungsten, p. 38, Oxford, Pergamon Press 1971
54. Evans, W.H., Read, J.I., Caughlin, D.: Analyst (London) *110*, 873 (1985)
55. Ficklin, W.H.: Anal. Lett. *15A*, 865 (1982)
56. Frank, R., Stonefield, K.I., Suda, P.: Can. J. Soil Sci. *59*, 99 (1979)
57. Förstner, U., Wittmann, G.T.W.: Metal Pollution in the Aquatic Environment, p. 16, Berlin, Springer-Verlag 1981
58. Förstner, U., Wittmann, G.T.W.: *ibid*, p. 95
59. Furr, A.K., Parkinson, T.F., Hinrichs, R.A., Van Campen, D.R., Boche, C.A., Gutenmann, W.H., St. John, L.E., Pakkala, I.S., Lisk, D.J.: Environ. Sci. Tech. *11*, 1194 (1977)
60. Galli, A.: Ann. Biol. Clin. Paris *24*, 165 (1966)
61. Galli, Z.A., Sheina, N.M., Polikarpova, N.V., Pleshakova, T.V.: Zh. Anal. Khim. *30*, 1148 (1975)
62. Garner, C.D., Buchanan, I., Collison, D., Mabbs, F.E., Porter, T.G., Wynn, C.H.: in Proceedings of the Climax Fourth International Conference on the Chemistry and Uses of Molybdenum (eds.) Barry, H.F., Mitchell, P.C.H., p. 163, Ann Arbor, MI, Climax Molybdenum Co. 1982
63. Ghe, A.M., Carati, D., Stefanelli, C.: Ann. Chem. (Rome) *73*, 705 (1983)
64. Ghe, A.M., Stefanelli, C., Tsintik, P., Veschi, G.: Talanta *32*, 359 (1985)
65. Gott, R.D.: American Mining Congress Convention San Francisco, CA, September 1977; Pollut. Abstr. *10*, 79-03136 (1979)
66. Greenwood, N.N., Earnshaw, A.: Chemistry of the Elements, p. 1167, Oxford, Pergamon Press 1984
67. Grigg, J.L.: Analyst (London) *78*, 470 (1953)
68. Gupla, U.C., Chipman, E.W., Mackay, D.C.: Can. J. Plant. Sci. *58*, 983 (1978)
69. Gürtler, O.: Fresenius Z. Anal. Chem. *285*, 259 (1977)
70. Gutteridge, S., Bray, R.C.: in Proceedings of the Climax Third International Conference on the Chemistry and Uses of Molybdenum (eds.) Barry, H.F., Mitchell, P.C.H., p. 275, Ann Arbor, MI, Climax Molybdenum Co. 1979
71. Hainberger, L., de Oliveira Andrade, W.: Mikrochim. Acta II, 1 (1982)
72. Hay, R.W.: Bio-Inorganic Chemistry, p. 20, Chichester, Ellis Horwood Ltd. 1984
73. Heinemann, H.: in Catalysis (eds.) Anderson, J.R., Boudart, M., vol. 1, ch. 1, Berlin, Springer-Verlag 1981
74. Hesse, P.R.: A. Textbook of Soil Chemical Analysis, London, John Murray Pub. 1971
75. Hettche, H.O.: Air Water Pollut. *8*, 185 (1964)
76. Heumann, Th., Stolica, N.D.: in Encyclopedia of Electrochemistry of the Elements (ed.) Bard, A.J., vol. 5, p. 136, New York, Marcel Dekker 1976
77. Himeno, S., Ueda, Y., Hasegawa, M.: Inorg. Chim. Acta *70*, 53 (1983)
78. Holten, C.H.: Acta Chem. Scand. *15*, 943 (1961)
79. Horak, O.: Öst. Stud. Atm. SGAE Seib. Br. 4019 (1980); Pollut. Abstr. *12*, 81-01944 (1981)
80. Huisingh, J.: Environ. Health Perspectives *10*, 265 (1975); Pollut. Abstr. *6*, 04050 (1975)
81. Ivanova, I.F., Ganago, L.I., Pushkareva, T.M., Ezerskaya, T.V., Kartashova, G.I.: Gig. Sanit. 57 (1983); Anal. Abstr. *45*, 5H78 (1983)
82. Jackson, D.R.: J. Environ. Quality *4*, 223 (1975)
83. Jarrell, W.M., Page, A.L., Elseewi, A.A.: Residue. Rev. *74*, 1 (1980)
84. Jeffrey, P.G.: Analyst (London) *82*, 558 (1957)

85. Johnson, J.L.: in Molybdenum and Molybdenum Containing Enzymes (ed.) Goughlan, M.P., p. 347, New York, Pergamon Press 1980
86. Kabata-Pendias, A., Pendias, H.: Trace Elements in Soils and Plants, p. 19, Boca Raton, FL, CRC Press 1984
87. Kabata-Pendias, A., Pendias, H.: *ibid.*, p. 57
88. Kabata-Pendias, A., Pendias, H.: *ibid.*, p. 60
89. Kabata-Pendias, A., Pendias, H.: *ibid.*, p. 204
90. Kamiya, S., Takutomi, M., Matsuda, Y.: Bull. Chem. Soc. Jpn. *40*, 407 (1967)
91. Kawabuchi, K., Kuroda, R.: Anal. Chim. Acta *46*, 23 (1969)
92. Kawabuchi, K., Kuroda, R.: Talanta *17*, 67 (1970)
93. Khan, S.U., Clouter, R.O., Hidiroglou, M.: J. Assoc. Off. Anal. Chem. *62*, 1062 (1979)
94. Khristova, R., Nomova, D.: Acta Chim. Acad. Sci. Hung. *81*, 433 (1974)
95. Killeffer, D.H., Lenz, A.: Molybdenum Compounds, New York, Wiley-Interscience, 1952
96. Kim, C.H., Alexander, P.W., Smythe, L.E.: Talanta *23*, 229 (1976)
97. Kim, C.H., Owens, C.M., Smythe, L.E.: Talanta *21*, 445 (1974)
98. Kim, Y.S., Zeitlin, H.: Anal. Chim. Acta *51*, 516 (1970)
99. Kiriyama, T., Kuroda, R.: Talanta *31*, 472 (1984)
100. Klement, R.: Fresenius Z. Anal. Chem. *136*, 17 (1952)
101. Koizumi, T., Saito, S., Yamane, Y.: Chem. Biol. Interactions *51*, 219 (1984)
102. Korkisch, J., Gödl, L., Gross, H.: Talanta *22*, 669 (1975)
103. Korkisch, J., Krivanec, H.: Anal. Chim. Acta *83*, 111 (1976)
104. Korkisch, J., Steffan, I.: Mikrochim. Acta 651 (1973)
105. Korkisch, J., Steffan, I., Arrhenuis, G.: Anal. Chim. Acta *94*, 237 (1977)
106. Kubota, J.: in Molybdenum in the Environment (eds.) Chappell, W.R., Petersen, K.K., vol. 2, p. 555, New York, Marcel Dekker 1977
107. Kulathilake, A.I., Chatt, A.: Anal. Chem. *52*, 828 (1980)
108. Kumar, A., Rana, S.V.S.: Ind. Health *20*, 219 (1982)
109. Kuroda, R., Torui, T.: Fresenius Z. Anal. Chem. *269*, 22 (1974)
110. Lal, F., Biswas, T.D.: J. Indian Soc. Soil Sci. *22*, 333 (1974); Chem. Abstr. *83*, 150548n (1975)
111. Lansdown, A.R.: MoS_2 Lubrication A Continuation Survey 1975–1976, NTIS N78-20345/2WK (1978)
112. Latimer, W.M.: The Oxidation State of the Elements and Their Potentials in Aqueous Solutions, New York, Prentice-Hall 1952
113. Lillie, E.G., Greenland, L.P.: Anal. Chim. Acta *69*, 313 (1974)
114. Lisk, D.J., Gutenmann, W.H., Pakkala, I.S., Churey, D.J., Kelly, W.C.: J. Agr. Food Chem. *27*, 1393 (1979)
115. Livingstone, D.A.: in US Geol. Surv. Paper 440-G (ed.) Fleischer, M., p. 8 (1963)
116. Lowe, R.H., Massey, H.F.: Soil Sci. *100*, 238 (1965)
117. Luckey, T.D., Venugopal, B.: Metal Toxicity in Mammals, vol. 1, p. 18, New York, Plenum Press 1977
118. Malik, W.U., Srivastava, S.K., Bansal, A.: Anal. Chem. *54*, 1399 (1982)
119. Manzoori, J.L.: Talanta *27*, 682 (1980)
120. Mariella, R.P.: Chem. Eng. News *60*(23), 43 (1982)
121. Martin, M.H., Coughtry, P.J.: Biological Monitoring of Heavy Metal Pollution, p. 180, London, Applied Science Pub. 1982
122. Masarky, N.H., Schwitzgebel, K., Wolbach, C.D.: Control of Air Pollution Emissions from Mo Roasting, NTIS PB83-264192 (1983)
123. Maziere, B., Gros, J., Comar, D.: J. Radioanal. Chem. *24*, 279 (1975)
124. McCharthy, J.P., Caruso, J.A., Wolnik, K.A., Frick, F.L.: Anal. Chim. Acta *147*, 163 (1983)
125. McLeod, C.W., Otsuki, A., Okamoto, K., Haraguchi, H., Fuwa, K.: Analyst (London) *106*, 419 (1981)
126. Mertz, W.: in Molybdenum in the Environment (eds.) Chappell, W.R., Petersen, K.K., p. 267, New York, Marcel Dekker 1976
127. Mertz, W.: *ibid.*, p. 275

128. Mills, C.F.: in Proceedings of the Climax Fourth International Conference on the Chemistry and Uses of Molybdenum (eds.) Barry, H.F., Mitchell, P.C.H., p. 134, Ann Arbor, MI, Climax Molybdenum Co. 1982
129. Miska, K.H., Semchyshen, M., Whelan, E.P., Kruzich, D.J. eds.: Physical Metallurgy and Technology of Molybdenum and Its Alloys Proceedings of A Symposium, Ann Arbor, MI, Amax Materials Research Center 1984
130. Monien, H., Bovenberk, R., Kringe, K.P., Rath, D.: Fresenius Z. Anal. Chem. *300*, 363 (1980)
131. Morre, J.A.: Amer. Zool. *25*, 514 (1985)
132. Morgen, E.A., Rossinskaya, E.S., Vlasov, N.A.: Zh. Anal. Khim. *30*, 1384 (1975)
133. Morris, A.W.: Deep Sea Research *22*, 49 (1975)
134. Müller-Vogt, G., Vendl, W., Pfundstein, P.: Fresenius Z. Anal. Chem. *314*, 638 (1983)
135. Nakata, R., Okazaki, S., Hori, T., Fujinaga, T.: Anal. Chim. Acta *149*, 67 (1983)
136. Neidermeier, W., Griggs, J.H., Webb, J.: Appl. Spectrosc. *28*, 1 (1974)
137. Newton, W.E., McDonald, J.W., Burgess, B.K.: in Proceedings of the Climax Fourth International Conference on the Chemistry and Uses of Molybdenum (eds.) Barry, H.F., Mitchell, P.C.H., p. 150, Ann Arbor, MI, Climax Molybdenum Co. 1982
138. Nikolaev, A.S.: Metallurgist *27*, 49 (1983)
139. Norhein, G., Waasjø, E.: Fresenius Z. Anal. Chem. *286*, 229 (1977)
140. Notton, B.A., Hewitt, E.J.: in Proceedings of the Climax Third International Conference on the Chemistry and Uses of Molybdenum (eds.) Barry, H.F., Mitchell, P.C.H., p. 280, Ann Arbor, MI, Climax Molybdenum Co. 1979
141. Olsson, J., Wallen, B.: Desalination *44*, 241 (1983)
142. Otto, M., Müller, H.: Talanta *24*, 15 (1977)
143. Pakhol'chuk, S.F., Andrianov, A.M.: Zh. Anal. Khim. *34*, 193 (1979)
144. Pargeter, J.K.: in Accomplishments in Waste Utilization, p. 52, Chicago, IL, ITT Research Institute 1982; Pollut. Abstr. *14*, 83-05092 (1983)
145. Pavelenko, L.I.: Zh. Anal. Khim. *15*, 463 (1960)
146. Pavelenko, L.I., Karyakin, A.V., Bert, F.: Zh. Anal. Khim. *36*, 1793 (1981)
147. Patel, K.S., Mishra, R.K.: Talanta *29*, 791 (1982)
148. Patel, K.S., Mishra, R.K.: Ann. Chim. *73*, 91 (1983)
149. Pepin, D., Gardes, A., Petit, J., Berger, J.-A., Gaillard, G.: Analusis *2*, 549 (1973)
150. Petrie, S.E., Jackson, T.L.: Agronomy J. *74*, 1077 (1982)
151. Pienkos, P.T., Shah, V.K., Brill, W.J.: in Molybdenum and Molybdenum-Containing Enzymes (ed.) Coughlan, M.P., ch. 11, Oxford, Pergamon Press 1980
152. Podshivalova, A.K., Chernyak, A.S., Karpov, I.K.: Zh. Neorg. Khim. *29*, 2554 (1984)
153. Pollock, E.N.: At. Absorpt. Newsl. *9*, 47 (1970)
154. Postgate, J.R.: The Fundamentals of Nitrogen Fixation, Cambridge, Cambridge University Press 1982
155. Prasad, A.S. ed.: Clinical Biochemical and Nutritional Aspects of Trace Elements, p. 415, New York, Alan R. Liss 1982
156. Přibil, R., Vesslý, V.: Talanta *17*, 170 (1970)
157. Rana, S.V.S., Kumar, A.: Ark. Hig. Rada Toksikol. *34*, 9 (1983); Pollut. Abstr. *15*, 84-00996 (1984)
158. Rancicova, M., Cuta, J., Malat, M.: Vodni Hospod. *31B*, 19 (1981); Chem. Abstr. *95*, 17506d (1981)
159. Rao, D.V.R.: Curr. Sci. *21*, 257 (1952)
160. Ratnasamy, P., Sivasanker, S.: Catal. Rev. *22*, 401 (1980)
161. Riley, J.P., Taylor, D.: Anal. Chim. Acta *41*, 175 (1968)
162. Robinson, W.O.: Soil Sci. *66*, 317 (1948)
163. Robinson, W.O.: J. Assoc. Off. Agr. Chem. *38*, 246 (1955)
164. Robitaille, D.R.: Chem. Eng. (NY) *89*(20), 139 (1982)
165. Sandell, E.B.: Ind. Eng. Chem., Anal. Ed. *8*, 336 (1936)
166. Sasaki, Y., Sillen, L.G.: Ark. Kemi *29*, 253 (1968)
167. Savvin, S.B., Chernova, R.K., Beloliptseva, C.M.: Zh. Anal. Khim. *35*, 1128 (1980)
168. Schachter, M.M.: Mikrochim. Acta III, 317 (1983)
169. Schofield, M.: Metallurgia *68*, 31 (1963)

170. Schweizer, V.B.: At. Absorpt. Newsl. *14*, 137 (1975)
171. Sergeev, G.M., Korenman, I.M., Stepanova, L.I.: Zh. Anal. Khim. *38*, 1816 (1983)
172. Sherrell, C.G.: N. Z. J. Agric. Res. *27*, 417 (1984)
173. Shock, C.C., William, W.A.: Agronomy J. *76*, 35 (1984)
174. Shriadah, M.M.A., Kataoka, M., Ohzeki, K.: Analyst (London) *110*, 125 (1985)
175. Sillanpää, M.: Micronutrients and the Nutrient States of Soil, A Global Study, p. 95, Rome, Food and Agriculture Organization of the United Nations 1982
176. Sims, J.L., Suchy, M.E., Cornelius, P.L.: Agronomy J. *75*, 239 (1983)
177. Sittig, M.: Toxic Metals Pollution Control and Worker Protection, Park Ridge, NJ, Noyes Data Corp. 1976
178. Sittig, M.: *ibid.*, p. 277
179. Sittig, M.: *ibid.*, p. 278
180. Smith, A.E.: Analyst (London) *98*, 65 (1973)
181. Sokolov, A.V. ed.: Agrochemistry of the Soils of the USSR, p. 169, Jerusalem, Israel Program for Scientific Translations 1974
182. Somer, G., Cakir, O., Solak, A.O.: Analyst (London) *109*, 135 (1984)
183. Spackov, A.: Collect. Czech. Chem. Commun. *30*, 1255 (1965)
184. Spence, J.T.: Coord. Chem. Rev. *48*, 59 (1983)
185. Srivaslava, S.K., Sharma, A.K., Jain, C.K.: Talanta *30*, 285 (1983)
186. Standard Methods for the Examination of Water and Waste Water 13th ed., p. 519, New York, American Public Health Association 1971
187. Stanton, R.E., Mockler, M., Newton, S.: J. Geochem. Explor. *2*, 37 (1973)
188. Stark, J.G.: J. Chem. Educ. *46*, 505 (1969)
189. Sterritt, R.M., Lester, J.N.: Analyst (London) *105*, 616 (1980)
190. Subba Rao, N.S. ed.: Recent Advances in Biological Nitrogen Fixation, New York, Holmes & Meier Pub. 1980
191. Sugawara, K., Okabe, S.: J. Tokyo Univ. Fish. Spec. Ed. *8*, 165 (1966); Chem. Abstr. *68*, 43063u (1968)
192. Sugimae, A.: Appl. Spectrosc. *28*, 458 (1974)
193. Swaine, D.J.: Spectrochim. Acta *19*, 841 (1963)
194. Swaine, D.J.: in Trace Substances in Environmental Health XI (ed.) Hemphill, D.D., p. 107, Columbia, MO, University of Missouri 1977
195. Swanson, V.E., Medlin, J.M., Hatch, J.R., Coleman, S.L., Hood, G.H., Woodruff, S.D., Hildebrand, R.T.: US Geol. Surv. Rept. 76-468 (1976)
196. Swedo, K.B., Enemark, J.H.: J. Chem. Educ. *56*, 70 (1979)
197. Tabor, E.C.: Health Lab. Sci. *7*, 149 (1970)
198. Taimni, I.K., Agarwal, R.P.: Anal. Chim. Acta *9*, 203 (1953)
199. Tan, K.: Fenxi Huaxue *11*, 433 (1983); Anal. Abstr. *46*, 6B130 (1984)
200. Tao, H., Miyazaki, A., Bansho, K., Umezaki, Y.: Anal. Chim. Acta *156*, 159 (1984)
201. Ternero, M., Gracia, I.: Analyst (London) *108*, 310 (1983)
202. Theisen, A.A., Pinkerton, A.: Soil Sci. Soc. Amer. Proc. *32*, 440 (1968)
203. Thompson, M.: Analyst (London) *110*, 229 (1985)
204. Tie, A., Yu, W., Li, C., Liao, W., Shi, Y., Liao, X., Liu, H.: Fenxi Huaxue *11*, 839 (1983); Anal. Abstr. *46*, 10B122 (1984)
205. Tominaga, M., Bansho, K., Umezaki, Y.: Anal. Chim. Acta *169*, 171 (1985)
206. Tompseff, S.L., Fitzpatrick, J.: Analyst (London) *75*, 279 (1950)
207. Truesdale, V.W., Smith, P.J., Smith, C.J.: Analyst (London) *104*, 897 (1979)
208. Turekian, K.K., Wedepohl, K.H.: Bull. Geol. Soc. Amer. *72*, 175 (1961)
209. Ure, A.M., Bacon, J.A.: Analyst (London) *103*, 807 (1978)
210. van den Berg, C.M.G.: Anal. Chem. *57*, 1532 (1985)
211. Verbeeck, J., Vanderborght, B., Van Grieken, R., Ex, G.: Anal. Chim. Acta *128*, 207 (1981)
212. Verstuyft, A.W.: in Analytical Techniques in Occupational Health Chemistry, American Chemical Society Symposium Series 120 (eds.) Dollberg, D.D., Verstuyft, A.W., ch. 14, Washington, D.C., American Chemical Society 1980
213. Vieux, A.S., Rutagengwa, N., Mpeti, N.: Analysis *4*, 134 (1976)

214. Walravens, P.A., Moure-Eraso, R., Solomons, C.C., Chappell, W.R., Bentley, G.: Arch. Environ. Health *34*, 302 (1979)
215. Wan Ngah, W.S., Sarkissian, L.L., Tyson, J.F.: Anal. Proc. *20*, 597 (1983)
216. Ward, F.N.: Anal. Chem. *23*, 788 (1951)
217. Webber, M.D., Monteith, H.D., Corneau, D.G.M.: Training and Technical Transfer Division (Water) Environmental Protection Service Environment of Canada Ottawa, Ont. Rept. K1A 1C8 (1981); Pollut. Abstr. *13*, 82-00491 (1982)
218. Weeks, M.E., Leicester, H.M.: Discovery of the Elements, Easton, PA, Chemical Education Pub. 1968
219. Wide, M.: Environ. Res. *33*, 47 (1984)
220. Willard, H.H., Diehl, H.: Advanced Quantitative Analysis, New York, D. Van Nostrand Co. 1943
221. Woodis, T.C., Kolmes, J.H., Ardis, J.D., Johnson, F.J.: J. Assoc. Off. Anal. Chem. *63*, 1245 (1980)
222. Yamazaki, H., Gohda, S., Nishikawa, Y.: Bunseki Kagaku *29*, 58 (1980)
223. Young, E.G., Smith, D.G., Langille, W.M.: J. Fish Res. Board Can. *16*, 7 (1959)
224. Zaguzin, V.P., Ksenzova, V.I., Pogrebnyak, Yu.F.: Zh. Anal. Khim. *35*, 1143 (1980)

Subject Index

GPSR Compliance

The European Union's (EU) General Product Safety Regulation (GPSR) is a set of rules that requires consumer products to be safe and our obligations to ensure this.

If you have any concerns about our products, you can contact us on ProductSafety@springernature.com

In case Publisher is established outside the EU, the EU authorized representative is:

Springer Nature Customer Service Center GmbH
Europaplatz 3
69115 Heidelberg, Germany

Batch number: 10620881

Printed by Printforce, the Netherlands